KUNO STELLBRINK

Composites-standards testing and design

Composites- standards testing and design

National Physical Laboratory
8/9 April 1974

Conference Proceedings

IPC Science and Technology Press

© Crown Copyright 1974. No material may be reproduced from these Proceedings without the written permission of the Director of the National Physical Laboratory.

Published by IPC Science and Technology Press Ltd, IPC House, 32 High Street, Guildford, Surrey, England GU1 3EW.

ISBN 0 902852 31 0

Printed by R. Kingshott & Co. Ltd., Aldershot, Hants.

Contents

Foreword — Sir Alan Cottrell — 7

Some scientific points concerning the mechanics of fibrous composites — A. Kelly — 9

Tension-compression experiments on a fibre reinforced composite of Cu-W — H. Lilholt — 17

Prediction of properties for engineering design with composites — U. Hütter — 20

On a constraint effect in steady-state creep of fibre composites — O. Bøcker Pederson — 33

Effective structural use of grp and cfrp — W. Paton — 36

Applications of advanced composites in aircraft structures — I.C. Taig — 40

The disproportionate weakening of composites by sub-millimetre defects — L.E Dingle, R.G. Williams and N.J. Parratt — 51

Applications of fibre reinforced composites in marine technology — C.S. Smith — 54

Pyrolytic surface treatment of graphite fibres — D.J. Pinchin and R.T. Woodhams — 70

Fatigue effects in carbon fibre reinforced glass — K.R. Linger — 73

Acoustic emission and fatigue of reinforced plastics — M. Fuwa, A.R. Bunsell and B. Harris — 77

Composite materials in civil engineering. Their current use, performance requirements and future potential — L.H. McCurrich and M.A.J. Adams — 80

Fibre reinforced cements — scientific foundations for specifications — J. Aveston, R.A. Mercer and J.M. Sillwood — 93

Some engineering considerations of cement-based fibrous composites and polymerised concrete — R.N. Swamy and C.U.S. Kameswara — 104

Tensile stress-strain characteristics of glass fibre reinforced cement — B.A. Procter, D.R. Oakley and W. Wiechers — 106

Modification of grc properties — A.J. Majumdar — 108

Ferrocement — a summary of research activity — R.G. Morgan — 111

The characterization of reinforced thermoplastics for industrial and engineering uses — C.M.R. Dunn and S. Turner — 113

Time-dependence and anisotropy of the stiffness of fibre-plastics composites   120
R.M. Ogorkiewicz

The characterization of glass-reinforced polycarbonate by means of creep-rupture and tensile experiments   123
J. Hugo

Characterization of fibre composites using ultrasonics   126
G. Dean

Anisotropy arising from processing of reinforced plastics   131
K. Thomas and D.E. Meyer

Dynamic testing and performance of unidirectional carbon fibre-carbon composites   140
L. Boyne, J. Hill and S.L. Smith

Techniques for measuring the mechanical properties of composite materials   144
P.D. Ewins

Testing and characterization of composite materials   155
R.D. Adams

The compression strength of unidirectional cfrp   158
N.L. Hancox

Charpy impact strength of uniaxial cfrp   160
M.G. Bader

Opening remarks

SIR ALAN COTTRELL, FRS

Mr Chairman, Ladies and Gentlemen: It is a great pleasure for me this morning to have this opportunity to open your conference on composites. I would like first of all to pay a tribute to our hosts, the National Physical Laboratory, for arranging the meeting and for inviting us here. They have chosen a subject of the greatest 'timeliness and promise', to use a favourite term of jargon in UK scientific policy-making. It is particularly right that we should be meeting here at this time to discuss the subject of composite materials because no laboratory in recent years has done more than the National Physical Laboratory to advance the science of composites, both for its own intrinsic interest and also as a potential source of new materials of great practical benefit to mankind. I think there is a good reflection of the NPL's contribution here in the very fine turn out of this conference of delegates from both this country and from abroad. Perhaps, may I also say at this point how very delighted all of us in the UK are to see so many of our friends and colleagues from abroad.

There is, of course, a great deal of interest at the present time in man-made composite materials of high strength and other mechanical properties. This interest ahs been stimulated through the fact that such materials can now be purposely designed to have particular combinations of properties: combinations which can be controlled and varied; if you like, from place to place in the same material in accordance with the structural designer's need. Because there is such great scope for varying the structures and compositions of the composites over an extremely wide range of variables; because the properties of the composites can thus be changed so considerably, for example, going from soft rubber-like materials right up to materials stiffer than steel; so design engineers are challenged by the problem of how to relate those properties to actual performance of the materials on the job and, even more than that, they have the problem of knowing, or of wanting to know, exactly what properties they require in their constructions. With these composite materials we have moved a very long way from the traditional engineering materials – the metals which, from the design engineer's point of view, are usually sufficiently well specified just by quoting their Young's modulus and their yield strength. But these new materials have a whole spectrum of properties to be handled by the designer and so one needs a much better rapport, between the design engineer and the materials man, than can be provided by a cryptic sheet of materials data. Because of all this, there is a special problem in setting standards and specifications for high performance composites and also in devising sufficiently searching tests and procedures for analysing the properties of the structures.

The purpose of this conference is to review in depth all the areas where work on standards and testing is needed at the present time: to serve the interests of the design engineers and to uncover any scientific stumbling blocks which may hold up progress towards the general use of these new materials. All this, of course, falls exactly into the remit of the National Physical Laboratory; and the organisation of conferences such as this is part of its general process of planning new scientific programmes. The conference, from its own programme, promises to be one of very great interest. I, personally, was pleased to see that two of the main papers are going to deal with advanced composites based on carbon or boron fibres, which are now so promising for aircraft and other means of transport. The carbon reinforced plastic, for example, is now being used increasingly in aircraft, for air brakes, wing tips, airliner flooring and things such as that. Then there are the glass reinforced plastics: these are now beginning to be used on quite a considerable scale in ships, including ships' hulls, and in other fields; also glass reinforced plastic is steadily capturing markets as a constructional material in civil engineering, where its use has been growing at about 15% per year in Europe and America.

Nevertheless, it is a rather difficult material to standardise for engineering construction and I hope that the conference will be able to pay some attention to this problem during the next two days. Fibre reinforced cement has become quite important recently and is now set ready for rapid commercial growth. I am very pleased to see that this is an area where the science and the practical development of the material have gone along hand in hand closely together. Then there is the fibre reinforcement of the thermoplastic materials, which is one of the large and rapidly growing areas of the whole subject at this time, as this fibre reinforcement not only builds up the mechanical properties but it also permits the use of thermoplastics at slightly higher temperatures than are otherwise possible and that is a really important benefit which is strongly influencing the application of these materials in motor vehicles, for examples, and in household hardware. But because of the flexibility and viscosity of the plastic matrix in these thermoplastic materials, there is, in this area, an even greater need for standards and specifications.

Finally, I was glad to see that the conference intends to pay particular attention to methods of testing, for we should never forget that the data that we can give to the design engineer and the data needed for quality control are, at best, only as good as the test methods that provide those data. Furthermore, those test methods have to be the right ones, to identify those properties that will be critical in the practical applications. All this sets a number of very interesting scientific and technological problems; and so to conclude I wish the conference every possible success. I shall look forward with the greatest pleasure to reading the published Proceedings.

Sir Alan is Master of Jesus College, Cambridge, and was formerly Chief Scientific Adviser, Cabinet Office

Some scientific points concerning the mechanics of fibrous composites

A. KELLY

Some simple points are made concerning attainable modulus to density ratios. The attainable strengths of fibres are discussed and the recent evidence that RAE fibres fail in shear is compared with similar evidence for glass. An upper limit to the strength of perfect graphite fibres is given assuming that they behave as does high strength glass. Methods of estimating the moduli of random planar mats of fibres are reviewed. The theories of constrained cracking of a brittle matrix are reviewed and it is shown how the theories of Romualdi and Batson and of Aveston, Cooper and Kelly are related by the concepts of fracture mechanics.

1 INTRODUCTION

Science without technology is useless, technology without science is blind. The blind may fall into a pit and so to prevent this, need the illumination of science. To meet this need the science needs to be developed enough to be firmly based and able to predict. It can then be used didactically for deduction of consequences and not still at the inductive or experimental stage of looking for the general principles. Of course, the two are never really separate and it's a rash scientist who arrogantly expects his predictions to be always correct. However, to assist the engineer, he needs to design his experiments so that as a result of them he is able to predict and, therefore, explain to an engineer why a particular difficulty arises or can be circumvented, instead of greeting his colleague with the unhelpful comment that 'such and such is interesting'.

The regions where fibre composites are being either very intensely researched (a spending say of more than £3M) or are being used are:
    (a)   as light weight constructional materials in the aircraft and marine fields;
    (b)   as reinforced plastics with many uses in civil fields;
    (c)   as new high temperature materials principally for gas turbines;
    (d)   as reinforcement for brittle materials, principally in an attempt to extend the range and variety of uses of that very cheap and easily fabricated material Portland cement.

In the regions (a), (b), and (d) applications of fibrous composites exist and will be dealt with at this conference. The research carried out under (c) has not yet made a new material used in service.

For some of these areas I will make some remarks of where the science can guide us.

Dr Kelly is the Deputy Director of the National Physical Laboratory, Teddington, Middlesex, England.

2 LIGHT WEIGHT CONSTRUCTIONAL MATERIALS

2a   Modulus and strength

Constructional materials used in aircraft, in high speed marine craft and in rapidly moving pieces of machinery need to be very strong and stiff per unit mass. Stiffness governs the deflections in tension and bending and usually controls failure in a compressed strut. Fig. 1a shows values of Young's modulus divided by density plotted against the strength divided by the density. Several of the fibres shown, eg boron and graphite, have much larger values (measured parallel to the fibre axis) than the high strength steel or aluminium alloys. Boron and graphite together with silicon carbide represent the largest values of specific modulus known and theory of the chemical bond shows that these values will not be exceeded (eg [1]). However, these large values are only found parallel to the axis of the fibre. Fig. 1 also shows values for an organic fibre developed by Du Pont and called PRD–49; it appears under several proprietary names, all of which (see, for example Black [2]) consist of chains of aromatic rings linked by non-flexible groups such as CO–NH and the specific modulus can be very large, eg 8 800 km as in Fig. 1a. Boron, graphite, PRD–49 and similar fibres have the highest specific stiffnesses because they represent solids with the largest areal density of strong directed bonds. Such cannot be formed solely between heavy elements and so the light elements with multiple valencies must give the stiffest solids per unit of weight. Bonds in a number of directions are needed and not less than three per atom are required and preferably more. The possible hybridization of s, p orbitals leads to tetrahedrally directed bonds with four bonds per atom, and this is the most that can be obtained. Hybridization of s, all three p and two d can give six bonding directions per atom but the inclusion of d orbitals means the larger atoms and hence a reduced areal density of bonds crossing any plane: s, 2p and d leads to four *coplanar* bonds.

With only three strong bonds an atom must have at least 2π of solid angle unlinked to another and so produce anisotropic

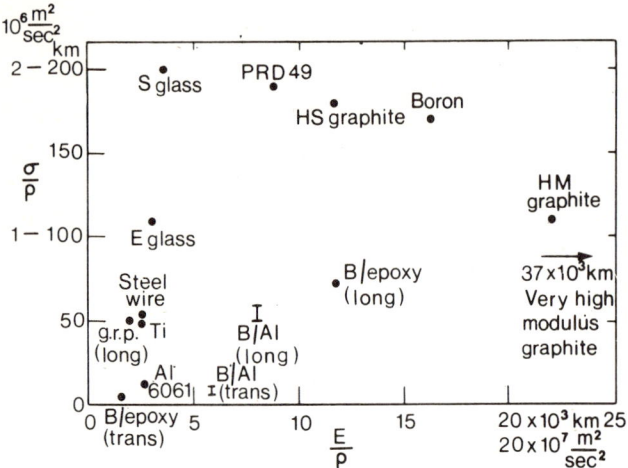

Fig. 1a Values of specific strength as a function of specific stiffness for various fibres and composites. Longitudinal and transverse Young's moduli for the composites are given.

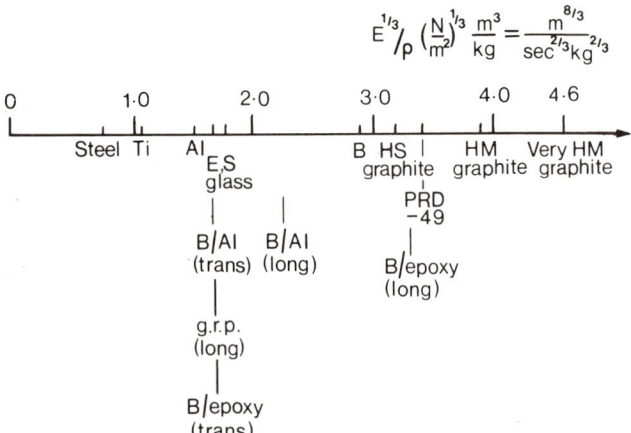

Fig. 1b Values of Young's modulus to the one third divided by density for various fibres and composites.

ementary theory of the thin solid disc of radius b spinning about its axis of revolution with angular velocity ω gives the maximum radial and tangential stresses as equal, located at the centre and given by

$$\sigma_r = \sigma_\theta = \frac{3 + \nu}{8} \rho \omega^2 b^2 \qquad (1)$$

where ν is Poisson's ration and ρ the density: $\omega^2 b^2$ is the linear velocity of the periphery of the disc; ν is usually between 0.2 and 0.3. The velocity, therefore, given as ordinates in Fig 1a represents about half the maximum linear velocity of the edge of such a disc.

If E/ρ is a minimum the increase in length of a rod under its own weight is a minimum for a given length of rod and the mean radial strain and hence the deflection of the outer edge of a spinning disc due to its own inertia is also a minimum. The velocity in the case of the abscissae in Fig. 1a represents the square of the velocity of sound in a thin rod of the material. Stiffness in compression of a strut or of a panel often governs the failure load of a structure. If lightness is sought, elastic modulus and density do not enter to the same power into the condition for least weight. Fig. 1b collects values of $E^{1/3}/\rho$ where E is Young's modulus and ρ the density. A maximum value of $E^{1/3}/\rho$ is the condition that a given load be supported by a member of least weight spanning a given length by the bending of a cantilevered panel with a given allowable deflection. It is also the condition for a panel of given height to support a given load with minimum thickness and hence with minimum weight.*

Fig. 1b illustrates the great advantage to be gained by reducing the density if buckling of a panel of fixed width is to be avoided with least weight. Comparing Figs. 1a and 1b steel appears much the least attractive material because of its high density. Boron graphite and PRD—49 emerge as clearly the leaders as they are on a plot of E/ρ but the striking fact is that the poorer transverse properties of the composite are no longer such a disadvantage in compression since the density is still low and, for instance, the transverse properties of a boron/epoxy composite are as good as that of a material with the properties of E-glass by itself.

2b Strengths of fibres

The actual strengths realized by the strong fibres are not thoroughly understood, in the sense that they can be predicted in the same way as can the yield strength of say a single crystal of a dispersion hardened alloy. Metallic wires have their ultimate strengths governed by a shear process and about ten years ago Marsh [4] suggested that the strength of glass is governed by a shear process also, even though the material cannot contain dislocations in the sense that dislocations in a crystal with discrete Burgers vectors are usually understood. Marsh's deduction rested (among other features) upon noting that the (time and humidity dependent) yield stress observed in compression under a hardness indenter was the same as the failure stress in tension of

stiffness — such is found in graphite fibres [3]. Boron does rather better and silicon much better as regards directionality, having a tetrahedral link. However, the Si—Si bond is not so stiff and the increase in stiffness obtained by interposing oxygen between the Si as in glass (or strictly silica) has the disadvantage of so increasing the distance between silicon atoms that the modulus is much reduced and reduced proportionately more per unit of weight. Science may still have something to offer by filling in the gaps in Fig. 1a provided it can be done cheaply.

Of course Fig. 1a is not the whole story: the fibres must be made into a composite and then transverse and longitudinal properties can differ more than in the component fibres. Fig. 1a also compares composites with the conventional materials and then I believe that we see that gap filling in Fig. 1a may be pointless unless combined with a really large reduction in cost. The units for both abscissae and ordinates in Fig. 1a are shown as either a length or as the square of a velocity. In the case of the strength, the first unit (a length) is useful if one is interested in the strength of objects stressed due to terrestrial gravity. The length given represents the greatest length of a uniform rod which could in principle be picked up from one end while lying on the surface of this planet. However, if one is interested in the construction of rapidly moving parts which are loaded by inertial forces, then the square of the velocity is the more convenient. The el-

---

* Buckling load of a strut and the deflection of a cantilever are controlled by bending forces only so long as the Young's modulus and the shear modulus are of the same order of magnitude. For individual carbon fibres and for carbon fibre composites this is not the case. The ratio of E to G is very large as for graphite single crystals [30]. For the carbon fibres in Table 1 the ratio of Young's modulus to longitudinal shear modulus is 30 and for the carbon fibre composite in reference [3] at $V_f = 0.50$ is 46. With these values *elastic* buckling of the composite could be controlled by shear in some cases and not by elastic bending as in the classic Euler formula.

a fine fibre. The same type of correlation exists for carbon fibres [5]. Marsh had first to convince that glass, the apparent paradigm of a brittle material could sustain plastic flow. This he did by noting that plastic furrows are produced by a stylus, that plastic flow obviously occurs beneath an indenter, that unloaded cracks do not close up and that the work of fracture is equal to many times the surface energy. If Marsh is essentially correct, as I believe, then it follows that the sudden fracture in tension is due to instability because the plastic flow is unstable. A marked instability is expected, because (a) no such thing as work hardening is expected in a glass and (b) in all covalent materials the stress to initiate plastic flow is much greater than that needed to maintain it. The possibility of plastic flow occurring in carbon fibres produces no initial disbelief when one considers the ease of basal slip in graphite crystals. Experiments by Ewins and Ham [5] were aimed to decide on the mode of failure in compression of carbon fibre reinforced epoxy resin composites. Collings [6] had found that the ultimate compressive strengths of composites compressed parallel to the fibres (longitudinally) were closely the same as the compressive strengths measured when an aligned composite was compressed in a direction normal to the fibres, with an experimental set-up such that Poisson expansion in a direction normal to both the fibre axis and the direction of the compressive axis was prevented. Furthermore, the failure mode observed in both cases was the same and involved failure of all of the fibres. Detailed observation of the fracture surfaces of composites, carried out by Ewins and Ham showed clearly that the failure of the composites in longitudinal compression occurs by a shear failure of the individual fibres on a plane at or close to that of the maximum shear stress. Composites failing in this way in compression fail at very similar loads if deformed in *tension* and this suggests (when coupled with the observation that a composite rod broken in flexure shows similar shear failures on both the sides of the neutral plane subject to compressive and to tensile forces) that the fibres fail in tension by a shear mechanism. The observed longitudinal failure stress in both tension and compression of a composite containing $V_f = 0.60$ of HT-S carbon fibres is $1.4 \, \text{GN/m}^2$. Since the compressive failure stress of the resin $\sim 200 \, \text{MN/m}^2$, then applying the rule of mixtures the compressive (and tensile) yield stress of the fibres is $2.2 \, \text{GN/m}^2$, corresponding to a yield stress in shear of $1.1 \, \text{GN/m}^2$. The value for the yield stress of high modulus fibre HM-S would be $1.2 \, \text{GN/m}^2$, corresponding to a yield stress in shear of $0.6 \, \text{GN/m}^2$. The shear modulus of a graphite fibre of Type 1 (HM) for shear on a plane at $45°$ to the axis (G') is $\sim 8.3 \, \text{GN/m}^2$ using the elastic moduli in reference [3], so that the shear stress corresponds to a value of $0.07 \, G'$. The corresponding figure for strong glass fibres at low temperature is $0.19 \, G'$ [1]. The highest shear strength likely, if graphite were like glass is $0.19 \, G'$. Taking Reynolds [8] values of the S_{ij} a figure of $3 \, \text{GN/m}^2$ is obtained for the ultimate shear strength and so $6 \, \text{GN/m}^2$ or close to 10^6 psi as the expected upper limit to the tensile strength for graphite fibres. Values close to this have occasionally been reported.

There have been reports that thin carbon fibres are *stiffer* than thick ones and this is related to the state of oxidation [7]. If this result is confirmed the possibility, therefore, exists of increasing the modulus without the very high temperature treatments presently necessary to obtain the high value of modulus. An upper limit to the modulus of graphite fibre is set by the value for a graphite crystal. This value is $\sim 1000 \, \text{GN/m}^2$ [8] which is not yet approached by fibres subject to hot stretching [9].

3 MODULI OF ALIGNED COMPOSITES

If the elastic moduli of fibres and of matrix are known then bounds for the moduli of the composite may be deduced for the case of aligned fibres [10] [11]. For the case of transverse isotropy, two are then well defined namely E_3** and ν_{13} but the other three moduli have rather widely separated bounds.

Hashin and Rosen [12] produced a convenient set of equations (based on what they call the composite cylinder assemblage) for calculating five independent moduli of the aligned continuous filament composite with transverse isotropy. Often the calculations of the composite properties are directed towards their use in design of laminated structures. In the simple theory of these only four elastic constants are needed for each lamella because normal stresses are neglected. These are the axial Young's modulus, transverse Young's modulus, axial Poisson's ratio and axial shear modulus. Hashin and Rosen's model gives good agreement with well defined experimental results for the first three of these but tends to give values lower than experiment for the axial shear modulus: the value could in principle be only $\tfrac{2}{3}$ the true value in the case of say glass in epoxy resin. Sometimes empirical interpolation formulae are recommended, eg those by Halpin and Tsai [13]. The recommended value of the parameter to be adjusted for the interpolation in the case of the axial shear modulus actually reduces the expression to one identical with that of Hashin and Rosen. A useful review of the present position has been given by Rosen [14].

Of course, exact expressions can be obtained in some cases for specific geometries. Since useful composites are not likely to possess a specific geometry, there seems to be little to be gained for the engineer or designer in pursuing these researches. The measurement of the elastic moduli of composites is however very important for two reasons. The first is in quality control, and the second arises because it is often not easy to determine accurately the elastic moduli of thin fibres. In this case the moduli of the composite are found and then using expressions for the composite moduli in terms of the component moduli, those of the fibre are deduced. A self consistent method must be used, in the sense that the experimental data, particularly those for values of the c_{ij} which are experimentally difficult to obtain, must be scrutinized with care and the values used only if they can be shown to correspond to reasonable general statements, eg that the bulk modulus is positive or that adding a stiff fibre to a very compliant matrix cannot in the absence of voids produce a composite more compliant than the matrix. If this is not done some rather strange figures may emerge [15]. Dean and Turner [3] have been able to deduce all five elastic moduli of carbon fibres in epoxy resin — assuming the fibres are transversely isotropic — by measuring the five independent moduli of the composite by ultrasonic methods. If axis 3 is the fibre axis, $c_{11} = c_{22}$, c_{33}, $c_{66} = \tfrac{1}{2}(c_{11} - c_{12})$ and c_{44} are deduced directly from the velocity of elastic waves. The variation of c_{33} and c_{66} is closely linear with V_f, so values for the fibre can be deduced by extrapolation to $V_f = 1$. c_{13} for the composite is hard to find since along no

---

** Axis 3 is parallel to the fibres.

direction does the elastic wave velocity depend only on c_{13}. Values of c_{13} can be found by expressing the other constants in terms of it in order to set limits to c_{13}. From the values for the composite and measured values for the matrix, values of the c_{ij} of the fibre can be found either by extrapolation, as above, or by forming combinations of the c_{ij} which should be linear when plotted against volume fraction. To do this, a theory must be used. Hashin and Rosens' was found to be adequate. For example to find c_{44} for the fibre one can form the expression

$$\left(\frac{c_{44c} - c_{44m}}{c_{44c} + c_{44m}}\right) = \left(\frac{c_{44f} - c_{44m}}{c_{44f} + c_{44m}}\right) V_f \qquad (2)$$

and a value of c_{44f} found by extrapolation to $V_f = 1$. c_{11} for the fibre is found in an analogous way. Fig. 2 shows the linearity of such a plot, and Table 1 the stiffness deduced for the fibres of the high modulus RAE type. The values so deduced can be compared with static measurements and if this be done very good agreement is obtained, eg the value of axial Young's modulus of the fibre from Table 1 is $1/S_{33} = 400\,GN/m^2$ and static values are between 400 and $410\,GN/m^2$. This comparison shows also that in most fibres there is little departure from linearity in the stress-strain curve since the ultrasonic values are obtained for vanishingly small strains.

4 LAMINATES AND THE RANDOM COMPOSITE

The application of high performance composites depends upon the assembly of unidirectional lamellae of plies into a laminated structure. The basic theory is available for calculating the elastic modulus, deflection, and hence the composite stress-strain curve and internal stresses prior to failure, of any interface or of any single ply — see eg [16]. The advanced use of these composites makes it necessary to abandon some of the simple hypotheses. This is particularly important if the layers are highly elastically anisotropic and there is only a small number of layers in the composite. An often quoted figure from Whitney [17] shows this – Fig. 3a and Fig. 3b. Here the maximum deflection of a square plate is given as a function of the angle of orientation of the fibres. The case shown is for a symmetric biaxial laminate and maximum deflection of the plate is plotted for a plate of fixed thickness containing different numbers of layers. For a small number of layers it is important to take into account coupling between bending and stretching. This is usually neglected in the simplest theory and such neglect is possible if there are so many layers that the laminate is symmetric with respect to the midplane. The figure shows that coupling must be considered for the two layer composite of graphite epoxy which is very highly anisotropic. For glass epoxy the difference is important but not to the same extent.

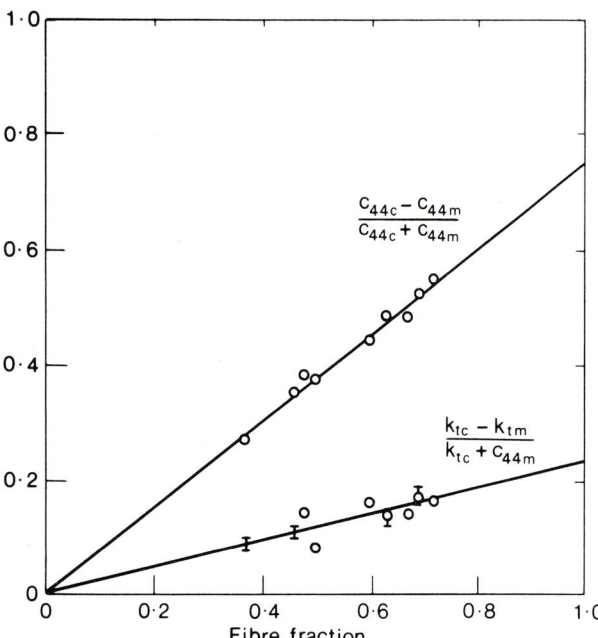

Fig. 2 From Reference [3] – a plot of two ratios of the elastic moduli versus volume fraction of fibres – see text. k_t is the transverse plane strain bulk modulus.

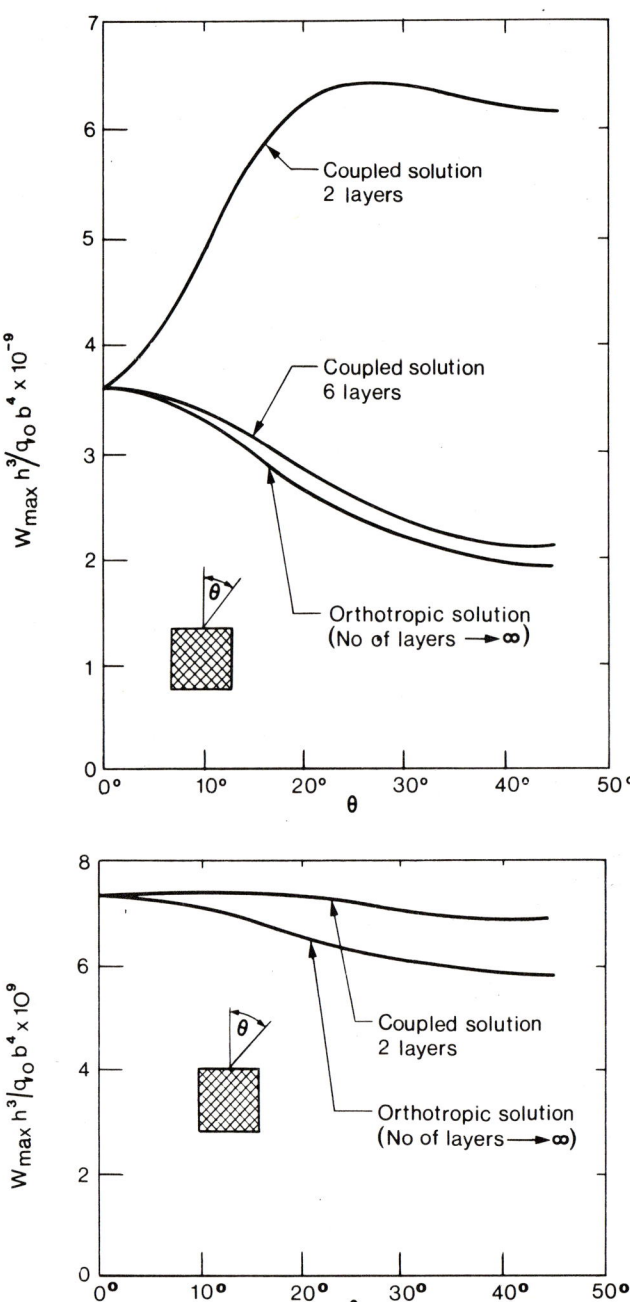

Fig. 3 Maximum deflection, W_{max}, as a function of fibre orientation for a simply supported square plate under transverse loading, characterised by the quantity q_0. h is the thickness and b the spin: (a, top) graphite epoxy (b, bottom) glass epoxy. These are theoretical results from [17].

Table 1. Elastic stiffness and compliance components for type 1 carbon fibres. c_{pq} GN/m² s_{pq} m²/GN deduced by Dean and Turner [3].

	$c_{11}=c_{22}$	c_{33}	$c_{44}=c_{55}$	c_{66}	$c_{13}=c_{23}$	c_{12}	ν_{13}	ν_{31}	ν_{12}
	$s_{11}=s_{22}$	s_{33}	$s_{44}=s_{55}$	s_{66}	$s_{13}=s_{23}$	s_{12}			
Type 1 fibres	12.1	410	13.7	2.8	6.5	6.5	0.35	0.01	0.53
	0.116	0.0025	0.073	0.357	−0.00086	−0.062			

A fibre array of importance in many reinforced plastics and which is becoming so in reinforced cement and plaster is that with fibres randomly arranged in a plane. In these composites the volume fractions are sometimes small – less than say 10% – because of the high cost of the fibres relative to the matrix. An early expression due to Cox [18] is often used which gives the modulus of the planar mat as

$$E_c = \frac{E_f V_f}{3} \qquad (3)$$

Application of this is restricted to the cases $E_f \gg E_m$ and it does not behave properly either as $V_f \to 0$ or as $V_f \to 1$. A simple extension to Cox can be made by supposing that the matrix occupies all of space (ie $V_m = 1$) and by considering in addition a set of fibres with modulus $(E_f - E_m)$ and volume fraction V_f. Since the fibres form a random planar array the application of netting analysis gives their effective modulus as $\frac{1}{3}(E_f - E_m)$ and application of the rule of mixture gives

$$E = E_m + \frac{1}{3} V_f (E_f - E_m) \qquad (4)$$

which behaves properly both as $V_f \to 0$ and $E_f \to E_m$.

A much more detailed analysis based on the results of [10, 11, 12] has been given by Christensen and Waals [20] who obtained a result by averaging exact expressions for aligned materials over all planar orientations. Fibre-fibre interaction, is, therefore, neglected. The result contains other moduli than E_f and E_m in the expression for the composite modulus. Christensen and Waals showed that their results could be approximated very well by

$$E = E_m + \frac{1}{3} V_f (E_f + 3E_m) \qquad (5)$$

for low volume fractions (< 0.2) and $E_f \gg E_m$. This expression clearly has the correct behaviour as $V_f \to 0$. The limit $E_f \to E_m$ is not permitted in the approximation leading to equation (5). Lee [21] measured the elastic modulus of a composite consisting of random planar glass fibres in polystyrene ($E_m = 3.24$ GN/m²) and in polystyrene-acrylonitrile ($E_m = 3.66$ GN/m²) and the results are shown in Fig. 4. Both equations (4) and (5) appear to be useful approximations. Christensen and Waals result is based on a better foundation than equation (4) but the real effect of fibre 'cross-overs' is not known. Lockett [22] has extended the approach leading to equation (4) by considering the matrix more accurately but again assuming the matrix fills all space and there are in addition fibres of modulus $(E_f - E_m)$. He finds that equation (4) is a good approximation provided $\frac{1}{3} < \nu_M < \frac{1}{2}$ but that it is less good when the Poisson ratio of the matrix is less than $\frac{1}{3}$. For very small values of ν_m (a rather unlikely case) (4) could overestimate the Young's modulus of the composite by a factor as large as 2. The Poisson's ratio of the matrices in Lee's experiments would be about $\frac{1}{3}$, but accurate values are not available.

5 CONSTRAINED CRACKING OF A BRITTLE MATRIX COMPOSITE

A little more than ten years ago Romualdi [23] suggested that the actual cracking strain of concrete could be increased by incorporating fine fibres. Since then Majumdar [26] and others have developed glass fibre reinforced cements and the prediction and control of the initial cracking strain has become an important engineering feature.

There are in the literature essentially two theories to account for the experimental fact that the introduction of fibres increases the strain at which cracks are first observed, ie increases the initial cracking strain. These are due to Romualdi and Batson [23] – RB theory – and Aveston, Cooper and Kelly [24] – ACK theory. The theories are thought to be distinct. The first regards the fibres as limiting the length of crack present in the matrix and the second as limiting the displacement of the surfaces of the crack. By developing fracture mechanics in a physically clear way it may be shown that the two theories are intimately connected [25]. This arises because the length and the displacement of an equilibrium crack under stress in a homogeneous material are necessarily related to one another. The two (apparently) different theories may be cast into forms which, within the accuracy of the quantitative development, are identical as we shall now show.

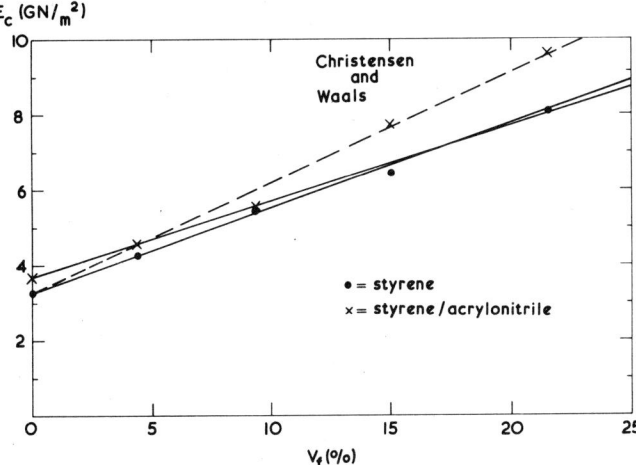

Fig. 4 Variation of in-plane Young's modulus with volume fraction measured by Lee [21] for random planar mats of glass in polystyrene and in styrene acrylonitrile copolymer. The full lines represent the predictions of equation (4) and the dotted line the prediction of Christensen and Waals for the composite with styrene as a matrix.

5a Theory of Romualdi and Batson

Romualdi and Batson choose to work in terms of a critical stress intensity K_c. They do not give explicit formulae for this since their calculations are carried out by computer methods. Their approach as stated in their paper is as follows.

'The basis for the appropriate analysis is the crack extension force concept advanced by Irwin as a measure of the influences that tend to extend a crack in a stressed medium. As a flaw in the concrete tends to enlarge to a crack, displacements develop in the material ahead of the crack as a result of the stress field singularity at the crack edge. The greater rigidity of steel reinforcement in the immediate vicinity of the crack, however, opposes these displacements, and forces are exerted by the reinforcement on the concrete matrix.

By means of equations describing compatibility between adjacent points on the reinforcement and in the concrete, it is possible to calculate these forces and, from suitable fracture mechanics applications to interpret them in terms of a reduction in the crack extension force.'

The ACK theory [24] assumes that the reinforcement limits the maximum opening displacement of the crack, while the RB theory views the reinforcement as limiting the crack extension force. These two approaches are related by the theory of fracture mechanics and, if both sets of authors have done their sums correctly, it must be possible to show that they are the same. [25] shows this in detail.

RB work in terms of a stress intensity factor K, and consider only one specific case, namely a penny-shaped crack of radius less than the separation of adjacent (parallel) fibres in a square array. They assume that in the absence of a crack the strains in fibre and matrix are the same. When the crack is present the additional displacements in the concrete caused by the extensional strains in the neighbourhood of the crack are resisted by the fibres which *are assumed infinitely stiff*. This resistance causes a distribution of shear forces along the wires which act to close the crack. The forces acting to close the crack are calculated as follows. If x is the displacement of a point at the surface of a fibre, in the y direction (normal to the crack and parallel to the wires) due to the presence of the crack under stress σ in an infinite (concrete) medium, then if the (rigid) wires are introduced forces P arise at the wire surface which reduce this displacement to zero. The effect of all of these factors at any one point is summed to produce an expression for the distribution of these forces in terms of the x due to the crack. The critical stress intensity factor K_c caused by the remote stress in the absence of the reinforcement is

$$K_\sigma = \frac{2\sigma}{\pi}\sqrt{c} \qquad (6)$$

for a penny-shaped crack under normal stress σ. Note σ will be the stress on the concrete. To find the value of K_c for the crack in the reinforced concrete K_T, say, the stress intensity factor due to the forces P, K_p say, is calculated and subtracted from the value of K given by equation (6). To find K_p RB note that the forces in the wires tend to close the crack, so in the absence of a crack their net effect must be to produce a compressive stress over the area which would be occupied by the crack. RB calculated the *average* compressive stress over the crack area and then assume that the effect of the crack is the same as that of a crack expanded by a pressure p numerically equal to the average compressive stress. (K_c has the same form for a crack under a remote tensile stress and one being expanded under internal pressure (see eg [27])). The final expression for K_T, ie the stress intensity in reinforced concrete, is

$$K_T = K_\sigma - K_p = \frac{2\sqrt{c}}{\pi}(\sigma - p). \qquad (7)$$

RB calculate P — the details being given in their paper — and then equate K_T to the measured value of the stress intensity factor for propagation of a crack in cement. This measured value of K_c is derived from measured values of g_c, the critical strain energy release rate. RB draw no distinction between the stress on the matrix and the stress on the composite. Their method of getting a theoretical result is clearly a complicated one and in addition a number of approximations are made in the course of the calculations, notably the fibres are taken to be rigid and a penny-shaped crack is considered, and matrix and fibre are considered to be elastic throughout, with no slippage at the interface. No analytical result is given, but a plot of the theoretical cracking stress against wire spacing for various volume fractions of wire and an assumed value of g_c of 3.5 J/m^2 (0.02 in lb/in^2) is presented; it is shown in Fig. 5. The cracking stress depends inversely on wire spacing to the one half power for a given V_f and depends very slowly on volume fraction; in addition the cracking stress is proportional to the modulus of the concrete times its work of fracture to the one half power.

5b ACK Theory

RB give no analytical result, so the accuracy is impossible to estimate but to convince of the identity between the two theories one way is to show that the ACK theory can be put in a form which also predicts a cracking stress proportional to the modulus of the concrete times the work of fracture of the concrete to the one half power, inversely proportional to the square root of the fibre spacing, and depending only slowly on V_f. To do this we note that the RB theory is developed for the fully elastic case, no non-elastic effects being considered in either fibres or matrix. This assumption is quite unrealistic and ACK theory has been developed for the fully elastic case (when it can be compared with RB) only for the sake of completeness [29]. It should normally be used in the non-elastic form [24]. Equation (24) of Aveston and Kelly [28] gives the cracking strain of the matrix in terms of the fracture surface work. From their equation (24), putting fracture surface work $g = 2\gamma$, ϵ_{mu} = cracking strain of matrix, so $\epsilon_{mu}^2 = \sigma_m^2/E_m^2$ and remembering $a = E_m V_m/E_f V_f$, so that $E_c = E_f V_f(1 + a)$, then

$$\sigma_m^2 = \frac{E_m g_m}{(1+a)r}\left(\frac{2G_m E_c}{\psi E_f E_m V_m}\right)^{1/2} \qquad (8)$$

where r is the radius of the fibres.

ψ should have the value appropriate to a square array, to compare with RB which, for instance with $V_f = 5\%$ is $\ln(\pi/V_f)^{\frac{1}{2}} \cong \ln 8 \approx 2$ and the spacing s of the fibres (centre to centre) is related to r as

$$r = s\left(\frac{V_f}{\pi}\right)^{1/2} \qquad (9)$$

and

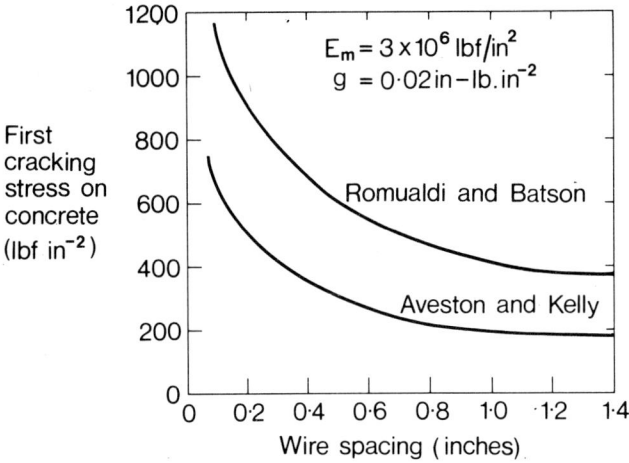

Fig. 5 Comparison of the prediction of the first cracking stress of concrete containing steel wires according to the theories of Romualdi and Batson [23] and of Aveston and Kelly [28].

$$G_m = \frac{E_m}{2(1+\nu_m)}. \tag{10}$$

Making these substitutions we have

$$\sigma_m = \left[\frac{\pi}{2(1+\nu)(1+a)V_m}\right]^{1/4}\left(\frac{E_m g_m}{s}\right)^{1/2} \sim \left(\frac{E_m g_m}{s}\right)^{1/2}$$

$$[V_m(1+a)]^{-1/4} \tag{11}$$

Equation (11) demonstrates that the cracking stress on the concrete according to ACK (elastic theory) is proportional to $(E_m g_m)^{1/2}$, inversely proportional to wire spacing and depends slowly on volume fraction. All of these features are those shown by RB theory. The two theories are compared in Fig. 5. In the fully elastic form they differ by a factor of about 2.0 and both yield very large values of the cracking stress or strain compared with experiment.

ACK theory and RB theory are fundamentally related by the concepts of fracture mechanics. RB is a fully elastic theory and a compeltely elastic analysis cannot apply to cracks in concrete. RB partly realised this since they calculate and comment on the large bond stresses between fibres and matrix which must be withstood if an elastic analysis is to have any meaning. In view of the uncertainty of applying a fully elastic analysis it seems much better to use the ACK formulation with measured values of parameters, such as x', the relaxation distance, derived from experimental measurements of crack spacing. One will then have a reasonably certain prediction based on experiment for the predicted limit of proportionality, the first cracking strain — see Aveston's paper (Paper 6) later in this Conference.

There are papers in the literature by Chan and Patterson [29] which state that the initial cracking strain of a reinforced cement is controlled by the fact that the presence of the fibres physically limits the possible size of initial flaws. It may or may not be so, that the presence of fibres limits the size of the initial flaws. If Chan and Patterson were right, it follows that the cracking strain or limit of proportionality would be altered by making a scratch on the surface of the specimen which is of length greater than the spacing of the fibres: experimentally this is not so. The important point is that cracks cannot extend through the matrix unless the stress on the matrix is sufficiently large to force the crack forward against the resistance offered by the fracture surface work of the matrix.

ACKNOWLEDGEMENTS

I am grateful to Mr. J. Aveston, Dr. G. Dean and Dr. F.J. Lockett for helpful discussions.

REFERENCES

1 **Kelly, A.** *Strong Solids, 2nd Edition.* Oxford University Press (1973)

2 **Black, W.B.,** Editor. *Symposium on High Modulus Wholly Aromatic Fibres. J Macromolecular Science, Chemistry* **7** No 1 (1973)

3 **Dean, G.D.** and **Turner, P.** 'The elastic properties of carbon fibres and their composites', *Composites* **4** No 4 (1973) pp 174–180

4 **Marsh, D.M.** 'Plastic flow and fracture of glass', *Proc Roy Soc A* **282** (1964) pp 33–43

5 **Ewins, P.D.** and **Ham, A.C.** 'The nature of the compressive failure in unidirectional carbon fibre reinforced plastics', *RAE Technical Report* 73057 (1973)

6 **Collings, T.A.** 'Transverse compressive behaviour of carbon fibre plastics', *Composites* **5** No 3 (1974) pp 108–116

7 **Johnson, D.J., Crawford, D.** and **Jones, B.F.** 'Observations of a three-phase structure in high modulus PAN-based carbon fibres', *J Mat Sci* **8** No 2 (1973) pp 286–290

8 **Reynolds, W.M.** *Physical properties of graphite.* Elsevier (1968)

9 **Watt, W.** 'Production and properties of high modulus carbon fibres', *Proc Roy Soc A* **319** (1970) pp 5–15

10 **Hill, R.** 'Theory of mechanical properties of fibre-strengthened materials. 1: Elastic behaviour', *J Mech Phys Solids* **12** (1964) pp 199–212

11 **Hashin, Z.** 'On elastic behaviour of fibre reinforced materials of arbitrary transverse phase geometry', *J Mech Phys Solids* **13** (1965) pp 119–134

12 **Hashin, Z.** and **Rosen, B.W.** 'The elastic moduli of fiber-reinforced materials', *J App Mech* **31** (1964) pp 223–230

13 **Ashton, J.E., Halpin, J.C.** and **Petit, P.H.** *Primer on composite materials: analysis.* Technomic Publishing Co. (1969)

14 **Rosen, B.W.** 'Stiffness of fibre composite materials', *Composites* **4** No 1 (1973) pp 16–25

15 **Goggin, P.R.** 'The elastic constants of carbon-fibre composites', *J Mat Sci* **8** No 2 (1973) pp 233–244

16 **Calcote, C.R.** *The analysis of laminated composite structures.* van Nostrand Reinhold (1969)

17 **Whitney, J.M.** 'A study of the effects of coupling between bending and stretching on the mechanical behaviour of layered anisotropic composite materials', *AFML–TR–68–330*

18 **Cox, H.L.** 'The elasticity and strength of paper and other fibrous materials', *British J App Phys* **3** (1952) pp 72–79

19 **Kelly, A.** 'Composites with brittle matrices', *Frontiers in Materials Science.* Muriel Dekker Inc., Murr, L.E. and Stein, C. editors (in Press)

20 **Christiansen, R.M.** and **Waals, F.M.** 'Effective stiffness of randomly oriented fibre composites', *J Comp Mat* **6** (1972) pp 518–532

21 **Lee, L.H.** 'Strength-composition relationships of random short fibre-thermoplastic composites', *Polymer Eng Sci* **9** No 3 (1969) pp 213–224

22 **Lockett, J.F.** Private communication

23 **Romualdi, J.P.** and **Batson, G.B.** 'Mechanics of crack arrest in concrete', *J Eng Mech Div, Proc American Soc Civ Engineers* 89 No EM3 (1963) pp 147–168

24 **Aveston, J., Cooper, G.A.** and **Kelly, A.** 'Single and multiple fracture', *The properties of fibre composites, conference proceedings of NPL conference*, IPC Science and Technology Press Ltd (1971) pp 15–24

25 **Kelly, A.** 'Constrained cracking of a brittle matrix', *NPL Internal Report* DD(A) 3 (1973)

26 **Majumdar, A.J.** 'Glass fibre reinforced cement and gypsum products', *Proc Roy Soc A* 319 (1970) pp 69–78

27 **Sneddon, I.N.** and **Lowengrub, M.** *Crack problems in the classical theory of elasticity.* John Wiley (1969)

28 **Aveston, J.** and **Kelly, A.** 'Theory of multiple facture of fibrous composites', *J Mat Sci* 8 No 3 (1973) pp 352–362

29 **Chan, H.C.** and **Patterson, W.A.** 'The theoretical prediction of the cracking stress of glass fibre reinforced inorganic cement', *J Mat Sci* 7 No 8 (1972) pp 856–860

30 **Baker, C.** and **Kelly, A.** 'The effect of neutron irradiation on the elastic moduli of graphite single crystals', *Phil Mag* 9 (1964) pp 927–951

QUESTIONS

1 On results obtained with metal matrix composites, by A.S. Wronski (School of Materials Science, University of Bradford, Bradford BD7 1DP)

Could Dr. Kelly comment on some results on a *metal* matrix composite (presented at the Second Liverpool Symposium on Metal Matrix Composites, 1972 — in press) of B.R. Watson-Adams, J.J. Dibb and myself, with reference to his proposal that carbon fibres fail in tension by a shear mechanism. The composite was 53% V_f type 2 Harwell carbon fibre/nickel prepared by RARDE technique of plating and hot compaction. Tensile specimens were tested in tension under superposed hydrostatic pressures ranging to 280MN/m². The mean strength of the individual fibres was $\sim 2.2\,\text{GN/m}^2$ and plating and hot compaction did not appear to degrade their strength. The tensile strength at atmospheric pressure of the composite was 0.73 ± 0.05 GN/m², ie $\sim 60\%$ of the law of mixtures value if allowance is made for the contribution of the nickel. Failure resulted from the fracture of the fibres which was followed by ductile failure of the nickel matrix. Under superposed hydrostatic pressure the net composite tensile strength remained approximately constant, as the maximum shear stress increased. Up to $\sim 140\,\text{MN/m}^2$ superposed pressure the composite failure mechanism resembled that at atmospheric pressure, at higher pressures the failure in the nickel matrix was predominantly intergranular. (When pure nickel was tested, it was, of course, the maximum shear stress at yield that was constant whilst the tensile stress decreased.) These results indicate to us that the critical stage of failure of this metal matrix composite is the tensile stress-operated fracture of the carbon fibres.

Comment by Dr Kelly

The view taken in my paper is that Ewins and Ham have shown that in careful experiments carbon fibres fail in shear. If that is the case (and in analogy with glass of high strength, it could be so) then I expect the upper limit to the strength of graphite to be a similar fraction of the (anisotropic) shear modulus as for glass. I evaluate this and find a value of 6GN/m² as the expected maximum value for the tensile strength of a highly perfect graphite fibre.

In Dr Wronski's interesting experiments, I doubt whether the carbon fibres can be considered as highly perfect; indeed their strength is quite low, less than that of fibres prepared under clean room conditions [**Moreton and Watt**, *Nature* 247 361 (1974)]. In the case of glass, failure initiated from surface flaws does not show the characteristics of a shear failure.

Tension-compression experiments on a fibre reinforced composite of Cu-W

H. LILHOLT

Specimens of single crystal copper with between 1 and 4 volume % of parallel tungsten wires of diameter 20 μm have been tested at room temperature in tension and compression through cycles of constant plastic strain amplitudes (Bauschinger tests). The data have been used to evaluate the (experimental) back stress of the composites; a preliminary comparison with recent theories of back stresses in two-phase materials indicates that the composite at room temperature behaves as a perfect memory solid (no relaxation of dislocation structures).

1 INTRODUCTION

Stresses in two-phase materials have been analysed theoretically [1,2] and measured in dispersion hardened metals like Cu-SiO$_2$ [3]. This report contains some preliminary measurements of stresses (back stresses, image stresses) in fibre reinforced composites of Cu-W.

2 THEORY

The analyses of the stresses in many types of two-phase materials of various geometries have been carried out by Brown and Clarke [4]. We shall use their result for cylindrical fibres in a cylindrical specimen. The image (shear) stress for a material containing fibres with the same elastic properties as the matrix is

$$\tau_{im} = \frac{11 - 8\nu}{8(1-\nu)} \times \mu \times V_f \times \epsilon_p \quad (1)$$

where ν is Poisson's ratio of the matrix, μ the shear modulus of the matrix, V_f the volume fraction of fibres, and ϵ_p the symmetrical plastic strain. Equation 1 refers to single slip of the matrix and is nearly the same as for multiple slip.

The necessary correction for the true elastic properties of the fibres has been worked out numerically [4] and is

$$\text{correction factor} = 0.88 \times \frac{\mu^*}{\mu}$$

where μ^* is the shear modulus of the fibres. The image stress for Cu-W is therefore

$$\tau_{im} = 0.88 \times \frac{11 - 8\nu}{8(1-\nu)} \times \mu^* \times V_f \times \epsilon_p \quad (2)$$

and is seen to be independent of the (shear) modulus of the matrix. This simplifies to a certain extent the use of single crystals of different orientations, although the conversion from shear to tensile properties still depends on the orientation. We shall take, as a first approximation, for the image stress in tension

$$\sigma_{im} = 2 \times \tau_{im} \quad (3)$$

In comparing the theoretical stresses with the experiments we use the method [3] of inverting the stress-strain curve, and measuring the difference in (tensile) stress $\Delta\sigma$ when the inverted curve and the forward (extrapolated) curve are parallel. We then get, taking $\nu = 1/3$ for the matrix,

$$\Delta\sigma = 2\sigma_{im} = 2(2\tau_{im}) = 4 \times 1.37 \times \mu^* \times V_f \times \epsilon_p \quad (4)$$

It should be noted that the theoretical stresses refer to the unrelaxed state of the material, ie no relaxation of dislocation structures has occurred.

3 EXPERIMENTAL

Single crystals of 99.999% Cu were cast with various contents of 20 μm long, parallel W-wires. The crystallographic orientation of the Cu was not the same for all specimens. All tests reported were carried out at room temperature in a special tension-compression rig mounted on an Instron machine. The strain was recorded with a strain gauge extensometer mounted on the gauge section of the specimen. The gauge section was given a length/transverse dimension ratio of about 3 to prevent buckling during the compressive part of the cycles. The cycling was carried out in tension and compression at constant plastic strain amplitude. Three or more cycles at a given plastic strain amplitude gave a good measure of $\Delta\sigma$ at the corresponding forward plastic strain. It was possible to continue with the same specimen to a higher plastic strain level and thus obtain $\Delta\sigma$ at various (forward) plastic strains ϵ_p.

4 RESULTS

Tests were carried out for volume fractions of fibres of about 1, 2 and 4 volume %. The stress/strain curves, as

Metallurgy Department, Danish Atomic Energy Commission, Risø, Denmark

exemplified in Fig. 1, show a work hardening rate which is nearly linear with forward or reverse strain, a cyclic hardening (at a given plastic strain amplitude) which is nearly linear with the cumulative plastic strain, and an image stress, which is linear with the (forward) plastic strain ϵ_p, see Fig. 2. The shape of the stress/strain curves is identical for the forward and reverse parts of the cycles at given plastic strain amplitude, except for the cyclic hardening contribution. This indicates a high degree of symmetry with respect to loading direction for the fibre composites; this is also borne out by the fact that the same image stresses are measured for specimens loaded first into compression.

The slopes of the curves of $\Delta\sigma$ versus ϵ_p (as Fig. 2) are plotted against the volume fraction in Fig. 3, and the slope is found to be

0.64×10^5 kp/mm^2.

In several specimens, which were taken to large values of the (forward) plastic strain ϵ_p, the plot of $\Delta\sigma$ deviates from the straight line, eg Fig. 2.

5 DISCUSSION

We shall limit the discussion to the image stresses. The theoretical correlation between $\Delta\sigma$ and ϵ_p (Equation 4) gives an estimate of the slope of Fig. 3 equal to

$4 \times 1.37 \times \mu^*$

Using the value of the shear modulus $\mu^* = 16 \times 10^3$ kp/mm^2.

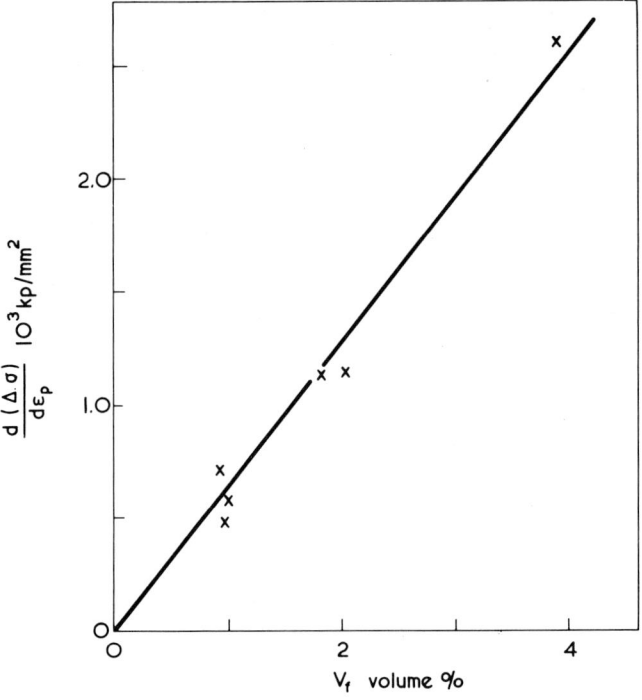

Fig. 2 The parameter $\Delta\sigma$, obtained as described in the text, as a function of (forward) plastic strain ϵ_p: the fibre composite is Cu with 1.8 volume % 20μm W and is tested at room temperature

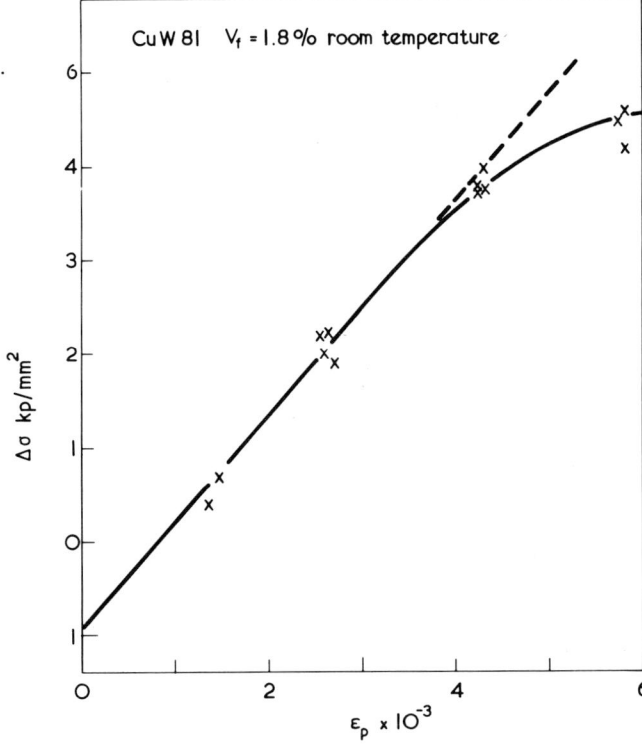

Fig. 3 The slopes (of the straight parts) of curves of $\Delta\sigma$ versus ϵ_p (like Fig. 2) for several fibre composites of Cu with various volume fractions of 20μm W.

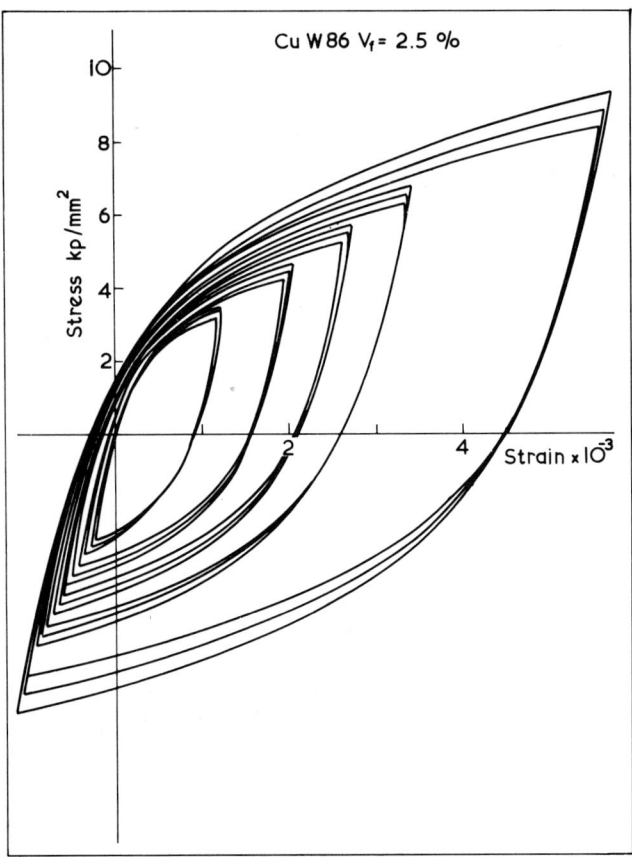

Fig. 1 Cyclic loading of a fibre composite of Cu with 2.5 volume % 20μm W: the test was started at low plastic strain amplitudes

of polycrystalline W we find the theoretical slope to be

0.88×10^5 kp/mm^2.

This is larger than the experimental slope by about 35%. This over-estimate, based on unrelaxed dislocation structures, could be caused either by inaccuracies in the estimate or by some recovery process taking place in the specimens during the test at room temperature.

The estimate for τ_{im} for single slip (Equation 1) is larger than that for multiple slip by about 14%. The conversion of shear stress to tensile stress was done with the assumption of a factor of 2 (Equation 3). This may cause an error in relation to the actual specimens, as well as in relating the experimental (tensile) strain to ϵ_p, the symmetrical plastic strain. The relation between $\Delta\sigma$ and σ_{im} (Equation 4) is not certain, but probably a good approximation. These errors may account for the discrepancy between the theory and experiments.

The experiments themselves may not be a true representation of an unrelaxed material. Experiments on dispersion hardened Cu-SiO$_2$ show recovery beyond a certain ϵ_p [3]; this process is understandable in terms of the dislocation manoeuvres which can take place at spherical particles [1,2]. It is difficult to imagine dislocation manoeuvres at fibres [5] which can be so effective and extensive as at spherical particles. A preliminary conclusion is therefore that the *straight* line of $\Delta\sigma$ versus ϵ_p indicates a true unrelaxed material, and that the discrepancy in the slopes is caused by inaccuracies in the theoretical estimate of τ_{im} and the correlation between τ_{im} and $\Delta\sigma$.

Some of the curves of $\Delta\sigma$ versus ϵ_p show a decreasing slope beyond a certain ϵ_p. This deviation from the straight line is not very well defined. This type of deviation indicates for dispersion hardened Cu-SiO$_2$ that relaxation processes start [3]. This is also a possibility for the fibre composites of Cu-W, although the necessary dislocation manoeuvres are not clear. It is also possible that the deviation from the straight line of $\Delta\sigma$ versus ϵ_p for Cu-W indicates a change in behaviour of the fibres. In dispersion hardened materials yield and fracture of the particles is not possible, while in fibre composites both these processes often occur. The deviation from the straight line occurs at $\epsilon_p \sim 4 \times 10^{-3}$; the total strain carried by the fibres is higher than this value by the elastic strain suffered by the composite; at the stress level in the composites this elastic strain is of the order 0.5×10^{-3}. As the fracture strain of the fibres is 1.1–1.2% [6] it is unlikely that fracture of fibres takes place. The yield strain of the fibres is of the order 0.4% [6], and it is therefore likely that the fibres yield in the composites during the cyclic loading to high plastic strains. Furthermore, a preliminary comparison of the shapes shows that the curve of $\Delta\sigma$ versus ϵ_p is very similar to the stress/strain curve of the fibres beyond the yield strain. This is partly expected because the properties of the fibres enter the expression for the image stress (Equation 2), while the image stress is rather unaffected by the matrix properties.

We therefore conclude that the deviation of the curve of $\Delta\sigma$ versus ϵ_p shows the change of behaviour of the fibres and not the onset of a relaxation process for the dislocation structures.

6 CONCLUSIONS

The experiments have demonstrated that fibre composites can be used to study the stresses in two-phase materials, especially the image stress. A comparison between theory and experiments indicates that Cu-W composites with 1–4 volume % fibres behave as unrelaxed materials, and that the behaviour of the fibres affects the image stress.

ACKNOWLEDGEMENTS

It is a great pleasure for the author to acknowledge the hospitality received during his stay with the Metal Physics Group in the Cavendish Laboratory, Cambridge, where the experiments were carried out.

REFERENCES

1 **Brown, L.M.** and **Stobbs, W.M.** 'The work-hardening of copper-silica: I', *Phil Mag* **23** (1971) pp 1185–1199

2 **Brown, L.M.** and **Stobbs, W.M.** 'The work-hardening of copper-silica: II', *Phil Mag* **23** (1971) pp 1201–1233

3 **Atkinson, J.D., Brown, L.M.** and **Stobbs, W.M.** 'Recovery and the Bauschinger effect in Cu-SiO$_2$', *3rd Int Conf on the Strength of Metals and Alloys* **1** (1973) pp 36–40

4 **Brown, L.M.** and **Clarke, D.R.** Work to be published

5 **Kelly, A.** 'Reinforcement of structural materials by long strong fibres', *Metall Trans* **3** (1972) pp 2313–2325

6 **Lilholt, H.** 'Work hardening and fibre reinforcement', Thesis, Cambridge University (1968)

Prediction of properties for engineering design with composites

U. HÜTTER

If it is required to predict the properties of composite structures so that engineering design calculations may be simplified, then it is necessary to question not only the reliability of analytical methods but also the interpretation of the many failure phenomena which influence the optimum utilisation of available fibre strength in advanced composites. Before discussing such problems, I would like to make a few quite personal remarks. Our team in the Stuttgart DFVLR research institute, as well as the group working in our university facilities on composite research, are looking at these problems from the standpoint of the direct application of results to engineering design and/or to tasks, derived from design problems. Thus we consider ourselves as engineers who are trying to find out concise answers to certain well defined questions on component applications which have to comply with stated requirements.

In pursuing these objectives quite unconventional methods proved to be helpful (see Fig. 1), such as the use of macroscopic models of microscopic structures, or the application of a ministrength test-machine fixed on the object carrier of a photomicroscope. The latter enabled us to observe, amongst other things, the phenomena of initial cracks in multi-layer composites under realistic conditions with loaded specimen and slowly increasing load (see Fig. 2). This is our way of looking at the problems, being sometimes a little suspicious of idealisations which may force the interpretation of the behaviour of complex multi-parameter, coupled composite-systems onto the Procrustes-bed of closed analytical solutions.

1 UNIAXIAL COMPOSITE EXTENSION STRENGTH SCATTER

Certainly everybody involved in composite research will consider the inevitable high scatter of any data from composite strength tests as an undesirable penalty in the applicability of data to structures. This holds even for results from a unidirectional composite test specimen stressed in simple tension [1,2], where the scatter is ± 6% to ± 8%. This is the case for the whole range of volume fractions reasonable for application in structures. Under bad conditions the scatter sometimes exceeds ± 12% or even more (see Fig. 5). These data may be correlated with the effect of fibre diameter on failure probability due to the statiatical distribution of flaws on the fibres [3], and thus to the evaluation of statistical strength data, gained by experiments with single fibres referring to fibre length as parameter. Such results show that the scatter of strength data of composites theoretically should be

Institut für Flugzeugbau der Universität Stuttgart, 7 Stuttgart 80 (Vaihingen), Pfaffenwaldring 31, W. Germany.

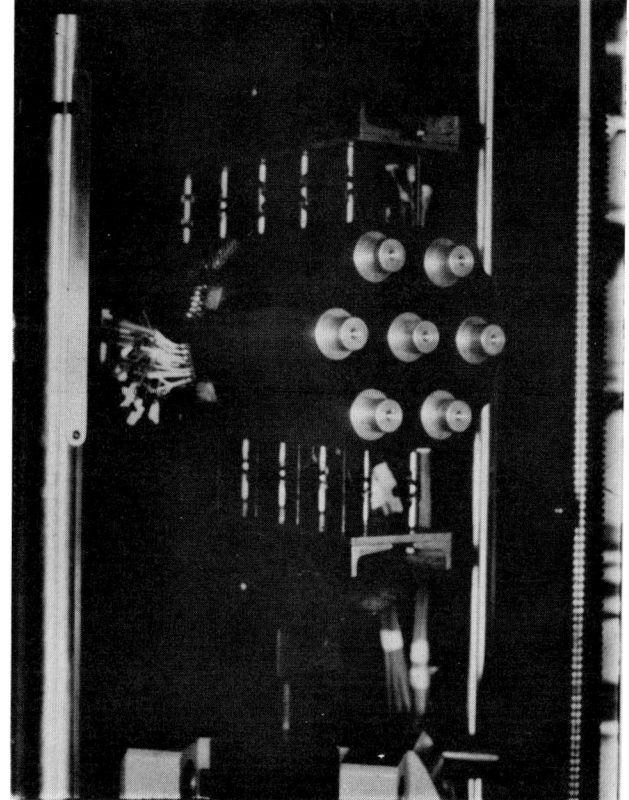

Fig. 1

smaller than that gained from single fibre tests (Fig. 3). Stellbrink [4] showed in an unpublished study that this holds independently of whether a regular hexagonal distribution with constant fibre diameter is assumed or a statistical distribution of fibres in the matrix with unequal fibre diameters (Fig. 4). Moreover the correlation of the modes, that is the maximum gradients of the probability versus fibre strength diagrams, related to fibre strength in the composite on the one hand, and single fibre strength on the other, gives values which are as good as the modes, related to single fibres having a length equal to the 'transfer length'. The transfer length is defined as the distance from the plane of crack of one individual fibre in the composite to the plane, where in the neighbouring fibres the difference from elevated stress to the average stress reaches zero (see Fig. 4). The actual values of the models measured in composites are still higher than those obtained from semi-theoretical analysis — based on single fibre strength statistics (see Fig. 3) [5,6].

Prediction of properties for engineering design with composites

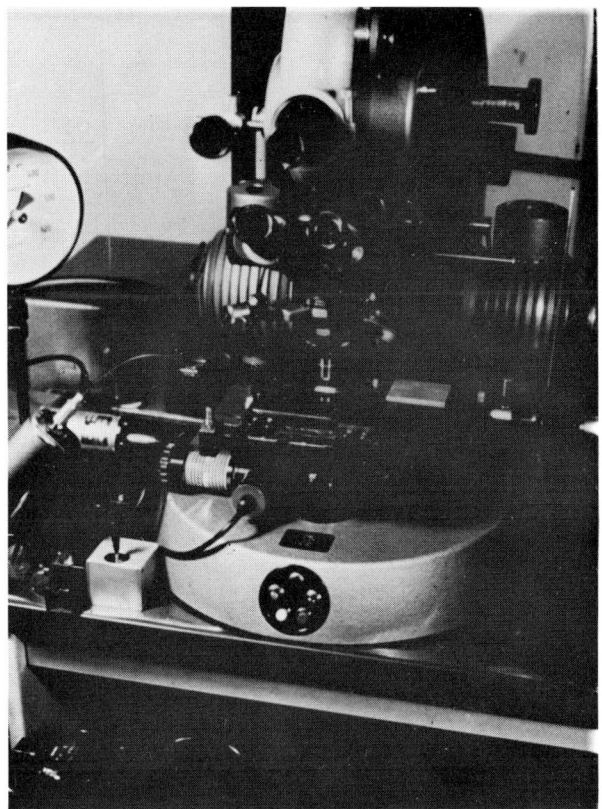

Fig. 2

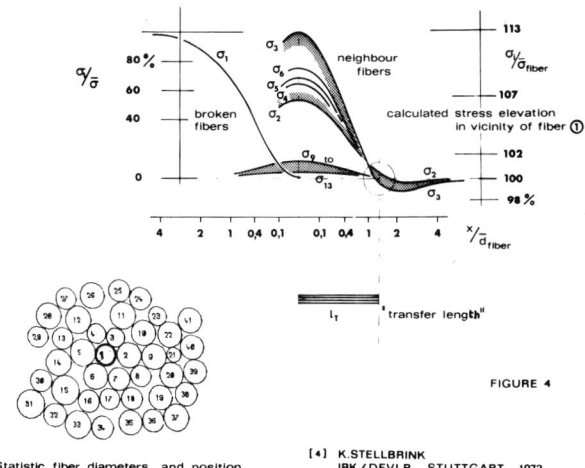

Fig. 4

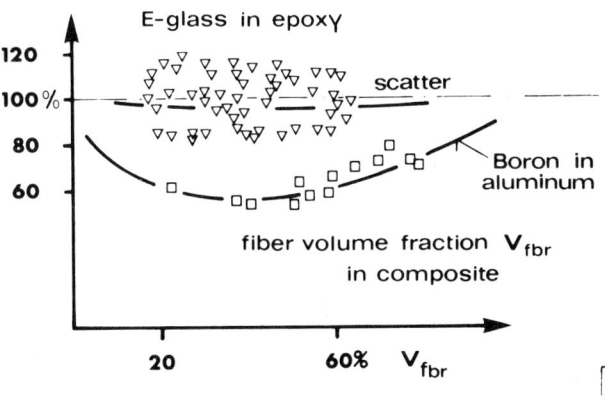

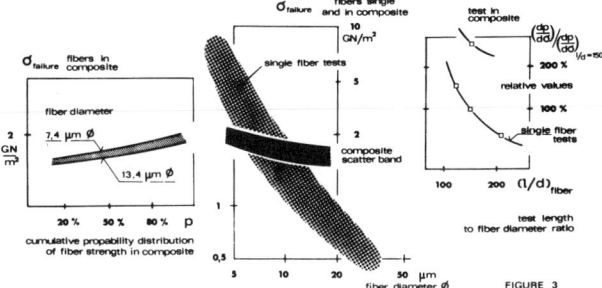

Fig. 3

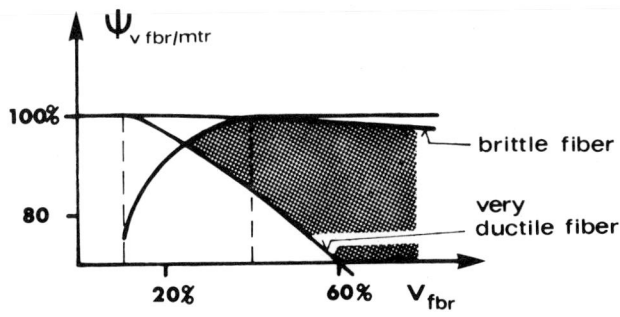

Fig. 5

Thus it seems in the light of much experience that the avoidance of a large scatter in composite material strength values is more a question of improvements in exact test-specimen dimensions, tight tolerances, clean fabrication, uniform quality of fibres and matrix materials, etc than a consequence of certain physical conditions.

2 IMPROVED MIXTURE RULE

Another question raised recently is that of the validity of the simple mixture rule for unidirectional, straight fibre composite elements (Fig. 5). An improved mixture rule formula takes into account, in addition, functions $\psi_{v(mtr/fbr)}$ of logarithmic measure of plastic deformation $\phi = \ln(1 + \epsilon)$ and a so-called strengthening coefficient $n_{mtr/fbr}$ [7]. The quotient of the total cross sectional areas unloaded and after plastic contraction must also be considered to distinguish between true stress and that which is applied in engineering design. Thus we obtain.

$$\sigma_{cmp}^{(f)} = \sigma_{mtr}^{(f)} (1 - V_{fbr}) \psi_{v(mtr)} + \sigma_{fbr}^{(f)} V_{fbr} \psi_{fbr}$$

Now ψ_{matrix} reaches values close to one – even with quite ductile matrices – if the fibre volume fractions are $V_{fbr} \geq 40\%$, whereas ψ_{fibre} is only close to one if brittle fibres are applied.

Only composites consisting of ductile fibres in ductile matrices deviate distinctively from the usual simple rule of mixtures in the range of fibre volume fractions which are of any interest for structures with high strength to weight ratios.

3 UNIAXIAL COMPOSITE COMPRESSION STRENGTH

In 1963 Kossira published, in Stuttgart, his thesis containing a careful investigation of glass fibre epoxy composite compressive strength. He used specimens, containing parallel straight fibres, loaded in the fibre direction (Fig. 6) (see [6]). In his investigation neither the ratio of the specimen length to the root of the cross-sectional area nor the shape of the cross-section showed any significant influence on the compressive strength of the specimen. At that time it was surprising to us that not even the fibre content in the composite — which was varied between 10 and 67% — proved to have noticeable influence on the compressive strength related to the cross-sectional area of fibres. We didn't expect this result, as the strength had been calculated assuming buckling of the straight fibres as elastic columns supported by the matrix.

Observations performed in 1962 on a macroscopic model consisting of transparent polymethylmethacrylate blades, representing the fibres and sheets of rubber, representing the matrix, (Fig. 7) have shown that — at least with fibre contents more than 20% — transverse buckling is not dominant. In addition it was shown that the wave-length of the shear buckling-mode is not significantly influenced by the fibre stiffness and the matrix Young's modulus, as is the case with the transverse mode but by certain boundary conditions in the specimen (see [6]). This is similar to the two dimensional solutions of Rosen and Schuerch, published in 1964 and 1966 [8,9]. These solutions indicate that the critical fibre stress to produce failure increases with the matrix shear modulus G_{mtr} and with the increasing volume fraction of fibres. This tendency has been confirmed by compressive strength tests with boron/epoxy composites performed by Lager and June in 1968 (see Fig. 6 [10]). Their specimens were laterally supported by an aluminium honeycomb core. These tests and those on which Levenetz reported 1965 [11] were performed with fibres of 50 to 130 μm diameter. Such large diameter fibres can be laid up and oriented into the matrix more easily and with more accuracy and regularity. This seems to be an additional reason why in these tests higher strengths have been reached under compression than under tension.

In 1967 Herrmann, Mason and Chan carried out a theoretical investigation on the response of reinforcing wire to a compressive state of stress. They took into consideration imperfections which occur in technical processes, assuming as the most critical state of initial imperfections that they have the shape of the buckling mode itself [12]. Any shear buckling analysis published so far is based on the ratio of the width

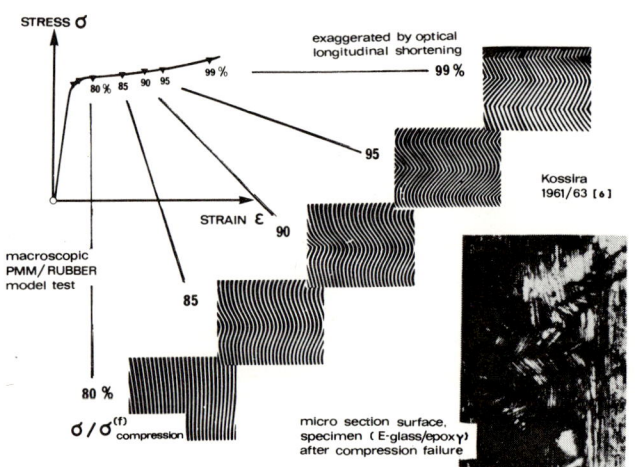

Fig. 7

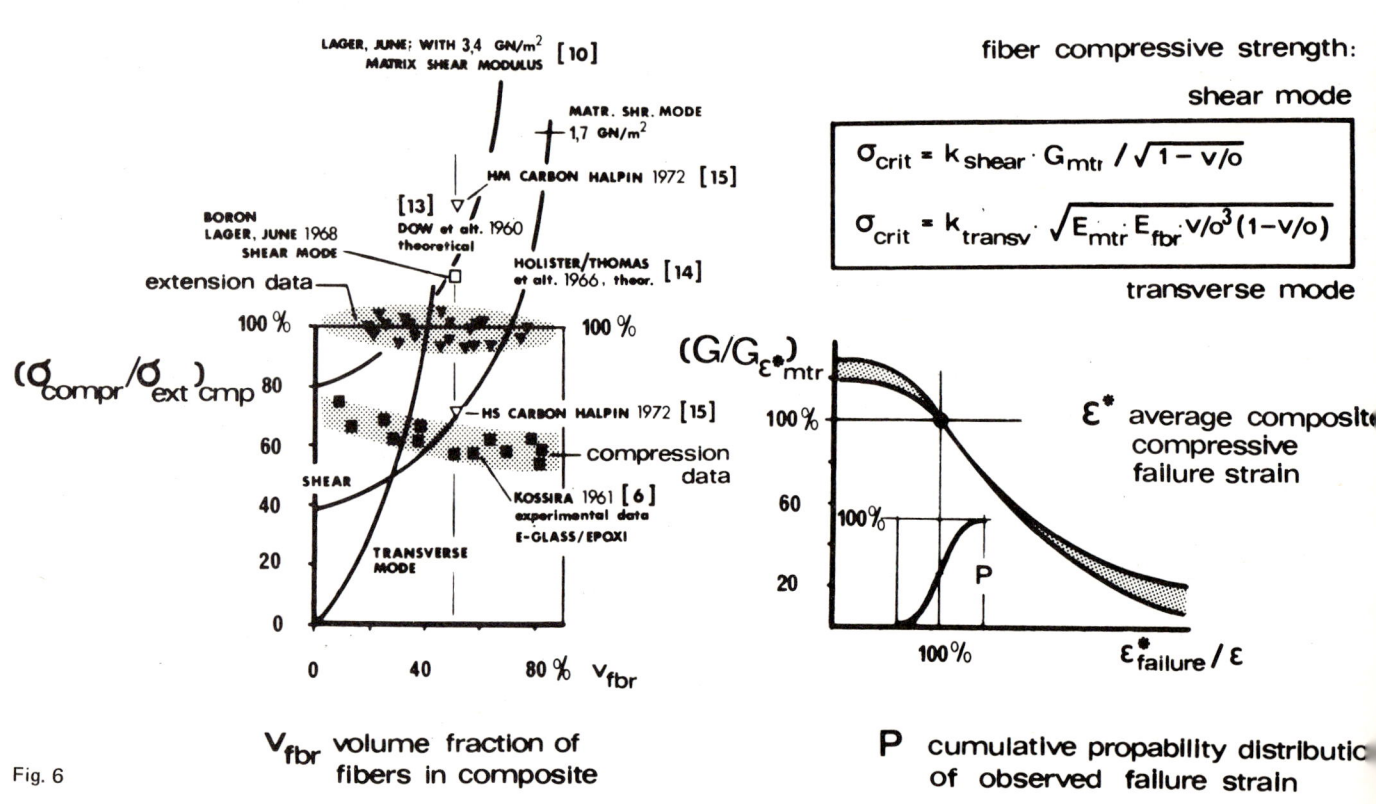

Fig. 6

fiber compressive strength:

shear mode

$$\sigma_{crit} = k_{shear} \cdot G_{mtr} / \sqrt{1 - v/o}$$

$$\sigma_{crit} = k_{transv} \sqrt{E_{mtr} \cdot E_{fbr} \cdot v/o^3 (1 - v/o)}$$

transverse mode

ε^* average composite compressive failure strain

v_{fbr} volume fraction of fibers in composite

P cumulative propability distribution of observed failure strain

of the narrowest matrix bridge between fibres to the fibre diameter, at a given volume fraction of fibres in the composite. All matrix shear deformations are caused by the angular deviations of fibres from the direction of the loading vector. At a given fibre deviation, these shear deformations are evidently bigger the narrower the matrix bridge and thus the higher the fibre volume fraction in the matrix [13–15]. As a consequence of this shear angle increasing effect, angles big enough to reach the yield limit of a matrix-material are expected with a hexagonal array at 0.7° of fibre deviation from the direction of the load vector at 40% volume fraction of fibres. At 60% volume fraction the limit-angle for matrix yield is less than 0.4 — all these values are related to epoxy/resin yield behaviour. This means that small deviations arising from practical imperfections must cause shear yield at higher fibre contents ab initio. Since any fibre distribution in a composite is a statistical one, we must expect yielding as a consequence of shear deformation within the matrix due to misorientation or angles of deviation even at lower values of fibre volume fractions. On the other hand we have to consider that many fibres within the cross-section of a composite may have on one side a relatively wide matrix band and on the other almost zero-distance from neighbouring fibres. Thus mixed buckling modes can be expected even at fibre volume fractions of technical interest, especially as the half-wave-length of fibre deviation under a transverse buckling mode is, at all reasonable fibre volume fractions, only a few fibre-lengths (Fig. 8), [16,17]. As a consequence individual fibre-deviations can cause comparatively big deviation angles locally. Thus the statistics of fibre-distribution and orientation in the composite have to be taken into account for any correct intepretation of composite failure phenomena under compression.

Fig. 8

Fig. 9

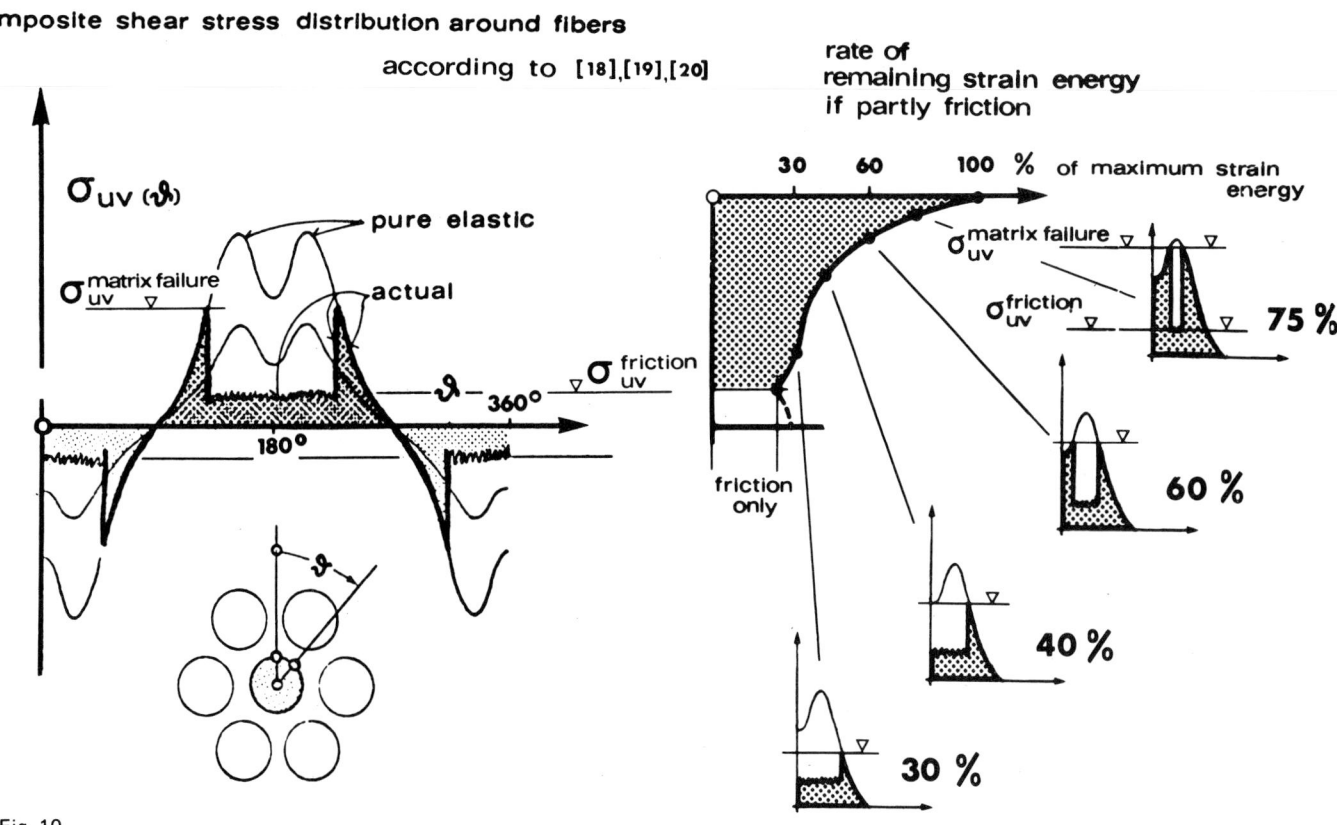

Fig. 10

We can summarize: if by whatever mechanisms somewhere in a composite body under compression a slight sinusoidal deviation mode is once initiated (Fig. 9), the shear angles there shift and remain far beyond the matrix yield point. Then the gradient of the shear modulus versus shear angle decreases, and finally the shear stress reaches a final constant value. Therefore the critical fibre stress is less influenced by fibre volume fraction than was earlier assumed — at least for fibre volume fractions in technical applications (Fig. 10) [18–20]. Additional observations are that failure under compression occurs mostly in limited regions within a structure, causing relatively large deformations locally, thus leading to post-buckling — failure phenomena other than those described above, including part or entire rupture and/or detachment of the matrix from the fibres.

4 INITIAL MATRIX CRACKS

To derive the optimum benefit from composites for structures we must accept the occurrence of initial cracks in certain layers of multilayer composite shells at certain loading conditions. As the transfer of considerable shear deformation in composite shells requires multilayer, crossply patterns, there always exist layers in such shells which are subjected to strain perpendicular to the fibre direction and to strains of a magnitude which cause initial cracks (Fig. 11). If such layers could not be supported by neighbouring lamina, any initial matrix cracks would, with increasing transverse strain, run through the whole panel to both edges and split the layer into a number of narrow strips only poorly connected together. The magnitude of the transverse strain which would produce such catastrophic through-running cracks is only approximately 10% of the failure strain in the fibre direction at about 70% fibre volume fraction in the composite and about 16% at fibre volume fraction of 50%. These figures are correct for glass fibres in epoxy resin and are little different for carbon or boron fibres in the same resin (see [20,21]). If neighbouring layers exist with sufficiently large fibre crossing angles and are plied together with the transversely strained layers, the development of cracks with increasing strain is less detrimental. Moreover the cracked layer retains a certain loadability in the direction of the indicated extension and there is a tendency to establish stable crack patterns under certain conditions. The reason is that the cross ply layers support the transversely extended ones preventing them from extending an unlimited amount.

Grüninger reports, in a still unpublished study on crack pattern stability, that, especially with a fully developed ideal diamond crack pattern, such layers can reach, without disintegration in the direction perpendicular to the fibres, values of extension even beyond the failure strain of fibres extended in the fibre direction [22]. The investigation of potential crack pattern stability is being performed by macroscopic structure models, by elasto/plastomechanic analysis, by controlled extension tests on specimens and by photoelasticity methods. A theoretical study showed that an unstable phase exists in 90° cross-ply composite laminates at the start of extension, where even at constant strain the cracks tend to run through the whole layer in a similar manner to that described earlier for single layer panels. The analysis indicates that the panel should be unloaded to a smaller extension than that which caused the first crack. After that further loading would be possible until the next crack appears — and so on — until the point of stability is reached. From that point on the cracks remain stable, which

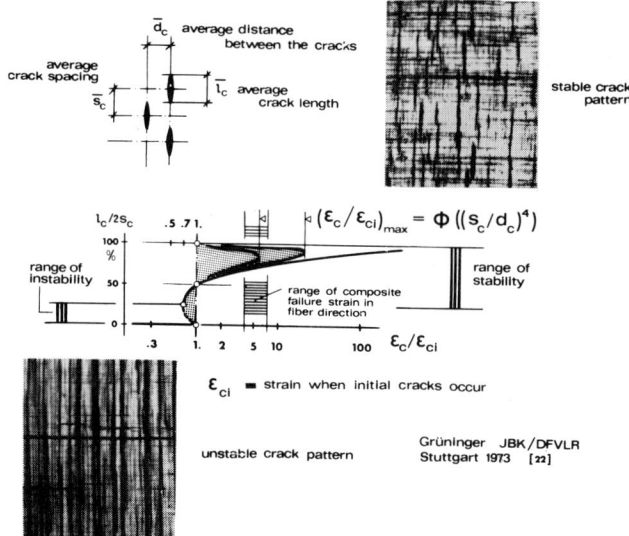

Fig. 11

means their tendency to run through the whole layer is considerably reduced. Tests with specimens of different size and wall thickness show that these two parameters influence the appearance and stability of the cracks. Moreover it seems that with better adhesion between matrix and fibres, the tendency to build up longer cracks is less than with poor adhesion.

It has been observed and reported earlier that the cracked layers keep, to a surprisingly high extent, their ability to bear compressive loads in the fibre direction, whereas the ability to bear a tensile load in the fibre direction remains almost entirely unchanged. In 1971 Weissinger investigated this question of the remaining compression strength of tranversely extended layers in a series of experiments with three layer 90° crossply specimens [23]. Concerning such specimens Puck et al reported in the early sixties on experimental findings related to the nature of an abrupt bend or 'knee' in the stress/strain diagram if transversely extended [24]. However, if these 90° unidirectional crossply specimens are tested in tension by cycling several times from zero-load up to a high percentage of maximum load, the abrupt bend disappears and the stress/strain diagram is almost reduced to that of a unilayer specimen extended in the fibre direction (Fig. 12). From this follow two consequences: the first is that the strength contribution of the layer perpendicular to the loading direction is not only basically small but tends to go to zero after a few load-cycles with relatively high load (see [21,24]). Secondly, that the initial contribution of the 90° layer can be extracted from the first cycle diagram. The compression behaviour perpendicular to the fibres is different: no marked abrupt bend or knee in the stress/strain diagram can be observed, and the reduction of the contribution of the 90° layer concerning the total strength of the two layer composite is reduced less by cycling than is the case if the specimen is extended (Fig. 13). Layers in a multi-layer composite with fibres at much smaller angles than 90° to the axis of symmetry — for example as shown on Fig. 13 — and loaded in the direction of the axis of symmetry, also endure extensions and compressions perpendicular to the direction of fibres.

Prediction of properties for engineering design with composites

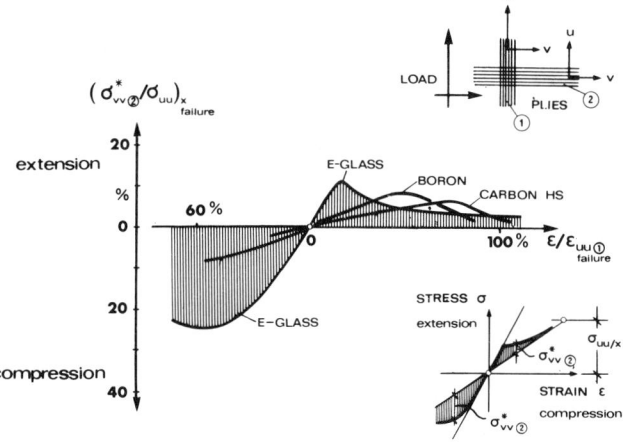

Fig. 12

5 ADVANTAGES OF APPLYING 'SUPPORTING LAYERS'

To have in the main load-bearing layers fibre angles related to the axis of symmetry — or main load — by just ± 20 – 30° seems to us to be a great advantage, especially if in addition a layer group exists which we call the 90° supporting layer. In this configuration there results a state of compression perpendicular to the load-bearing fibres in the case of extension in the main load direction, and just a slight extension of them, if the main load is compression. Theoretical findings and a large test programme with tubular and flat specimens indicated, as reported in [20,25,26] that the above mentioned supporting layers result in a considerable improvement of strength to mass ratio for light structure composite shells (Fig. 14). Optima do exist for the volume fraction of supporting layers in a three layer shell — these optima are functions of the angle ω of the fibre directions in the bearing layers to the axis of symmetry.

As flat specimens are much less expensive than tubular ones, the discrepancies between the results gained from flat and from tubular specimens have been calculated and tested. A discrepancy arises from the fact that over the width of a flat specimen of the three layer-type under symmetric, uniaxial load, the stress is not constant but decreases to both edges, where the interlaminar shear which causes the compatibility of both layers, reaches its maximum. Experimental evidence showed (see Fig. 14) [27,28] that clamping of the edges by a number of independent small clamps gave better strength results than when unclamped. A group of tests dealing with this phenomenon was aimed at the practical problem of assembling structural bodies from separately produced composite panels.

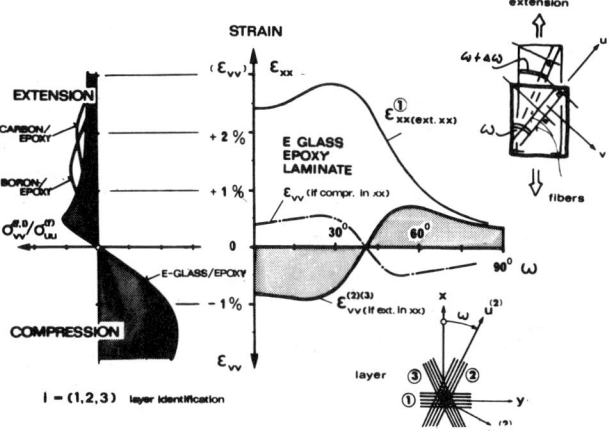

Fig. 13

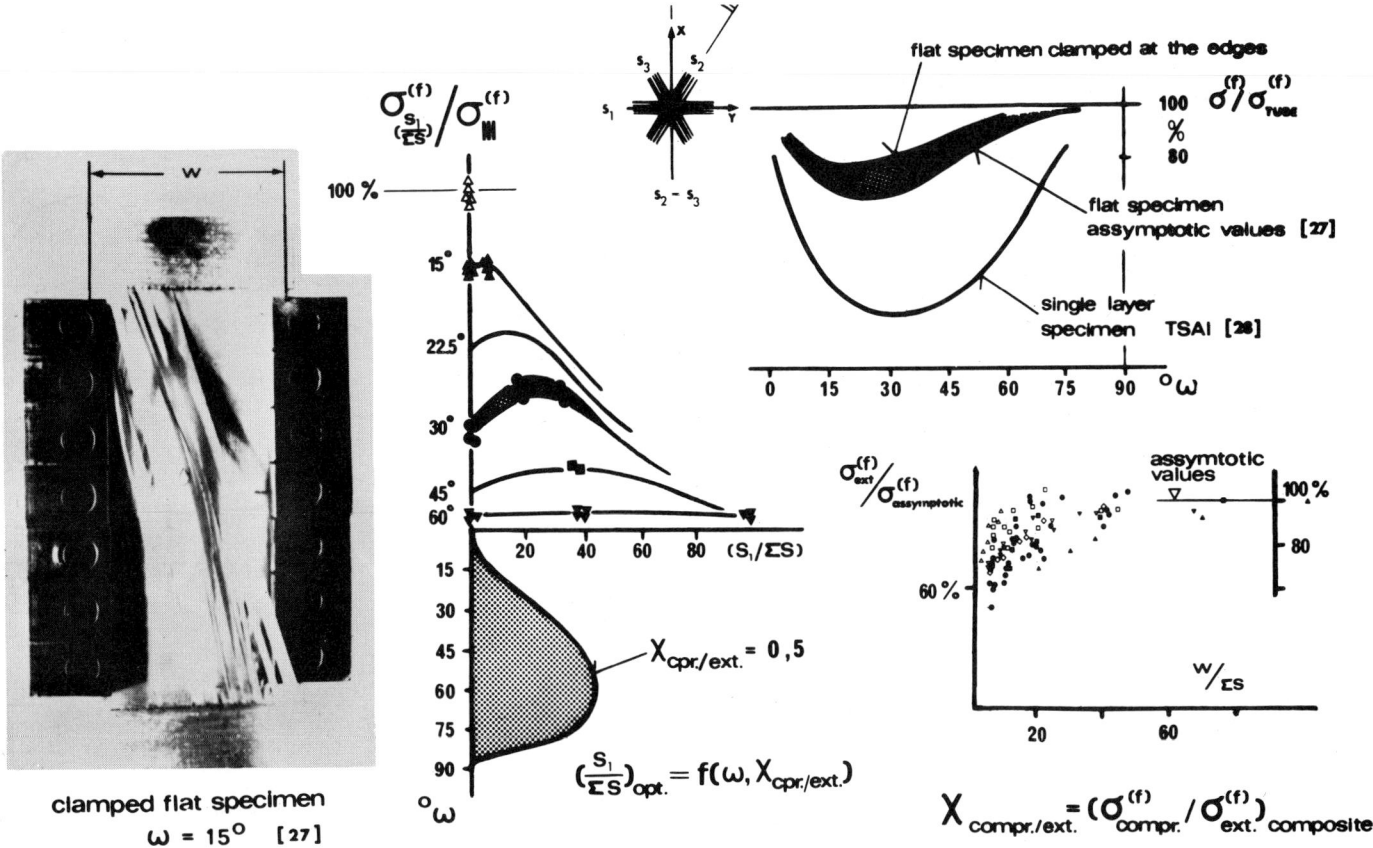

Fig. 14

6 LAYER COUPLING EFFECTS

The same type of basically three-layer composite shells have been subjected to some other tests, among which the influence of the magnitude of load steps on the elastic constants have been of particular importance [29]. Experiments with relatively large specimens, with slenderness ratios beyond 120 (see Fig. 14) and optimal volume fractions of supporting layer with the load-bearing layer angle as parameter showed that the Young's moduli in the specimen symmetry axis was considerably influenced by the magnitude of load between 32° and 55°, with the greatest influence between 40° and 45°. The influence on the Poisson's ratio had a maximum in the vicinity of 30° degrees (Fig. 15) [29,30]. A plausible explanation of this phenomenon can be derived from the assumption that with increasing load in certain interlayer matrix-lamina — especially at the edges of a multilayer shell — the shear stresses which effect the strain compatibility between the layers of different fibre direction surpass the yield point. Thus a slight plastic slip occurs with the effect that one of the layers seems to behave as if it had a lower Young's modulus than it in fact has.

This physical explanation can be checked by an elastomechanic analysis in which a coupling coefficient ψ_c is inserted into the Lamé [31], Voigt [32] equations, where the quasi-diminution of the respective layers' Young's modulus is indicated. If that coefficient is defined as a function of the specific load on the composite, a quite good correspondence with the experimental results can be obtained.

If we assume that

$$\nu_{uv} \approx \nu_{vu} \cdot E_{vv}/E_{uu}$$

we get, for the coefficient matrix C_i defined by $\sigma_i = C_i \cdot \epsilon_{ji}$

$$C_i = \begin{bmatrix} \dfrac{E_{uu} \cdot [\psi_c]}{1 - \nu_{uu}^2 \cdot E_{vv}/E_{uu} \cdot [\psi_c]} & \dfrac{E_{vv} \nu_{vu}}{1 - \nu_{vu}^2 \cdot E_{vv}/E_{uu} \cdot [\psi_c]} & 0 \\ \dfrac{E_{vv} \cdot \nu_{vu}}{1 - \nu_{vu}^2 \cdot E_{vv}/E_{uu} \cdot [\psi_c]} & \dfrac{E_{vv}}{1 - \nu_{vu}^2 \cdot E_{vv}/E_{uu} \cdot [\psi_c]} & 0 \\ 0 & 0 & E_{uv} \end{bmatrix}$$

$$\sigma_{ki} = T_{\sigma,\omega_i} \cdot \sigma_{ji}$$

and $\epsilon_k = \epsilon_{ki}$

as well as

$$\sigma_k = i \sum_1^n (\sigma_{ki} \cdot s_i / i \sum_1^n s_i)$$

we get

$$\sigma_{ji} = C_i \cdot T_{\sigma,\omega_i}^t \cdot D^{-1} \cdot \sigma_k$$

where

$$D = i \sum_1^n (s_i / i \sum_1^n s_i) \cdot T_{\sigma,\omega_i} \cdot C_i \cdot T_{\sigma,\omega_i}^t$$

$[\psi_c]$ is the layer coupling coefficient, and it appears only in the matrix C_i. See also Fig. 16.

We assume that the description of experimental results by such a procedure not only avoids a more complicated plasto-

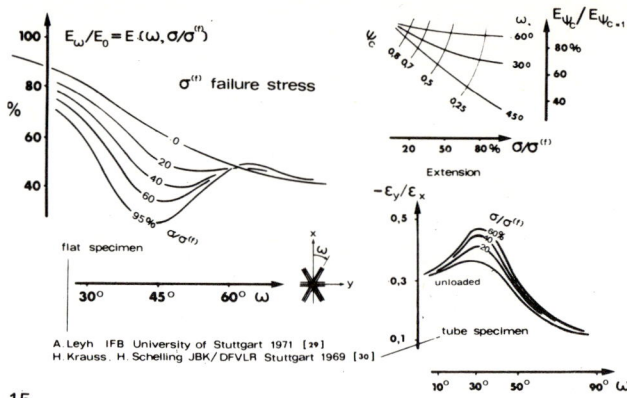

Fig. 15

Fig. 16

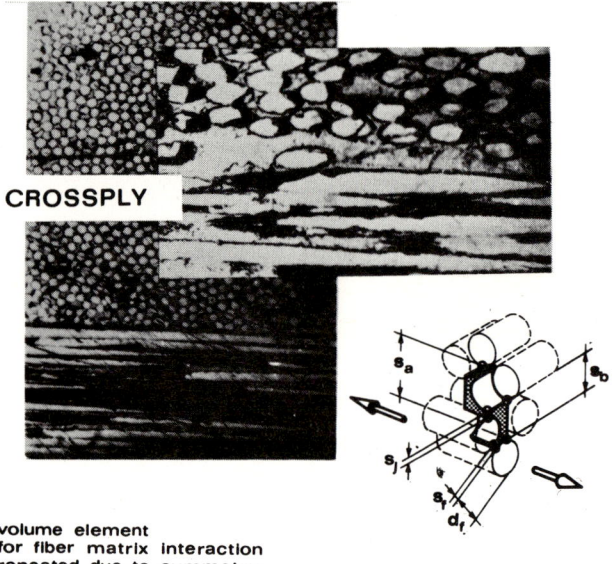

volume element for fiber matrix interaction repeated due to symmetry in joint surface between crossing layers

Fig. 17

mechanic calculation requiring many iterations but also leads to a better understanding of the coupling mechanism itself — a mechanism which is based on an inter-layer composite lamina which in fact is a sort of a cross ribbed plate (Fig. 17).

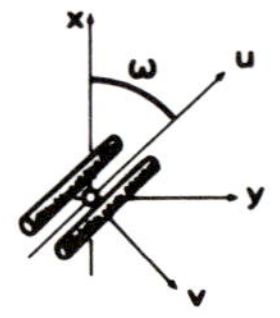

7 SYNOPTIC PRESENTATON: DESIGN CHARTS

If we now try a synoptic presentation (Fig. 18) of the different types of loading conditions and the failure behaviour of a multilayer composite shell by plotting the respective stress/strain curves and the Young's modulus strain curves for the main directions of the fibre axis within the layers, we see inter alia that we can either calculate the shell behaviour as

Fig. 18

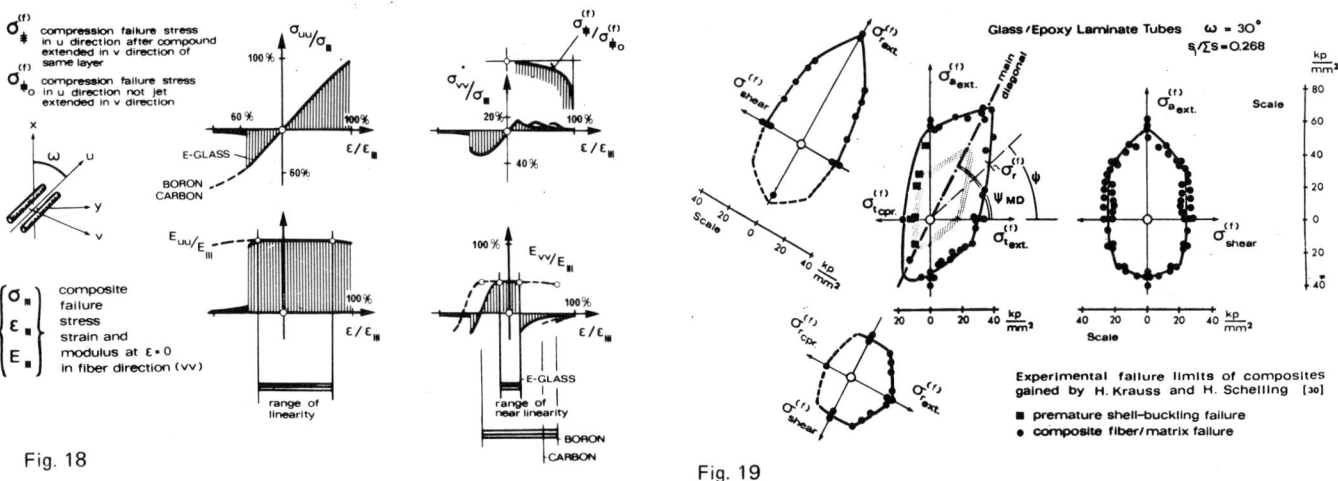

Fig. 19

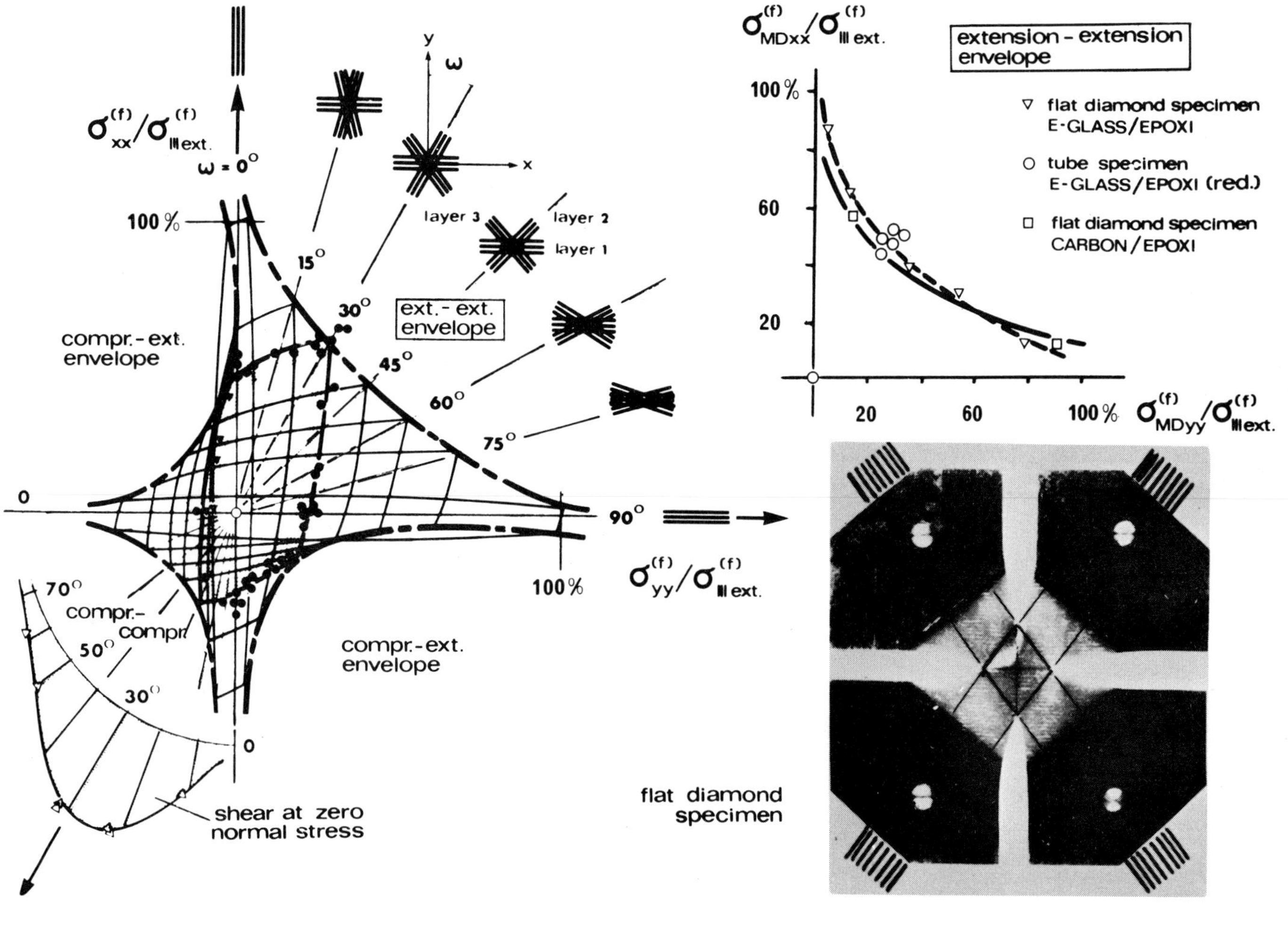

Fig. 20

strictly elastic within more or less narrow limits, or we have to take into consideration all the special nonlinear effects, thus performing the calculations of the expected failure limits by iteration, accepting the higher amount of calculation required. We believe that it would be helpful for industrial applications to give engineers — at least for primary design studies — a sort of design chart which describes the failure limits of biaxial stressed three- or four-layer composite shells (Fig. 19 (see [30]) and Fig. 20). But the experimental checking of such design charts for a broad variety of geometric and material parameters is a problem.

8 THE 'DIAMOND SPECIMEN'

A reliable experimental investigation of failure limits for composite shells using tubular specimens extended or compressed in the axial direction with internal or external pressure to stress them in the tangential direction requires a large number of specimens, process calculations and time. We have therefore looked for a simpler way to present the essential points of the failure envelope, if possible using flat specimens. Our experimental evidence, as well as the results of the Puppo/Evensen 'System of equations' [33], indicates that, for a two-layer-group composite shell, some especially important points exist in the envelope. We mean those points where the Puppo/Evensen ellipses cross each other, making 'corners': these are of extraordinary significance for the shape of the failure envelope. Four of these corners exist in the type of envelopes we have investigated so far. The connection of the corners to a central point of the diagram we call the 'main diagonals'. We call a layer group those sets of layers which are symmetric in any respect to the layers' axis of symmetry in the sense that the plus/minus layers are.

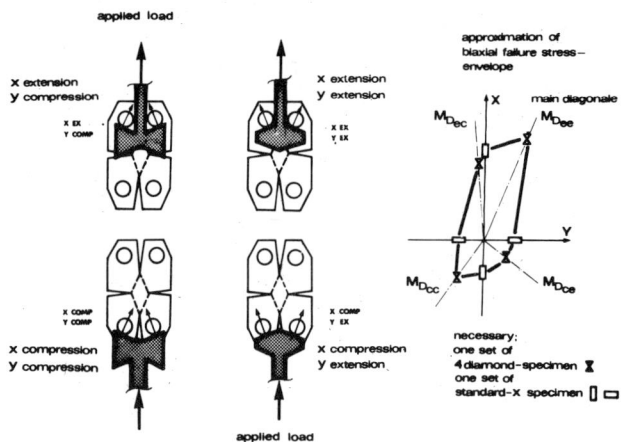

Fig. 23

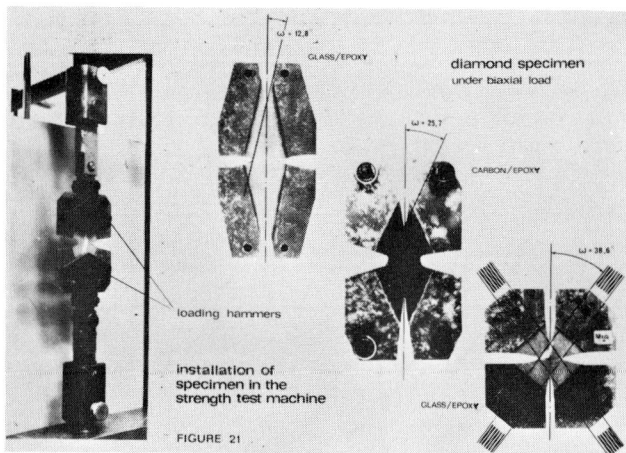

Fig. 21

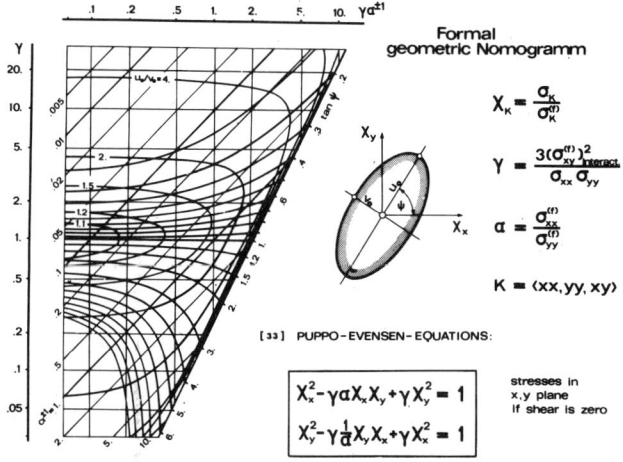

Fig. 24

Fig. 22

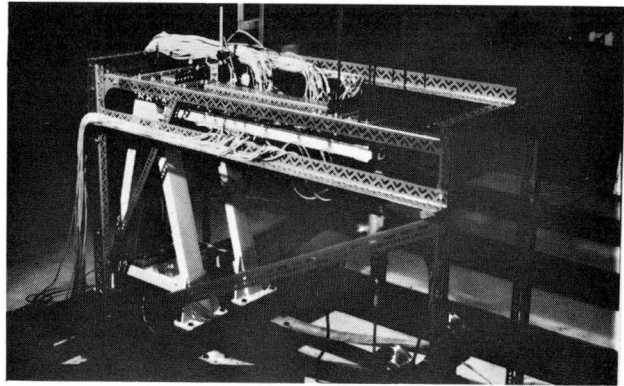

Fig. 25

Thus a type of specimen has been developed which takes advantage of the fact that for composite shell elements of the two-layer-group type, the main diagonal direction is easy to calculate and for the extension/extension case quite close to the direction of the fibres in the main bearing layers (Fig. 21). Sets of ball bearings connected by simple bolts to the grips of the specimen are able to apply, by simple hammer-shaped or butterfly-shaped tools, loads to the specimen whose load vector directions differ widely from the direction of the extension or compression which can be applied directly from an ordinary strength test-machine (Fig. 22). As the area of nearly constant stress in the specimen is shaped like a rhombus, we call this type a 'diamond specimen'. Pilot tests with them proved that it is possible to obtain values from them which fit quite well to the calculated main diagonal points envelope for a given type of symmetric composite shell (Fig. 23). The main advantage of this technique is that only four sets of diamond specimen are needed and in addition four sets of standard specimen of the type usually used for unidirectional strength tests, to get within a fairly good approximation to the whole failure envelope of a certain type of composite shell. The relation of the strength-tests results obtained by this method to the Puppo/Evensen equations can be easily found from nomographic methods (Fig. 24) (see [33]).

9 JOINT HARDWARE DESIGN AND TEST PROGRAMMES

Recently two joint programmes, concerning carbon/epoxy structures for advanced aircraft, have been performed by Dornier-Werke Friedrichshafen, DFVLF/Stuttgart and University of Stuttgart. One has been a study on the applicability of carbon primary-structure shells to slender wings (Fig. 25). The second was a fatigue life study of an airbrake for the Alpha-jet, where in addition to the load, temperature and humidity were varied within a wide range to simulate the whole environmental envelope of certain flight programmes [34].

Fig. 26

The main intention in the slender wing study has been to perform a comparison between elastomechanic and strength calculations based on finite element methods and experiments on the deformation of that piece of hardware under laboratory conditions (Fig. 26). The design of the wing box and the loading tests on structure components was done in cooperation between Dornier-Werke Friedrichshafen and our

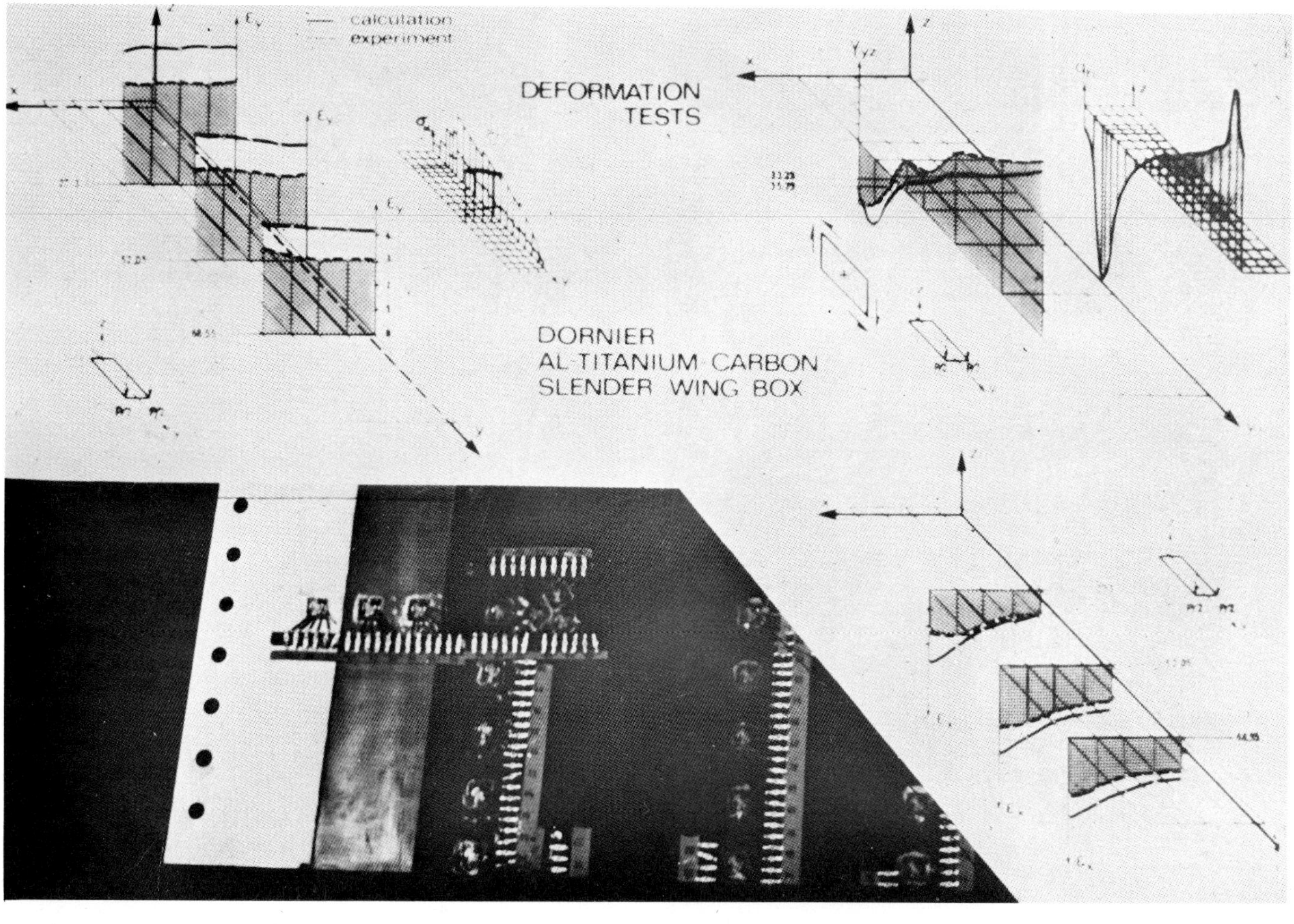

Fig. 27

DFVLR-Institute. The construction of the slender wing box was performed under ordinary factory circumstances at Dornier-Werke Friedrichshafen, where the programming and the computer calculation were also carried out. It was the task of our DFVLR-Institute in Stuttgart to perform the deformation tests. The intended load/level was such that almost any layers of the carbon/epoxy shells were well within linearity limits (Fig. 27). There appeared to be a certain discrepancy in the magnitude but not the tendency of deformations. We believe that this is due to the underestimation of certain interaction effects between the partly aluminum substructure and the carbon shells.

The test results with the Alpha-jet airbrake after 2×10^5 load cycles, which corresponds to 2.5 times the lifetime of the brake, proved to be so successful that flight tests will follow quite soon (Fig. 28) (see [34]). The combined climate and load cycling was performed in the climate chamber of our University-Institute. Load was applied by a hydraulic cylinder within a frame which contained the whole system and which transferred the reaction loads (Fig. 29).

10 COMPOSITE STRUCTURES UNDER FATIGUE

Generally it looks as if the fatigue behaviour (Fig. 30) ([35–38]) of advanced composites is no worse, if not better, than that of classical metallic materials. Moreover there are certain advantages, such as the less serious notch sensitivity if certain general design rules which concern the special qualities of composites are observed. In addition certain hybrid materials, formed by inter-connecting fibre bundles by wrapping around fibres of smaller diameter, seem to result in considerably better strength at high load cycle numbers [39]. Load cycling tests on small tubular composite specimens recently performed over a broad variety of frequencies have also shown that the frequency does not significantly

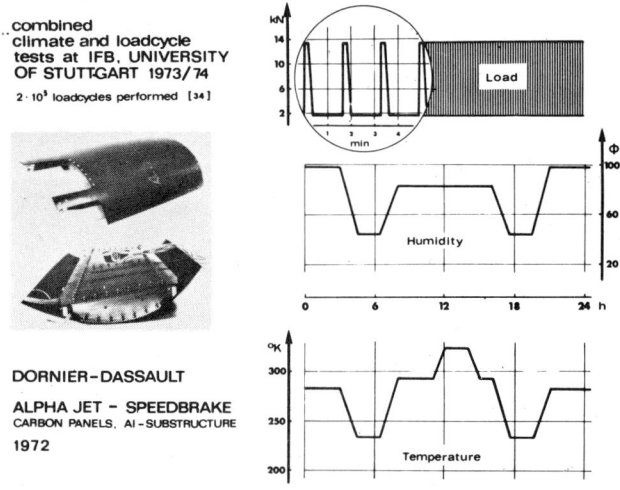

Fig. 28

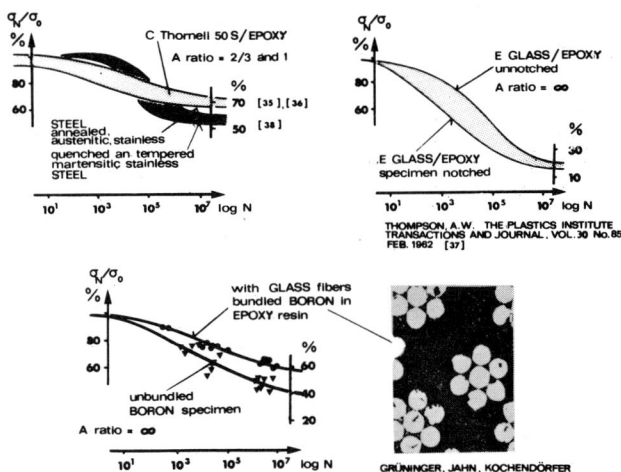

Fig. 30

Fig. 29

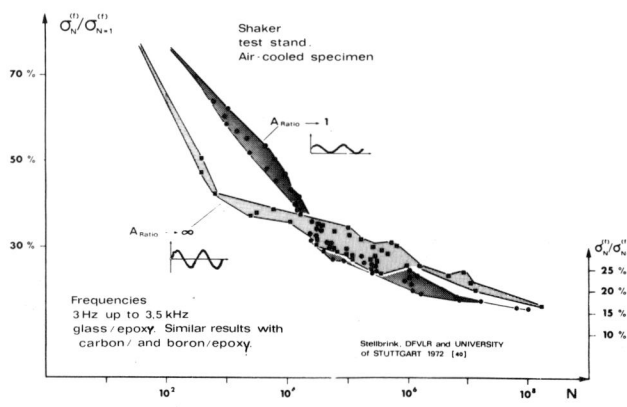

Fig. 31

influence the fatigue behaviour of those composites which have been tested, and that the fatigue behaviour of composites does not essentially differ from that of other metallic materials, such as aluminium alloys, up to frequencies beyond 10^8 (Fig. 31), ([40]). The only condition is to keep the specimen temperature constant, by sufficient airstream cooling, at all frequencies applied.

Thus it seems that the threshold for the safe application of engineering design to advanced composites has been passed, in spite of the difficulties, making it possible to solve geometric problems and optimize composite structures. This optimistic statement is made with a full awareness of the education barrier, which should be overcome in the near future.

ACKNOWLEDGEMENTS

The author would like to thank Mr P.G. Grüninger and Mr K. Stellbrink of DFVLR Stuttgart for essential contributions, Mr D. Muser for the preparation of the Diamond test specimen as well as for performing the tests. The author is especially thankful to Mr. H. Dörner, who contributed much to displaying the results presented by the excellent slides and figures, and to Mrs H.L. Faber who applied the utmost care to shape the concentrated text into legible English. Grateful thanks are also due to Dr A.F. Johnson (NPL) for further help in this regard.

REFERENCES

1. **Zweben, C.** 'Tensile failure of fiber composites', *AIAA Journal* 6 No 12 (1968)
2. **Zweben, C.** and **Rosen, B.W.** 'A statistical theory of material strength with application to composite materials', *J Mech Phys Solids* 18 (1970) pp 189-206
3. **Schmitz, G.K.** and **Metcalf, A.G.** 'Characterization of flaws on glass fibers', *Proc Tech Conf, SPI Reinforced Plastics Division* Section 3A (1965) pp 1–14
4. **Stellbrink, K.** *Unpublished study*, IBK/DFVLR Stuttgart (1973)
5. **Hütter, U.** 'Technologie von Composite-Bauweisen', *Vortrag auf der 4. Jahrestagung der DGLR, Baden-Baden* (1971) Jahrbuch 1971 der DGLR, pp 143-161
6. **Kossira, H.** 'Der Einfluss von Faserdurchmesser und Glasgehalt auf die Festigkeit harzgebundener Glasfaserstränger', *Dissertation, Technische Hochschule Stuttgart*, Stuttgart (March 1963)
7. **Troost, A.** 'El-Magd, Zur Brauchbarkeit der Mischungsregel für faserver stärke Metalle; *Zeitschrift für Metallkunde*, Bd. 64
8. **Rosen, B.W.** 'Mechanics of composite strengthening fibre composite materials', American Society for Metals (1964)
9. **Schuerch, H.** 'Prediction of compressive strength in uniaxial boron fibre-metal matrix composite materials', *AIAA Journal* 1, No 1 (1966)
10. **Lager, J.R.** and **June, R.** 'Compressive strength of boron-epoxy composites', *J of Composite Materials* 3 (1969)
11. **Leventz, B.** 'Compressive applications of large diameter fibre reinforced plastics', *Proceedings 19th Annual Meeting SPI Reinforced Plastics Division,* Section 14D, New York (1964)
12. **Herrmann, L.R., Mason, W.E,** and **Chan, S.T.K.** 'Response of reinforcing wires to compressive states of stress', *J Comp Mat* 1 (May 1967)
13. **Dow, N.F. et al.** *GE–TIS* 60 SD 389 (June 1960)
14. **Holister, G.S.** and **Thomas, C.** 'Fibre reinforced materials', Elsevier Publishing Co. Ltd (1966)
15. **Halpin, J.C.** 'Structure-property relations and reliability concepts', *J Comp Mat* 6 (April 1972)
16. **Timoshenko, S.** 'Theory of elastic stability', *Engineering Societies Monographs* McGraw-Hill Book Company, Inc., New York (1936)
17. **Bader, M.G.** and **Johnson, M.** 'Fatigue strength and failure mechanisms in uniaxial carbon fibre reinforced epoxy resin composite systems', *Composites* 5 No 2 (March 1974)
18. **Haener, J., Puppo, A.** and **Feng, M.Y.** 'Oblique loading of unidirectional fiber composites; shear loading', *USAAVLABS Technical Report* 68–81. US Army Aviation Material Laboratories, Fort Eustis, Virginia (January 1969)
19. **M. Weist,** Unpublished Study, Institute für Flugzeugbau University of Stuttgart (1968)
20. **Hütter, U.** 'Present and future possibilities of high strength and stiffness-to-weight ratio composites in primary structures', *AGARD Conference on impact of composite materials on aerospace vehicles and propulsion systems,* Toulouse (September 1972)
21. **Roth, S.** and **Grüninger, P.G.** 'Beitrag zur Deutung des Querzugversagens von Stranglaminaten', *Kunststoffe* 59 (1969) S 967–974.
22. **Grüninger, P.G.** Unpublished study, IBK/DFVLR Stuttgart (1973)
23. **Weissinger, H.** 'Die Restdruckfestigkeit von Stützlagen nach Zugbelastung von Dreilagenstranglaminaten', Unpublished Study, *IBK/DFVLR* Stuttgart (1971)
24. **Puck, A.** 'Zur Beanspruchung und Verformung von GFK-Mehrschichten-Verbund-Bauelementen', *Kunststoffe* Heft 12 (1967) S 965–973
25. **Hütter, U.** Optimization of shell structures under bending and torsion loads', Paper presented at the International Conference on the Mechanics of Composite Materials, Philadelphia, Pennsylvania (May 9, 1967) *Proceedings of the Fifth Symposium on Naval Structural Mechanics* Pergamon Press
26. **Hütter, U.** 'Weight saving by composite primary structures', *The Eighth Congress on the International Council of the Aeronautical Sciences* ICAS–Paper No 72–38, Amsterdam (September 1972)
27. **Günther, W.** Unpublished study, Institute für Flugzeugbau, University of Stuttgart (1972)
28. **Tsai, S.W.** and **Azzi, V.D.** 'Strength of laminated composite materials', *AIAA Journal* 4 No 2 (1966)
29. **Leyh, A.** 'Messung der Querdehnungszahlen und Elastizitätsmoduli an GFK-Platten mit verschiedener Faserorientierung', *Studienarbeit* Institute für Flugzeugbau, University of Stuttgart (1971)
30. **Krauss, H.** and **Schelling, H.** 'Mehrachsig beanspruchte Drei-Richtungs-Wickelrohre aus verstärkten Kunststoffen', DFVLR Stuttgart, *Kunststoffe* 59 (1969) S 911–917
31. **Lamé, G.** 'Leçons sur la théorie mathématique de l'élasticité des corps solides' (Paris 1852)
32. **Voigt, W.** 'Theoretische Studien über die Elastizitäts-Verhältnisse der Kristalle', *Abhdlg Akad d Wissenschaften* Bd 34 (Göttingen 1887)
33. **Puppo, A.H.** and **Evensen, H.A.** 'Strength of anisotropic materials under combined stresses', *AIAA Journal* 10 No 4 (April 1972) pp 468–474
34. **Muser, D., Carl, U. Günther, W.** and **Molly, J.P.** 'Dynamischer Klimatest der Alpha-Jet-Klappe', Unpublished report, Institute für Flugzeugbau, University of Stuttgart (1973)
35. **Preuss, T.** 'Kohlenstoffaser Wickelrohre unter Torsionswechselbelastung', *Interner Bericht* 14/70, IBK/DFVLR Stuttgart (1970)
36. **Preuss, T.** 'Messungen an Kohlenstoffaser Wickelrohen unter statischer Zug-, Druck- und Torsionsbelastung und unter Torsionswechselbelastung', *Interner Bericht* 11/71, IBK/DFVLR Stuttgart (1971)
37. **Thompson, A.W.** 'The fatigue and creep properties of reinforced plastics', *The Plastics Institute Transactions and Journal* 30 No 85 (February 1962)

38. Madayag, A.F. 'Metal fatigue: theory and design', John Wiley and Sons, Inc., New York (1969)
39. Grüninger, P.G., Kochendörfer, R. and Jahn, H. 'Verbundwerkstoffe mit neuartigen Faserwerkstoffen unter dynamischer Beanspruchung', IBK/DFVLR Stüttgart, *Kunststoffe* Bd 60 Heft 12,(1970) S 1029–1036
40. Stellbrink, K. 'Vergleich der Ermüdungsfestigkeit von GFK im Zug-Druck-Wechsel-und Zug-Schwell-Versuch', *Institutsbericht* 029 72/15, IBK/DFVLR und Institut für Flugzeugbau/Universität Stuttgart (1972)

QUESTIONS

1 On comparison of results with diamond and clydinrical test specimens, by I.C. Taig (British Aircraft Corporation).

Have the results of multidirectional tests using diamond specimens been compared with tests on cylinders and is there any significant effect on the results due to the stress concentrations between the arms of the specimen?

Reply by Professor Hütter

The diamond specimens have been developed especially to facilitate the strength tests under certain two dimensional loading conditions. They serve as indicated only to gain the most significant failure strength values at the 'main diagonal points' of the failure envelope.

The results gained so far from pilot tests show no essential discrepancies between tube tests and diamond tests. For the two most essential envelope — main diagonal points — extension/extension and compression/compression stress concentrations at the diamonds' corners seem to have no remarkable influence if matrix strength is by far smaller than fibre strength. Evidently this is true only for the type of symmetric three layer composite panels.

For the main diagonal-failure envelope points with extension/compression and compression/extension a slightly different shape of the specimen wings where the load is induced is necessary, to avoid undesirable additional stresses due to bending moments. However, the same procedure and type of gear can be applied.

2 On long term creep and creep rupture, by L. H. McCurrich (Chemical Building Products).

The work Professor Hütter is doing on compressive failure in grp is particularly important with regard to building applications. I should like to have his comments on the long term creep and creep rupture characteristics in compression as instability of the fibres could occur prematurely due to matrix creep.

Reply by Professor Hütter

As the compressive failure of undirectional composite panels or rods — loaded in the direction of fibres — is a stability failure up to loads quite close to the failure limits, no preliminary local fibre deviations have to be expected. This is true only if the fibre volume fraction is sufficiently high and if the quality of the composite structural element meets certain standards of homogeneity — within the structure as well as during production from part to part.

We never experienced critical long time or creep failure phenomena related to fibre shear buckling — except at loads close to experienced compression failure.

As I understand, the safe load requirements for building applications are such that long time composite (grp inter alia) compression failure certainly can be avoided.

COMMENTS

1 By J. R. Bowcock (Senior Engineer, Technical Services Department, Production Engineering Research Association, Melton Mowbray, Leics)

Professor Hütter is correct when he says "it might be an essential help to engineers if composite material design charts were available which contained fibre-volume content, fibre direction angles of layers, layer thicknesses and wall thickness".

The information readily available at the present time, mainly in the 'glossy type' of technical bulletins, in general relates to the properties of the individual materials and not the 'composite' ie resins, fibres (glass, carbon, boron etc) and the testing of small laboratory samples.

Over the last decade many papers have been presented and/or written on the techniques for measuring and testing the mechanical properties of composite materials. The results given in these papers are extremely variable and without qualification, obtuse and meaningless to the composite structure design engineer. Very often the learned gentlemen carrying out and analysing these test programmes do not know what use the results are when they have derived them.

Is it not now time that the people and organisations with the necessary capability initiate programmes for testing samples of composites which bear a relation to those structures that can be produced on the 'shop floor'.

What the design engineer working on composite structures requires is data charts that give the various relevant mechanical properties in relation to various combinations of thickness of laminate, number of fabric layers, resin/fibre volume ratios, resin types and type of make-up (eg hand lay-up, pressure moulded etc). The test samples should be manufactured under factory conditions.

Engineers in the Technical Services Department at PERA are working on the designs of very high strength composite structures where the wall thicknesses range from 6.35 mm to 76.2 mm and where the glass/resin volume ratio is very high (in certain cases 1:1).

Data charts of the type mentioned would provide an invaluable guide for work of this kind.

On a constraint effect in steady-state creep of fibre composites

O. BØCKER PEDERSEN

The tensile creep strength of a unidirectional discontinuous fibre composite is discussed on the basis of an analysis of the equilibrium conditions in the matrix. This and previous theories predict that a volume fraction V_f of non-creeping fibres of aspect ratio l/d strengthens the matrix by the factor $a\, V_f\, (l/d)^{1 + 1/n}$ where n is the stress exponent in the matrix creep law. Each theory yields an approximately constant a near unity. The magnitude of a seems to reflect the importance of the constraint effect. Preliminary experimental data is presented.

1 INTRODUCTION

The fibre diameter is being recognized as an important parameter in the description of many of those properties of fibre composites that involve an interaction between the components. One of these properties is the strength under conditions of steady-state creep of a unidirectional discontinuous fibre composite containing non-creeping perfectly bonded fibres of equal length l and diameter d. As the matrix extends plastically its transverse contraction is constrained by the non-creeping fibres. This constraint gives rise to high tensile stresses in the matrix and is likely to depend upon fibre spacing, and hence upon fibre diameter for a given fibre content.

Several authors have recently analysed the steady-state creep of discontinuous fibre composites. Mileiko [1] neglects the contribution of the matrix to the load carried by the composite and Kelly and Street [2] include a contribution from an unconstrained matrix. These analyses therefore do not take the constraint effect into account. McLean [3] considers the rate of energy dissipation in the matrix during creep and derives a creep strength which is roughly three times as great as that derived by Kelly and Street. The reason for this discrepancy could be that McLean's analysis implicitly takes the constraint effect into account. We derive the creep strength by analyses [4,5] which explicitly include the full constraint effect and also by an analysis which ignores it [5]. The results of these analyses indicate that the constraint effect can amount to roughly a doubling of the composite creep strength. Preliminary experimental investigations of the Cu-W system are described and results are discussed in the light of these theories.

2 THEORY

Many metals follow a power relation between steady-state creep rate and applied stress over a fairly wide range of temperature and stress [6]. For uniaxial loading the relation takes the form

$$\sigma_{zz} = \sigma_0 \left(\frac{\dot{\epsilon}_{zz}}{\dot{\epsilon}_0} \right)^{\frac{1}{n}} \quad (1)$$

where σ_{zz} is the tensile or compressive stress, $\dot{\epsilon}_{zz}$ is the steady-state uniaxial creep rate, and $\sigma_0, \dot{\epsilon}_0$, and n are constants at a given temperature. The expressions for the composite creep strength, σ_c, derived on the basis of Equation (1) by Mileiko, Kelly and Street, and McLean may all be rearranged to form the general relation

$$\sigma_c = a\, V_f \left(\frac{l}{d} \right)^{1 + \frac{1}{n}} \sigma_0 \left(\frac{\dot{\epsilon}}{\dot{\epsilon}_0} \right)^{\frac{1}{n}} \quad (2)$$

in which $\dot{\epsilon}$ is the steady-state creep rate of the composite, $\sigma_0, \dot{\epsilon}_0$, and n are the constants appearing in Equation (1) l/d is the fibre aspect ratio, V_f is the fibre volume fraction, and a is a very slow function of V_f and n, as shown by Bøcker Pedersen [4]. At large values of n these functions are approximately given by the constant values

$a_1 = 0.3$ Mileiko
$a_2 = 0.5$ Kelly and Street
$a_3 = 1.5$ McLean

We obtain similar results by considering the equilibrium conditions in the matrix. We select a cylindrical system of coordinates with origin, r = z = 0, at the fibre midpoint and z-axis parallel to the fibre axis. Assuming axial symmetry the integral form of one of the equilibrium equations is

$$2r\tau_{rz} - \tau_i d + 2 \int_{d/2}^{r} \rho \frac{\partial \sigma_{zz}}{\partial r} d\rho = 0 \quad (3)$$

For large values of n the interface shear stress, τ_i, is according to several analyses [2,3,5,7,8] almost constant along the major portion of the fibre. Integration of τ_i from the fibre end therefore results in an approximately linear axial distribution of tensile stress in the fibre. A point made by McLean

---

Metallurgy Department, Danish Atomic Energy Commission, Research Establishment Risø, 4000 Roskilde, and the Department of Structural Properties of Materials, The Technical University of Denmark, 2800 Lyngby, Denmark

is that the tensile stress in the matrix may be calculated by a similar integration of τ_i, resulting in a nearly linear distribution of stress in the matrix, too. McLean further concludes that the load on the composite is approximately equally divided between fibres and matrix, as long as there is flow in the matrix. The large values of σ_{zz} in the matrix associated with this load distribution are possible without excessive creep-rates because the constraint generates a transverse stress, σ_{rr}, which nearly balances σ_{zz}. The corresponding triaxial stress state is described by a general relation between the stress tensor and the strain-rate tensor [9], rather than by Equation (1). However, Equation (1) does apply when the constraint effect is negligible, ie when $\sigma_{rr} \ll \sigma_{zz}$. Following these ideas we derive the composite creep strength on the basis of three different sets of assumptions:

(a) τ_i is approximately constant and any value is allowed for σ_{zz} [5];

(b) Fibres and matrix share the total load equally and the z-dependence of σ_{zz} is linear [4];

(c) The integral term in Equation (3) is negligible [5].

The first set of assumptions leads to the load distribution assumed under (b) (equipartition of load), which we take to represent the maximum constraint effect. Assumption (c) is taken to represent the case of no constraint effect in which σ_{zz} is given in terms of $\dot{\epsilon}_{zz}$ by Equation (1). As in previous analyses we obtain a creep strength which may be expressed by Equation (2). The appropriate a-coefficients are shown as functions of V_f for $n = 9$ in Fig. 1. The results for maximum constraint effect a_a and a_b are very similar although they are obtained by slightly different approaches. Since the dependence of a is quite small the coefficients are approximately given by the constant values

$a_a = 1.2$

$a_b = 1.2$

$a_c = 0.5$

The agreement between the results of Mileiko and Kelly and Street (a_1 and a_2) and a_c is fairly good as is the agreement between the result of McLean (a_3) and a_a and a_b.

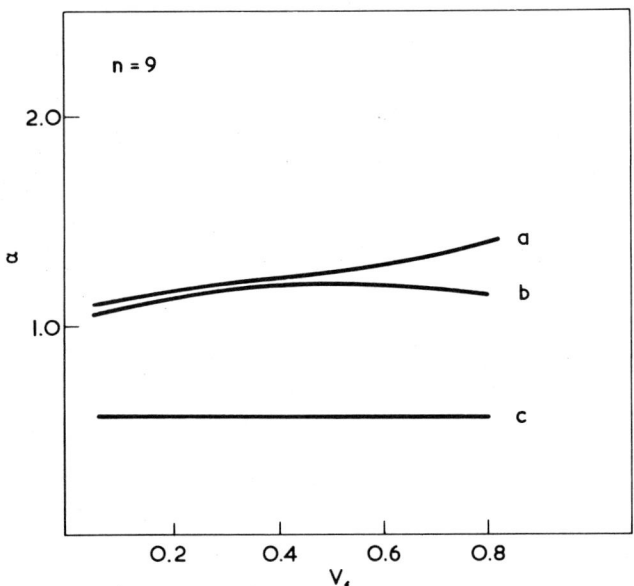

Fig. 1 Creep strength coefficients, a, derived on the basis of assumptions a, b and c

Table 1 Creep tests of discontinuous Cu-W fibre composites

Specimen	σ [MN/m^2]	$\dot{\epsilon}$ [10^{-5}h^{-1}]	A_f
1	122.6	12.59	0.22
2	103.0	20.15	0.17
2	78.5	1.89	0.17
3	103.0	1.98	0.25
4	122.6	2.78	0.22
5	137.3	13.71	0.27
5	147.2	22.27	0.27
5	157.0	44.16	0.27

The importance of the constraint effect is indicated by the difference between these two sets of results which is a factor of two or three. The constraint is likely to depend upon the mechanical properties of the components and upon geometrical factors, in particular the fibre spacing. In the region where the analyses are valid we therefore expect a to be a function of these variables with maximum and minimum given roughly by 1.5 and 0.5.

3 EXPERIMENTAL

The Cu-W system was investigated. Polycrystalline 0.5 mm W-wire was chopped into equal lengths and embedded by vacuum infiltration in a single crystal matrix of spectrographically pure Cu. Composite specimens of gauge length 45 mm containing a volume fraction $V_f = 0.28$ of fibres with an aspect ratio $l/d = 25$ were creep tested in air using a 13:1 lever type apparatus. Extension was magnified 200X and continuously recorded on a revolving drum. All creep tests were conducted at a temperature of 500°C ± 1°C, which was measured with a thermocouple in contact with the specimen.

4 RESULTS AND DISCUSSION

Values of applied stresses and the resulting minimum creep-rates are given in Table 1 together with the fibre area fractions, A_f, observed on the rupture surfaces. The values of A_f are all less than $V_f = 0.28$ indicating, as predicted by Street [10], that rupture occurred at cross-sections of low fibre content. Fig. 1 shows a double logarithmic plot of the data in Table 1. If we disregard the measurements on specimen 2 (data points in parentheses) the plot indicates a linear relation between $\log \sigma$ and $\log \dot{\epsilon}$. The comparatively high creep rates exhibited by specimen 2 could result from the extremely low area fraction observed on this specimen, $A_f = 0.17$.

In calculating the least-squares fit represented by the line in Fig. 2 we have used only one data point per specimen (and left out the results for specimen 2), since the experimental scatter primarily results from differences in composition of the specimens. This average is appropriate when determining $\sigma_{0c} = \sigma_c / (\dot{\epsilon}/\dot{\epsilon}_0)^{1/n}$, whereas a series of measurements on a single specimen (say those on specimen 2 or 5) appears to be more suitable when determining the composite stress exponent. The present data yield roughly the same exponent from both approaches. The creep strength determined by the least-squares method is

$$\sigma_c \text{ [MN/m}^2\text{]} = 130 \left(\frac{\dot{\epsilon} \text{ [h}^{-1}\text{]}}{10^{-4}} \right)^{\frac{1}{7}} \quad (4)$$

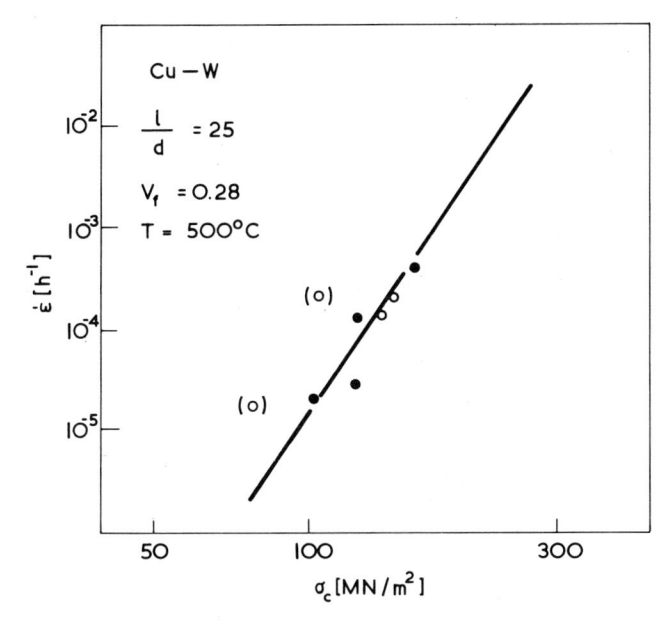

Fig. 2 Double logarithmic plot of the stresses and minimum creep-rates listed in Table 1

Table 2 Matrix properties at 500°C

Reference	σ_0 [MN/m^2]	$\dot{\epsilon}_0$ [10^{-5} h^{-1}]	n	a
11	13	10	6	0.8
12 extrapolated	25	10	5	0.4
13 extrapolated	23	10	5	0.4
14 extrapolated	16	10	5	0.6

We have not yet measured creep curves for the unreinforced matrix, but creep data on nominally pure Cu are available in the literature. Table 2 lists the parameters σ_0, $\dot{\epsilon}_0$, and n for Cu at 500°C. Only one work [11] contains results obtained at 500°C, but the parameters were extrapolated by means of the generalized creep equation [6] from three other works [12–14]. The values of a required in order that Equation (2) predicts Equation (4) on the basis of these data are also shown in Table 2.

Experimental evidence for the constraint has usually been found in tensile experiments where the fibre spacing was considerably smaller than the average spacing $\Lambda \sim 500\,\mu m$ in our specimens. Kelly and Lilholt [15] observed a constraint effect for fibre spacings less than $\Lambda \sim 20\,\mu m$, Cheskis and Heckel [16] for $\Lambda \sim 25\,\mu m$, Stuhrke [17] for $\Lambda \sim 100\,\mu m$, and Garmong and Shepard [18] for $\Lambda \sim 2.5\,\mu m$. For this reason we expect the constraint effect to be fairly unimportant in the present experiments, and most of the a-values in Table 2 are indeed close to the value predicted for the case of no constraint effect. However, the unextrapolated data from the work of Pahutova et al perhaps gives the most reliable value of a, and this value is consistent with some constraint effect.

ACKNOWLEDGEMENTS

It is a pleasure to acknowledge the continued interest of Professor R.M.J. Cotterill and Dr. N. Hansen. Thanks are due to Drs H. Lilholt and J.B. Bilde-Sørensen for several helpful discussions and to Messrs J. Kjøller, J. Larsen, P. Nielsen, and O. Olsen for skilful experimental assistance.

REFERENCES

1. Mileiko, S.T. 'Steady-state creep of a composite material with short fibres', *J Mat Sci* **5** (1970) pp 254–261

2. Kelly, A. and Street, K.N. 'Creep of discontinuous fibre composites II. Theory for the steady-state', *Proc Roy Soc A* **328** (1972) pp 283–293

3. McLean, D. 'Viscous flow of aligned composites', *J Mat Sci* **7** (1972) pp 98–104

4. Bøcker Pedersen, O. 'Creep strength of discontinuous fibre composites', *J Mat Sci* (in press)

5. Bøcker Pedersen, O. Unpublished work

6. Mukherjee, A.K., Bird, J.E. and Dorn, J.E. 'Experimental correlations for high-temperature creep', *Trans ASM* **69** (1969) pp 155–179

7. Kelly, A. and Tyson, W.R. 'Tensile properties of fibre reinforced metals II. Creep of silver-tungsten', *J Mech Phys Solids* **14** (1966) pp 177–186

8. de Silva, A.R.T. 'A theoretical analysis of creep in fibre reinforced composites', *J Mech Phys Solids* **16** (1968) pp 169–186

9. Odquist, F.K.G. 'Multi-axial state of Stress. Mathematical theory of creep and creep rupture', *Oxford University Press* (1966) pp 19–21

10. Street, K.N. 'Steady-state creep of fibre-reinforced materials', *The Properties of Fibre Composites. Conference Proceedings,* National Physical Laboratory, 4 November 1971, IPC Science and Technology Press Ltd (1971)

11. Pahutova, M., Cadek, J. and Rys, P. 'High temperature creep in copper', *Phil Mag* **23** (1971) pp 509–517

12. Feltham, P. and Meakin, J.D. 'Creep in face centred cubic metals with special reference to copper', *Acta Met* **7** (1959) pp 614–627

13. Barrett, C.R. and Sherby, O.D. 'Steady-state creep characteristics of polycrystalline copper in temperature range 400°C to 900°C', *Trans AIME* **230** (1964) pp 1322–1327

14. Monma, K., Suto, H. and Oikawa, H. *J Japan Institute of Metals* **28** (1964)

15. Kelly, A. and Lilholt, H. 'Stress-strain curve of a fibre-reinforced composite', *Phil Mag* **20** (1969) pp 311–328

16. Cheskis, H.P. and Heckel, R.W. 'In situ measurements of deformation behaviour of individual phases in composites by X-ray diffraction', *ASTM STP* **438** (1968) pp 76–91

17. Stuhrke, W.F. 'The mechanical behaviour of aluminium-boron composite material', *ASTM STP* **438** (1968) pp 108–133

18. Garmong, G. and Shepard, L.A. 'Matrix strengthening mechanisms of an iron fiber-copper matrix composite as a function of fiber size and spacing', *Metall Trans* **2** (1971) pp 175–180

Effective structural use of grp and cfrp

W. PATON

Anisotropy, lack of ductility and material variability introduced during processing present special problems in composite design for load-bearing applications. In some cases it is possible to use a duplex approach in which efficient composite elements are used in conjunction with other materials to produce an overall effective design. In this way, for example, metal can be retained in regions of complex stress. The merits of this approach are discussed with reference to several specific examples.

1 INTRODUCTION

An interest in composites can be justified only if we expect to gain some overall advantages in design situations either in terms of functional performance, economics or perhaps in the utilisation of material or energy. Effective methods of incorporating composites into designs should therefore be such that advantage is achieved without loss of reliability. Much of the research effort on composites of the fibre/plastic type has been directed at problems associated with the maximum utilisation of their favourable structural characteristics, on the assumption of a 100% composite approach. The difficulties arising from anisotropy, variability in properties and lack of ductility and toughness are well known. In many practical design situations a duplex approach, in which simple but efficient composite elements are combined with conventional metals, can be worthwhile and, in fact, avoid many of the difficulties and uncertainty of an all-composite design. Each material can be utilised in its most effective role and provided that the problems of combining dissimilar materials are taken into account, overall reliability can be achieved.

2 STRUCTURAL ELEMENTS

Microstructural parameters which influence composite properties are most easily controlled in simple geometric forms with unidirectional fibre alignment. Filament wound rings, for example, can be reproducibly manufactured with a high volume content of fibre, and good mechanical properties. They are therefore an ideal reinforcement for certain types of structure. Similarly, pultruded bar and strip [1,2] can be made to accurate specifications and permit highly efficient utilisation of fibre properties under uniaxial loading conditions.

Force transmission in structures involves both tension and compression members. Consequently, the buckling characteristics of elements in compression must be established in addition to purely material strength considerations. Because of this the buckling instability mechanism of composite elements requires special consideration, in order to establish the best type of construction and material design parameters. Optimum element design is likely to vary with structural index, P/L^2, which is the measure of loading intensity for a load P, and transmission path L.

For simple pin-ended columns the well-established equations for predicting the critical buckling load (P_{cr}) are:

Euler equation for elastic instability

$$P_{cr} = \frac{\pi^2 EI}{L^2} \qquad (1)$$

or Engesser equation for inelastic behaviour

$$P_{cr} = \frac{\pi^2 E_R I}{L^2} \qquad (2)$$

where E_R is the reduced modulus.

The bending rigidity $D\,(=EI)$ is the controlling parameter and therefore account must be taken of factors which can reduce this when anisotropic tubular sections are used. If the shear rigidity of the column is low, as may well be the case with cfrp, the buckling load will be reduced below the values predicted by Equation (1), since lateral deflection by shear becomes significant. The effect of shear on the critical load can be accounted for by an equation of the type obtained by Timoshenko [3]:

$$P_{cr} = \frac{P_E}{1 + n\dfrac{P_E}{AG}} \qquad (3)$$

where P_E is the Euler load $= \dfrac{\pi^2 EI}{L^2}$

and A/n is the effective area in shear.

National Engineering Laboratory, East Kilbride, Glasgow, Scotland

The critical load is therefore reduced by

$$\frac{1}{1 + n\frac{P_E}{AG}}$$

For solid columns, and when $G(A/n)$ is much greater than P_E, the effect is small but in composite tubular columns it may be of some significance.

The critical buckling stress (σ_{cr}) derived from Equation (1) is given by

$$\sigma_{cr} = \frac{\pi^2 E}{(L_e/\rho)^2}$$

where ρ is the radius of gyration, and L_e is the effective column length, which takes account of end constraints.

For a tubular column the relationship with the diameter/wall thickness ratio (d/t) and structural index can be shown to be

$$\sigma_{cr} = \pi E^{½} \left(\frac{1}{8\pi}\frac{d}{t}\right)^{½} \left(\frac{\rho}{L^2}\right)^{½} \quad (4)$$

Column efficiency is therefore increased with higher d/t ratios. Obviously 'cut-off' values exist for permissible column stresses, based either on the maximum compression stress of the material or the local buckling instability for columns of high d/t. Curves constructed in Fig. 1 from representative composite data show the relative effectiveness of various composite types in column performance. Maximum compression stress cut-offs were used. On this simple basis the most effective column should be of essentially unidirectional construction. However the implications of Equation (4) on optimum column design indicate that very high d/t ratios may be necessary and that local buckling modes will impose a serious limitation.

The results of experimental testing of unidirectional cfrp tines are presented in Figs. 2 and 3. The agreement between experimental values and the corrected Euler prediction is reasonable for high tensile (HT) and high modulus (HM) composites although the latter exhibit some divergence at low slenderness ratios.

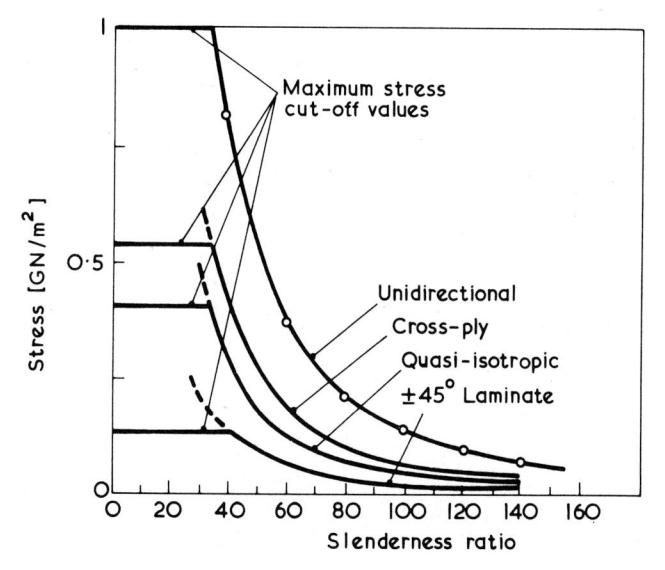

Fig. 1 Theoretical column curves for cfrp

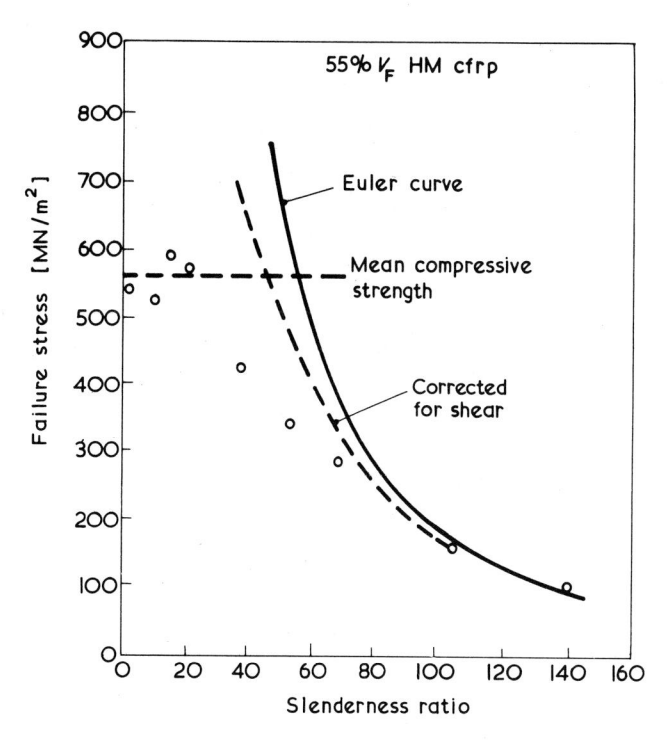

Fig. 2 Column results for 55% high modulus cfrp

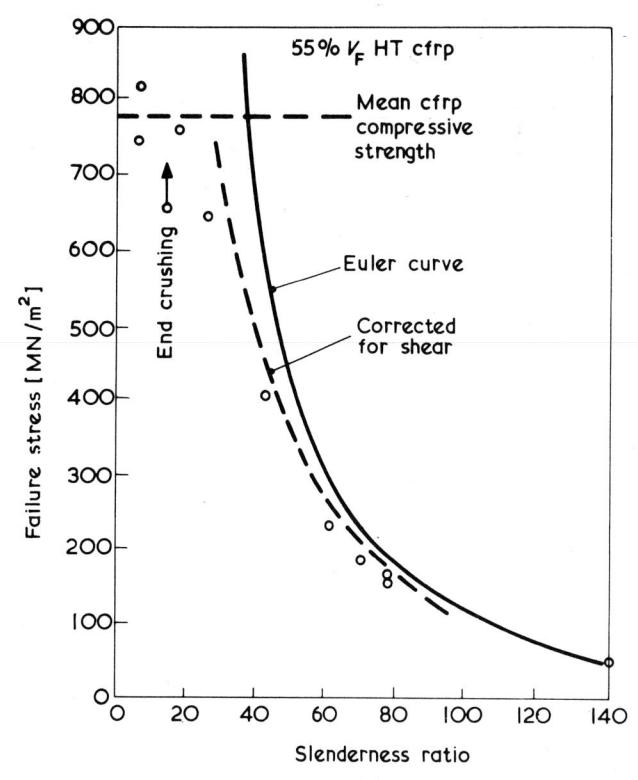

Fig. 3 Column results for 55% high strength cfrp

3 METAL-COMPOSITE COMBINATIONS

3.1 Beam structure

A structure of the type shown in Fig. 4 has essentially uniaxial member loads and its load carrying capacity will be determined by elastic stability consideration as discussed above, rather than tensile strength. The transmission of load between members is achieved by locating an isotropic and ductile metal in the region of complex stress at the joints.

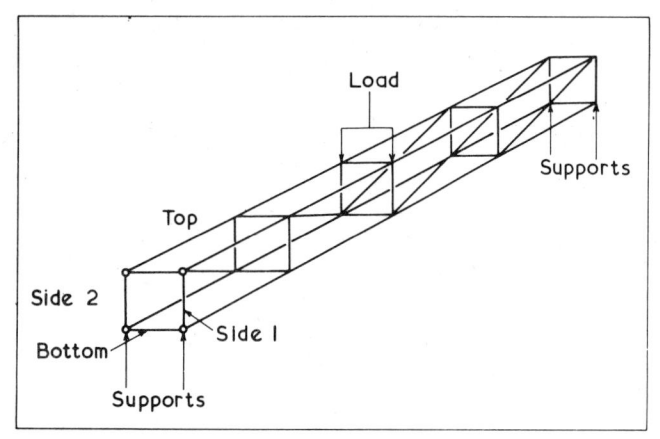

Fig. 4 Beam structure

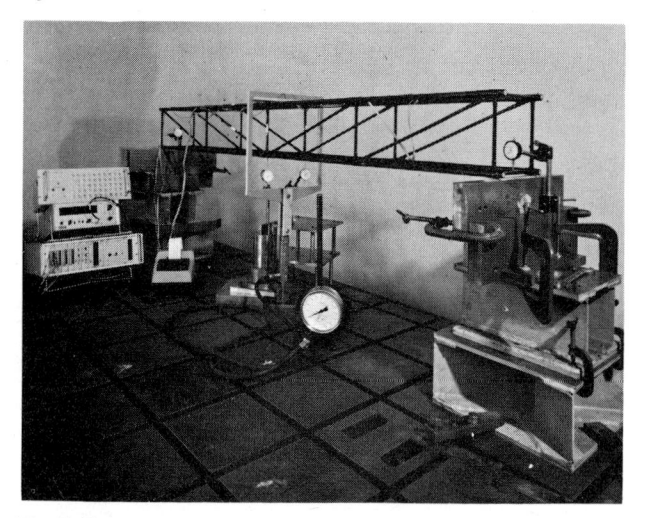

Fig. 5 Experimental cfrp test structure under test

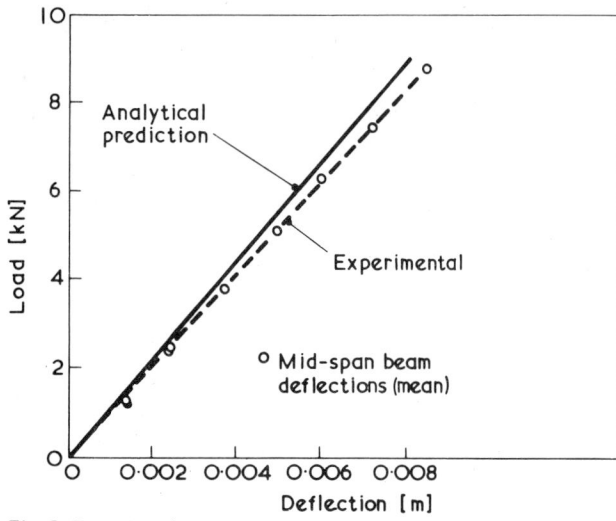

Fig. 6 Beam load/deflection curve

Fig. 7 CFRP tie-bar

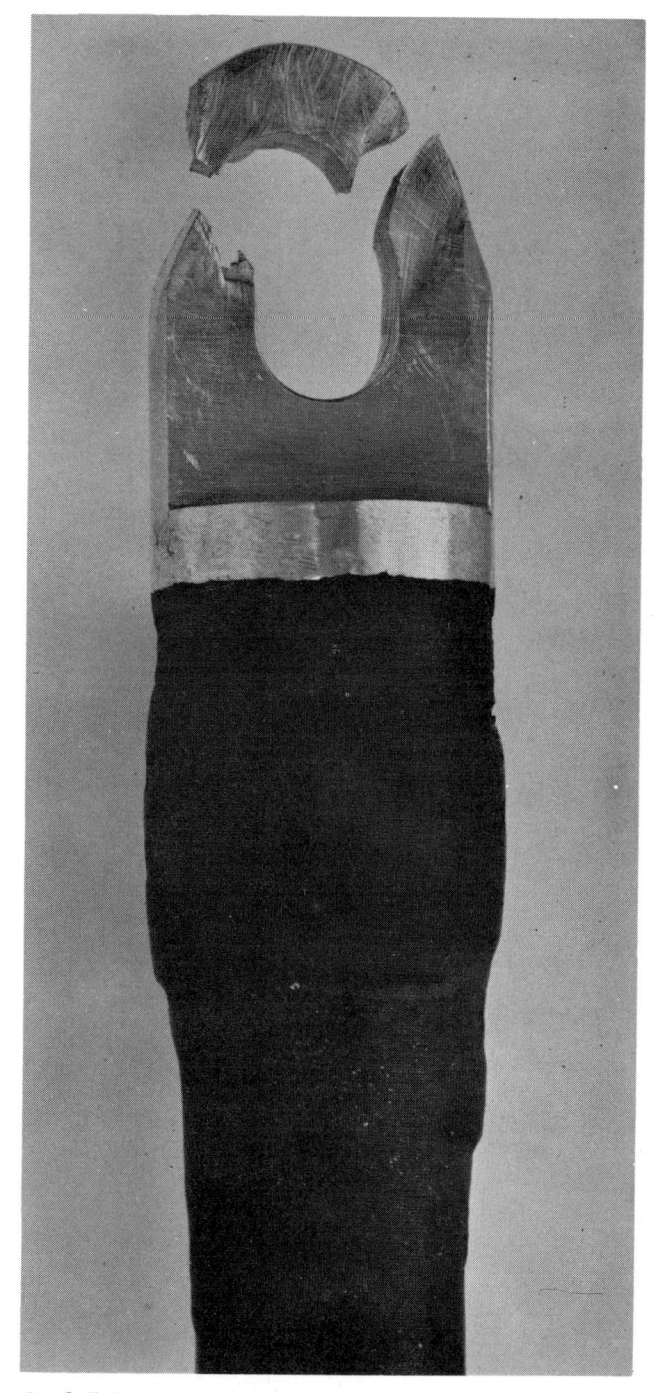

Fig. 8 Failure of cfrp/metal test bar

The cfrp/steel joint test structure shown in Fig. 5 was designed and constructed as part of an experimental evaluation programme [4]. The experimental test results were related to analytical predictions and the mechanical behaviour under load established up to buckling instability. The load/deflection relationship (Fig. 6) is in reasonable agreement with the analytical predictions. Uniform diffusion of load was observed and the structure failed by compression buckling in critical members at a stress consistent with the experimental column results. Thus, this was a reasonable basis on which to design and build structures which would perform in a reliable and predictable manner.

Whilst some weight penalty was incurred at the metal joints, it was still possible to gain a factor of 3 weight advantage over the equivalent steel structure.

3.2 CFRP tie-bar

In most cases the transmission of primary load into composite elements and their connection to sub-structure can only be achieved through metal attachments. In the example shown in Fig. 7 a high degree of joint reliability under uniaxial tension in an adverse environment was required. Provision had to be made for a fail-safe mechanism in the event of accidental overload or cfrp/metal de-bonding. This was achieved by fabricating the bi-directional cfrp tube with integrally bonded metal ends mechanically interlocked to the composite. This type of structural component can thus behave in a non-brittle manner and fail-safe — even although about 75% of the load path comprises a completely elastic material. A typical behaviour of a pin-loaded member is shown in Fig. 8.

3.3 Shell reinforcement with grp

The stiffness and strength of thin metal shells is greatly increased by the addition of distributed stiffeners of relatively simple form. When low mass is an important factor, for example, in dynamic operation, the use of low-density composite stiffeners can be advantageous. A hoop-reinforced conical shell (Fig. 9) consists of a steel shell reinforced by grp rings, filament wound in situ. Since differential straining or thermal effects could result in structural degradation, the adhesive bond between grp and steel is supplemented by integral mechanical fastening. Reinforced shells of this type can be analysed by replacing the stiffened section by an equivalent monocoque section with modified values of extensional and flexural rigidity. Dynamic characteristics can be tailored to meet particular requirements and a high degree of damping has been observed in this type of construction.

REFERENCES

1   Spencer, R.A.P. 'Advances in pultrusion of carbon fibre composites', *Proc Int Conf on Carbon Fibres* London (1971) paper 21

2   Paton, W., Lockhart, A.H. and Montgomery, I. 'The use of drawing methods in cfrp fabrication', *Proc Int Conf on Carbon Fibres* London (1974) paper 22

3   Timoshenko, S.P. and Gere, J.M. 'Theory of elasticity', McGraw-Hill, New York (1961) p 133

4   Paton, W. and Lockhart, A.H. 'The structural use of carbon fibre composites', *NEL Report* No 555, National Engineering Laboratory, East Kilbride, Glasgow, Scotland (1973)

Fig. 9 GRP ring-stiffened conical shell

QUESTIONS

1 On apparent emphasis being given to joints; stress concentration effects and space frame (cfrp) structures, by W.M. Banks (Department of Mechanics of Materials, University of Strathclyde, Glasgow)

In using duplex structures of the kind described by Mr Paton it would appear that the emphasis in design is being shifted from the basic composite system, whether grp or cfrp, to the joints. Would Mr Paton care to elaborate and in particular give some quantitative assessment of stress concentration effects?

Also, have the space frame type cfrp structures been tested to failure? If so, I would be interested in having some comments on the mode of failure.

Reply by Mr Paton

In many composite design situations the primary problem is indeed the transmission of load into the composite material, and therefore joint design is a more critical factor than the basic composite material properties. Load transmission into composites by shear through an adhesive layer is a preferred means which does not introduce material discontinuity and minimises stress concentrations. The magnitude of the stress concentration, which inevitably arises at joints, can only be evaluated quantitatively in relation to a particular geometry and material combination. CFRP space frame structures have been tested to buckling failure. Failure occurs through elastic instability of critical compression members and in the post buckled condition the structure continues to support a substantial proportion of the load. On removal of load it is possible for the structure to completely recover elastically, in contrast to the permanent plastic deformation observed in equivalent metal structures.

Applications of advanced composites in aircraft structures

I.C. TAIG

The paper deals with some present and potential applications of high strength and modulus fibrous composites, particularly carbon fibre/resin materials, in aircraft structures. Their justification is considered from the twin viewpoints of economic efficiency and structural integrity. A simplified appraisal is given of the structural efficiency of various structural elements employing carbon/epoxy material, in relation to the efficiency targets set by aircraft economics. In considering integrity, particular attention is given to detail design features such as joints, end fittings, reinforcements and discontinuities which have a most important influence on material characterisation and structural performance.

1 BROAD OBJECTIVES

The airframe designer is interested in the new composite materials primarily because of their combination of high specific strength and stiffness. These properties offer the prospect of substantial savings in structure weight which is a factor of high value to the aircraft designer and operator. This value can be quantified according to the type of aircraft and the stage during its design at which weight-saving technology can be exploited. Typical values are indicated in Table 1 as the monetary equivalent V of unit reduction in structural mass directly attributable to the new technology.

In the early stages of exploitation of a new material, particularly one involving radical changes of design, manufacture and safety assurance, it is introduced gradually on a substitution basis, that is by replacement of individual conventional components by composite items. The simplest economic assessment of such replacement is obtained by comparing the installed cost of the new component with the conventional item cost plus the value of the weight saved. Thus if a conventional component has mass M_1 and costs C_1 per unit mass and a composite item has mass M_2 and costs C_2 per unit mass, the application is economical, based on first cost, if,

$$C_2 M_2 \not> C_1 M_1 + V(M_1 - M_2) \tag{1}$$

For many simple situations a more useful formulation of the cost criterion can be written in terms of the structural efficiency ratio R defined as:—

$$R = \frac{\text{Mass of conventional structure replaced}}{\text{Mass of composite in component}}$$

Chief Structural Engineer, British Aircraft Corporation Limited, Military Aircraft Division, Warton Aerodrome, Preston, Lancs, England

The application is economic in terms of first cost if

$$R \not< \frac{C_2 + V}{C_1 + V} \tag{2}$$

or

$$C_2 \not> RC_1 + (R - 1)V \tag{2a}$$

Equations of the above form can be made more comprehensive by including within C_1 and C_2 some measure of service and maintenance costs. They can be used to give a first order assessment of the worth of various alternative materials and also to indicate those applications of a particular material which are most likely to prove cost effective.

Costs of the currently available composite materials are difficult to compare on a meaningful basis. Raw material costs are not stable in time and depend on the form in which the material is supplied and on the quantity ordered. Table 2 gives figures for five classes of fibre-resin composite which relate to material in preimpregnated tape form and are expressed as the cost of material as supplied, for unit mass of material in its cured condition.

To the material costs must be added the costs of fabrication, inspection and maintenance of the materials and as yet there are virtually no reliable data available for the advanced composites. Even where components are in fairly large scale production, the cost data are unpublished and in any case they are unlikely to reflect the potential of the composite fabrication process when quantity production techniques are properly developed. The figures used in setting efficiency targets below are believed to be conservative.

To assess the potential value of a material in a specific application some relatively simple calculations of structural efficiency can be made and compared with target values based on cost data [1].

Allowing for the expected drop in material prices in the

Table 1 Weight saving values for different aircraft types

Aircraft type	Exploitation stage	Value of weight saving, V (£/kg saved)
Subsonic commercial transport	Project study	130–170
	Detail design	75–110
Supersonic transport	Project study	450
	Detail design	200
Supersonic combat	Project study	100–159
	Detail design	40–60

Table 2 Costs of different classes of prepreg tape materials

Fibre type	1973–74 costs, £/kg		1980 costs, £/kg (estimated): all batch sizes
	Up to 5 kg	1000 kg	
E-glass	7	7	6–7
Carbon, type 2	110–140	70–135	20–30
Carbon, type 3	90–120	50–110	15–20
Boron	200	100	80
'Kevlar' 49	55–62	53	40*

*This figure, quoted by a distributor, is probably high.

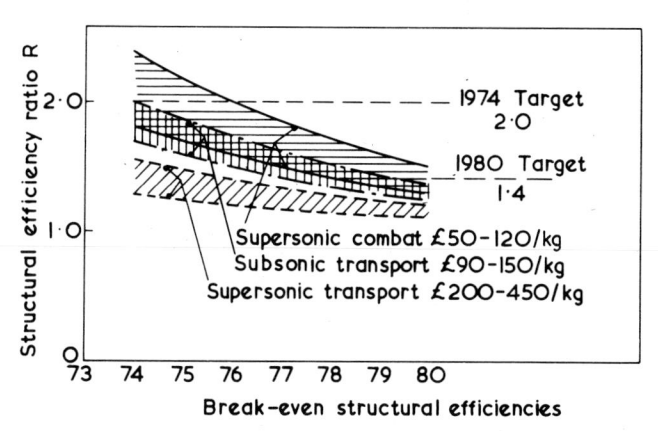

Fig. 1 Break-even structural efficiencies

next few years, the target efficiencies for various aircraft types are shown in Fig. 1 which updates the information in [1].

In some instances quite different economic arguments may be advanced for using composites, for example where a particularly efficient production process is feasible or where substantial modification costs can be saved by the addition of material to components in service [2]. Further examples will be quoted later.

Whilst structural efficiency and improved cost effectiveness are the driving forces behind the introduction of advanced composites, structural integrity is equally the concern of the designer and this aspect is receiving the greatest attention from research workers at the present time. It is necessary to appreciate from the outset that the over-glamorised laboratory properties of unidirectional test specimens are not the data on which realistic design can be based. In aircraft structures the number of components in which well-defined uniaxial load paths can be identified is relatively small (though important amongst the early application areas) and the number in which the static longitudinal strength in a laboratory environment is a sufficient criterion for integrity is nil. Some characteristics of the real environment are given in Table 3.

The useable material properties for assessment of efficiency are those which apply during or after exposure to this real environment. The designer's objective as regards integrity is that the structure shall have at least as high a probability of survival after exposure to this real environment as a conventional structure. To achieve this, with a material which is more difficult to characterise, because of its extreme anisotropy, than conventional materials, demands analysis and testing in much greater depth than has been necessary for metallic structures.

These twin aspects of structural efficiency and material characterisation for integrity are highlighted in the subsequent discussion of some particular airframe applications of composites. These are mainly confined to carbon fibre/epoxy resin materials but many of the points made will apply equally to other fibres and matrices.

2 SOME TYPICAL APPLICATIONS OF COMPOSITES

2.1 Unidirectional load bearing applications

Some typical mechanical and physical properties of carbon fibre/epoxy composites in their as-manufactured state are given in Table 4. The highly anisotropic nature of the material makes it obvious that the most efficient applications, from a structure weight standpoint, must be those in which most of the fibres can be aligned with a well-defined unidirectional load.

Two examples are tension (and stable compression) members in a framework or mechanism and the flanges of beams. Both illustrate immediately one of the principal design prob-

Table 3 Characteristics of 'real' environments

Environmental feature	Value or nature
Loadings	Fluctuating and often reversing
Operating temperatures	$-60°C$ to $+80°C$ (subsonic) $-60°C$ to $+150°C$ (supersonic)
Atmosphere	Clean, industrial and saline; varying humidities and temperatures (including freezing)
Fluids	Water, fuels, oils, hydraulic fluids, solvents, cleaners
Electrical hazards	Lightning strikes, static electricity
Erosion	Rain, hail, dust, etc at high relative velocities
Impact	Birds, stones, projectiles, handling
Corrosion	Galvanic effects in contact with metals.

Table 4 Anisotropy of carbon fibre composites

Property	Material		
	Aluminium alloy	Unidirectional carbon fibre composite	
		Type 1 fibre	Type 2 fibre
Strengths (MN/m^2)			
σ_{t1}	432	1 160	1 560
σ_{t2}	432	38	50
σ_{c1}	*371	940	1 200
σ_{c2}	*371	200	270
τ_{12}	242	85	110
Stiffnesses (GN/m^2)			
E_{11}	72	183	130
E_{22}	72	8.0	10.0
G_{12}	28	5.74	5.0
ν_{12}	0.3	0.3	0.31
ν_{21}	0.3	0.013	0.023
Strains (m/m)			
ϵ_{t1}	0.08	0.006	0.012
ϵ_{t2}	0.08	0.005	0.005
ϵ_{c1}	—	0.005	0.009
ϵ_{c2}	—	0.025	0.027
Coefficients of thermal expansion (per °C)			
a_1	22 × 10^{-6}	−0.5 × 10^{-6}	−0.5 × 10^{-6}
a_2	22 × 10^{-6}	25 × 10^{-6}	25 × 10^{-6}

Suffices: t = tension; c = compression; 1 = longitudinal; 2 = transverse;

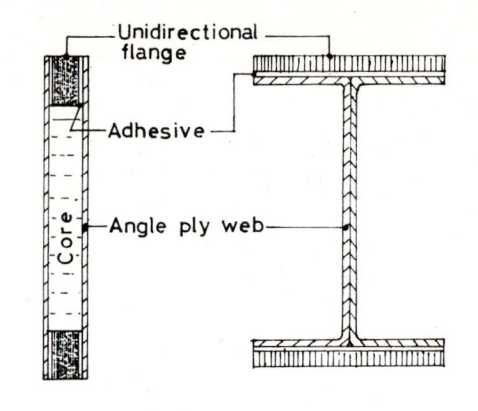

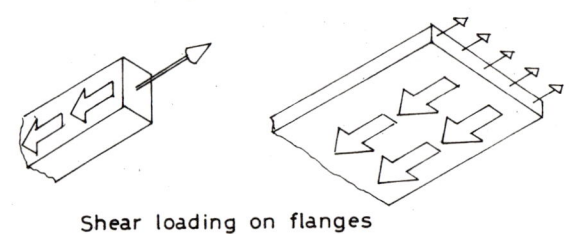

* 0.2% proof stress.

Fig. 2 Load introduction — beam flanges

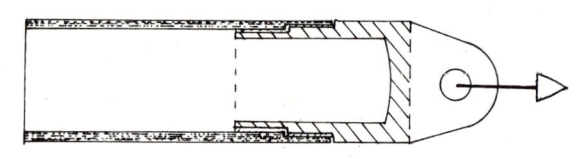

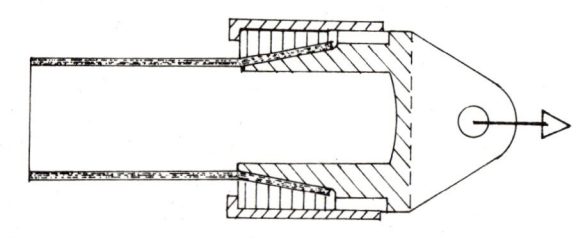

Fig. 3 Load introduction — rods and bars

lems in the exploitation of these materials. Whilst the minimum amount of material in the composite cross section is determined by the longitudinal load to be carried by the member and the material strength, the feasibility of realising the potential efficiency of the material also depends on the ability to introduce the load into the composite material which in most instances demands the transmission of shearing forces through the thickness or necessitates the introduction of tension and compression normal to the fibres. In the case of a beam flange as shown in Fig. 2 load transfer must involve considerable shearing. The ends of a rod or framework member as in Fig. 3 must either be designed to introduce loads by shearing or by more complex actions such as the wedging, lug or pulley types shown, which involve very high transverse stresses and probably shear as well.

As the dimensions of struts and beams increase and as the loads to be transmitted increase elastic stability becomes a more prominent criterion. The longitudinal stiffness of unidirectional material is high and so the overall Euler-type of strut stability is not a particular problem but the extremely low transverse modulus makes local instability (eg buckling of beam flanges or tube walls) occur much sooner than in comparable isotropic members. As loading index rises, therefore, an increasing amount of composite material must be provided in other directions than longitudinal or a separate material such as metal must be introduced to stabilise the composite. This in turn introduces problems of dimensional compatibility and may complicate the load transfer problem.

The former is well illustrated by the case of a metal beam with composite bonded to the flange. Here materials of different stiffnesses must strain compatibly, whilst thermal stresses are present because of their very different thermal expansions. The two materials will be unstrained at or near the bonding temperature (usually 100° to 170°C) whilst the normal operating temperature range for aircraft structures is from −60°C to 80°C (up to 150°C for supersonic aircraft). The thermal stresses are maximum at the lower temperature limit and are usually tensile in the metal and compressive in the composite (in the longitudinal direction). In addition there is a substantial shear stress in the adhesive near the ends of the composite reinforcement.

The effect of thermal stresses is to cause one or other material to reach its limiting stress prematurely. In the above example at low temperature the metal will yield in tension and the composite fail in compression at lower applied loads than if thermal stresses were absent.

In beam flanges comprising type 2 carbon/epoxy and high strength aluminium alloy it has been shown [1] that the structural efficiency ratio R varies with the proportion of composite from 1 or less to nearly 4. A very useful range of beam designs with about 50% composite in the flanges shows efficiency ratios of up to 3 according to loading direction and working temperature.

For these very simple applications of composites the following information is required for unidirectional material in order to design components effectively:—

Longitudinal tensile and compressive strength;

Interlaminar shear strength;

Longitudinal and transverse coefficient of thermal expansion;

Longitudinal and transverse elastic modulus and Poisson's ratio;

Strength and stiffness characteristics of an adhesive joint between the composite and adjoining materials.

These properties are required over the full working temperature range and before and after exposure to realistic environment. In addition to static strength data we also require fatigue information — not tests on composite tensile or flexural coupons, but realistic data from specimens which represent the critical load transfer mechanisms.

2.2 Two-dimensional panels, tubes and shells

The major part of an airframe structure is made up of relatively thin shells with local reinforcement which may be of the unidirectional type previously discussed or itself be two-dimensional. It is characteristic of such structures that whilst a predominant type of loading may exist there will also be other loading components or significant variations of magnitude and direction of the principal loadings which necessitate more complex laminate design. In a limited number of situations, such as pure shear members and pressure vessels, a two-directional fibre layup may suffice — either in the form of a filament wound shell or in a multilayer sheet with two sets of unidirectional layers identified by the fibre orientations. Cylindrical pressure vessels with longitudinal and circumferential fibres and torque tubes with fibres at ±45° to the axis are well known examples. The small control surface — a Jet Provost rudder tab, shown in Fig. 4 — is a less obvious example. This is attached by a continuous hinge along its length and apart from a local pressure loading it is subject to almost pure torsion. A ±45° skin layup is adequate to carry all load components.

In general more than two such layer directions are required and these may in theory be optimised to suit a particular situation. In practice it is often convenient to restrict the orientations to combinations of 3 or 4 basic directions (usually 0° ± 45° and 90° to the reference direction). Composites built up from such standard units can be characterised by means of a finite and standardised programme of tests.

The stiffness properties and built-in thermal stresses in laminated sheets are readily calculable, on the basis of plain strain theory, by simple superposition and through-thickness integration of individual layer properties. This process, adequate for definition of behaviour away from the edges, section changes or joints in a sheet, is not capable of dealing with the interlaminar and translaminar stress systems which occur near such discontinuities. These stresses, which are usually considered 'secondary' in conventional isotropic materials, are of primary importance in considering the integrity of composites because of their extreme anisotropy. Analysis of these stress systems is a very difficult undertaking because of the small scale and fine detail which are necessary if it is to be meaningful. The best that can be done with present day computing facilities is to analyse global behaviour of laminates on a plane strain basis and to use finite element or other numerical methods applied to a two dimensional slice through the material in the region of discontinuities (as illustrated in Fig. 5).

An added complexity whose significance may not be too great, but which varies according to the laminate layup is that the properties of the individual layers may change during the service life as a result of partial cracking of the matrix. The influence of matrix cracking on integrity and the determination of strength envelopes of degraded materials will be considered later; it is sufficient to note here that a rigorous analysis of laminated composite behaviour should reproduce the gradual degradation of the material. This is clearly not practical in anything other than typical model situations but it will often suffice to check the influence of

Fig. 4 Jet Provost rudder tab — a torsion tube

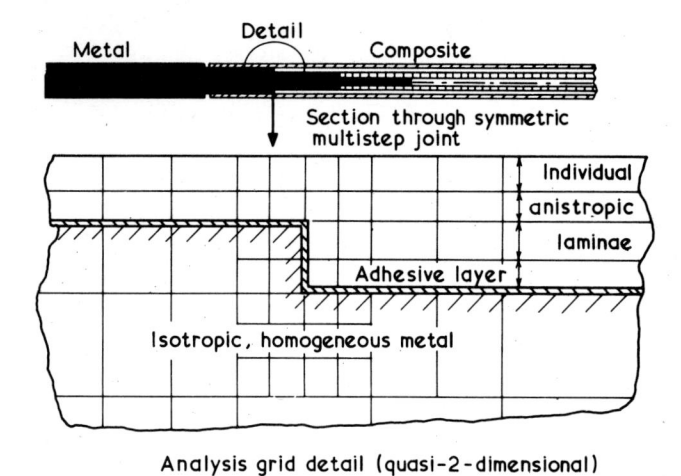

Fig. 5 Finite element analysis of stepped joint

the most severe likely (or tolerable) degradation and if necessary perform detail analyses before and afterwards. It would be highly desirable to avoid this problem altogether by choosing a sufficiently ductile matrix material, but the present-day epoxy, polyester and polyimide systems are brittle and the choice of laminate layup to avoid layer cracking is in most cases too restrictive to be economical.

To stabilise composite sheets either of the conventional methods of sandwich construction or discrete stiffening may be used and in both cases the designer has rather more flexibility in the choice of material layup to meet specific requirements than is the case with isotropic materials. For example, in sandwich panels, conventional practice suggests that the skins should be designed to carry all components of in-plane loading and a multidirectional layup with appropriate edge supports as in Fig. 6(a) is indicated. In many aircraft skin panels the primary loads are introduced as distributed shears along discrete longitudinal spars and in a conventional sandwich these shears must be diffused into the skin thereby adding a substantial shear loading to the primary longitudinal load. A wide panel must be designed to be stable under these loadings. An alternative design as in Fig. 6(b) concentrates the longitudinal loading into unidirectional composite strips within the core of the sandwich at those positions where the longitudinal shear loading is introduced. This greatly reduces both the longitudinal and shear loading in the skins and in many cases a simple angle-ply layup will suffice. The stiff longitudinal strips prevent excessive strains from developing in such skins. Both types of panel have been satisfactorily developed and tested at BAC and the choice between them depends on the detailed design requirements.

A special type of sandwich construction can be used for some of the thin flying surfaces and control surfaces typical in aircraft construction. This is the so-called full-depth sandwich in which the core not only stabilises the load-bearing skins but also carries most of the normal shear forces. Many examples of such structures exist — notably the boron/epoxy skinned tailplanes in service on the US Navy's F-14 aircraft [3]. This form of construction is not theoretically very efficient because of the very high weight of the core material, but it is essentially rather simple and lends itself to economical fabrication. Indeed it has been claimed that the fabrication cost of a composite skin including all edge members and reinforcements is lower than that for a corresponding all-metal skin with a similar degree of refinement in thickness graduation. Furthermore, a composite skin manufactured on a mould which defines its inside contour with precision is easier to match to a machined honey-comb core and framework than is a metal skin.

A general indication of the relative efficiencies of composite sheet structures and their conventional metal counterparts is provided by comparing a number of box beams typical of aircraft wing and tail surfaces. Fig. 7 shows such a comparison in which structural efficiency measured as skin load intensity/box mass per unit surface area is plotted against an appropriate loading index, for box structures with composite skinned sandwich and aluminium skin-stringer face panels and for full depth aluminium honeycomb core sandwich structures with composite skins and aluminium alloy skins. Over most of the loading range the composite sandwich skinned box structure is nearly twice as efficient as the conventional metal-skinned construction and the full depth honeycomb sandwich about 25% more efficient than the metal-skinned box. The actual efficiencies achievable in practice are greatly influenced by the design of edge members and attachment fittings which can in some cases weigh more than the whole

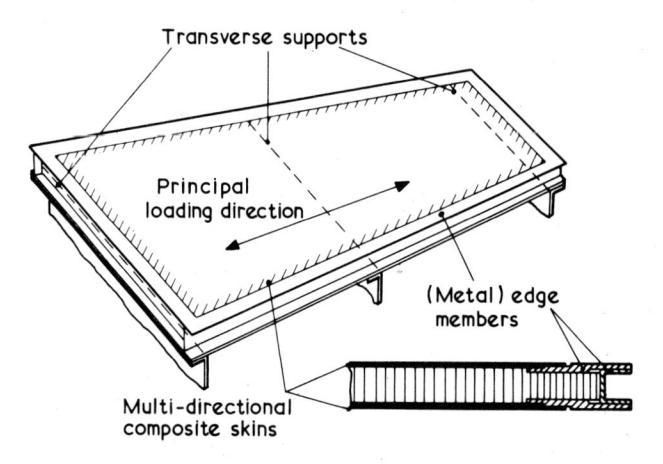

Fig. 6a Thick sandwich panel — 'picture frame' edges

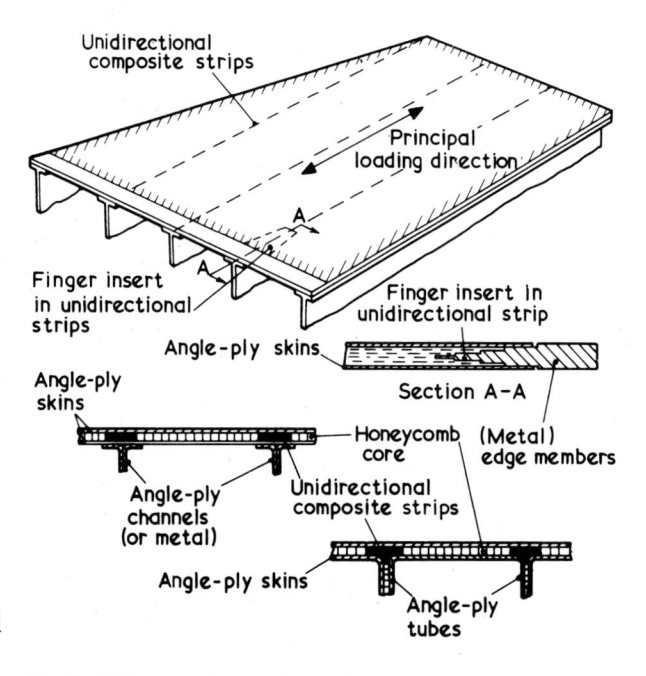

Fig. 6b Thin sandwich panel — multi-spar support

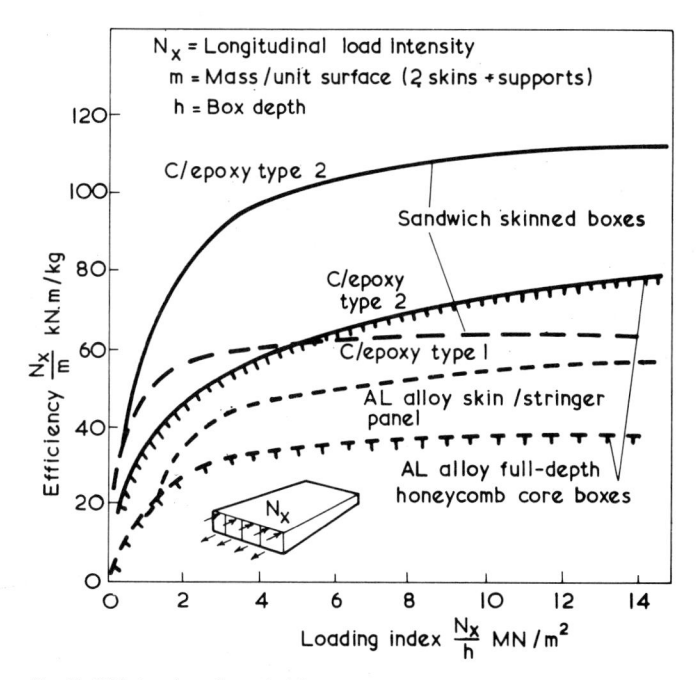

Fig. 7 Efficiencies of sandwich structures

of the composite skins. However, the full depth honeycomb sandwich type of construction has been produced with at least 20% overall weight saving compared with conventional construction.

2.3 Lattice structures

An attractive form of ultra-light construction which lends itself to composite construction using filament winding techniques is the open lattice structure. This has been used in spacecraft structures and is being developed for shear members in aircraft: Fig. 8(a) shows a small development component after testing at BAC. This concept derives its efficiency from the fact that material is concentrated into strips of considerably greater depth (and hence more stable) than the equivalent thin sheet. Substantial loads can be carried without the need for sandwich core or stabilising members. There are obvious problems associated with edge load introduction and with composite consolidation and fibre damage near the cross-over points in the lattice but these seem to be soluble with reasonable efficiency. Fig. 8(b) compares the theoretical efficiency of the lattice web with a plain sheet construction and the practically realisable efficiency depends on the ability of the designer to integrate the edge attachment into the overall panel design.

2.4 Solid members

There are not many examples known to the author of solid members made up from fibre reinforced composites except for various turbine and impeller blades used in the aero-engine and accessory fields. These centrifugally loaded components make good use of the anisotropic material but are notoriously difficult to design to deal with impact loadings. A fundamental difficulty in designing three-dimensional components is the difficulty of providing three-dimensional reinforcement or alternatively of avoiding delamination in components built up of two-dimensional layers.

However, an aircraft structure contains a number of solid members which can contribute quite a substantial proportion of the total weight, eg undercarriage components and wing and tail surface support members. There is a potential payoff from the use of composites in these applications but the difficulties of their design for adequate integrity will make it likely that their realisation will be delayed until a greater body of experience has been built up with simpler members.

3 JOINT AND FITTINGS

3.1 Joints

The ideal composite structure is a monolithic member with all structural features laid up in situ and cured in a single operation. Quite substantial structures such as grp helicopter blades and tailplane wings have been made by such methods and where possible this approach should be the ultimate objective or at least the standard beside which less ideal structures are measured. A typical airframe is far from being a pure structure in the sense that its sole function is the transmission of loads. Indeed the smooth external shape belies the fact that the airframe is structurally discontinuous, with doors, access panels, windows, control surfaces, inspection and maintenance hatches covering a large part of the surface. Internally the modern aircraft, in addition to crew and pay load, houses a vast and complex array of electronic and mechanical equipment, pneumatic and fluid systems interconnected by a labyrinth of wires and tubes.

Fig. 8a Lattice panels after test

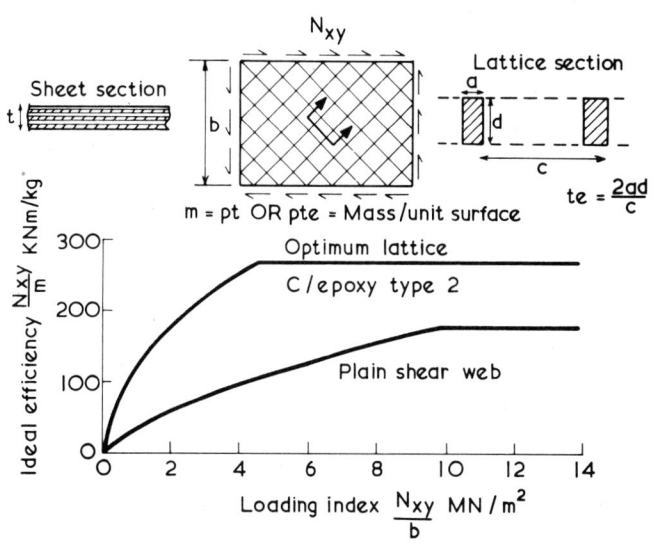

Fig. 8b Lattice and plain sheet shear webs

Installation, inspection, maintenance and replacement of structure and systems demand that many assembly joints must be present, some of which are permanent but many more require disconnection with varying frequency throughout the service life. In short, major structural discontinuities and joints are an essential feature of the complex aircraft structure and the successful exploitation of composite materials will stand or fall by the design and performance of the joints and fittings employed as well as by the designer's ingenuity in providing continuous load paths. Three principal types of joint are used in the assembly of carbon fibre reinforced composite structures:—

(i) bonded composite—composite or composite—metal assembly joints;

(ii) joints comprising bonded edge members, usually of metal, with conventional mechanical joints through the edge member;

(iii) direct mechanical joints through the composite, with or without reinforcement.

Joints of the first type have been widely studied and reported, usually for the simple case of unidirectional composite strips subjected to tension loads. The simple lap joint is often tested but rarely found in practice — more practical configurations are butt-strap joints with various refinements in geometry such as tapers or steps to improve joint efficiency. Fig. 9 shows some of the specimens tested in a recent programme which, together with supporting theoretical analysis, is improving our understanding of the mechanical behaviour of these joints. For example, the influences of adhesive strength, ductility and geometrical parameters are illustrated in Fig. 10 derived from this work. This knowledge is guiding the development of improved adhesives as well as geometrical design.

Considerable sophistication in design has been introduced into end fitting joints as illustrated in Fig. 11. The taper and step configurations play an important part in reducing stress concentrations and enable joints of extremely high strength to be made using adhesives of quite modest shear strength and low ductility. The joint illustrated in Fig. 12 can transmit tension and compression loads of over 4MN/m (23 000 lbf/in) with an average adhesive stress of 20MN/m^2 (3 000 lbf/in^2), whilst a symmetric double lap joint with the same materials would be limited to about 1.5MN/m^2. The manufacture of stepped lap and shimmed joints to the necessary tolerance is relatively simple if the principle of co-curing during layup

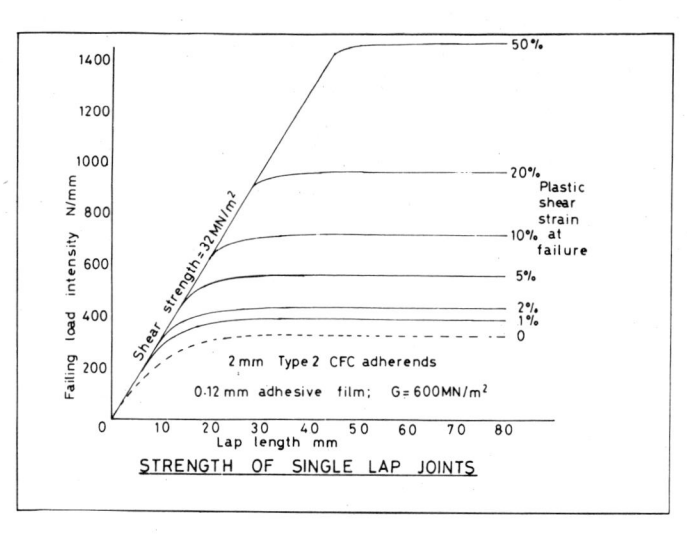

Fig. 10 Strength of single lap joints

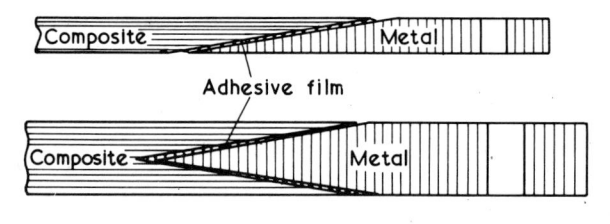

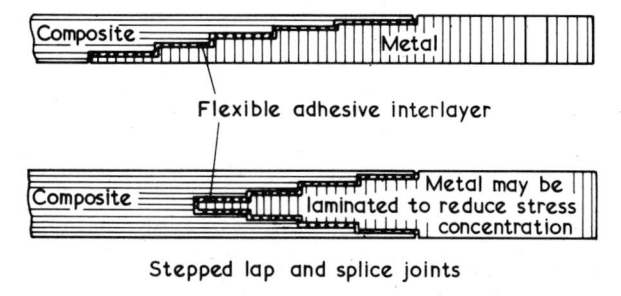

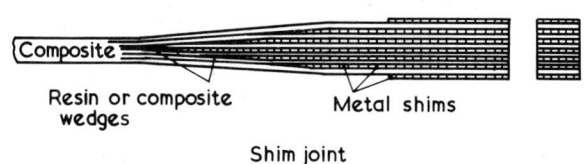

Fig. 11 Edge member concepts

Fig. 9 Composite bonded joint test specimens

is used. For example, the joint illustrated is made up of metal shims carefully matched in thickness to preimpregnated carbon fibre tape and laid up in a curing fixture at the same time as the composite. Interleaved layers of film adhesive are cured at the same time as the composite. This is an economical way of making accurate and strong joints even if the component is curved. The symmetrical layup minimises distortion after cure.

The major proportion of joints in airframe structures, particularly at junctions between skins and internal members, are shear joints. Similar basic theories can be used to design them but the layup of the composite skins being joined in such instances tends to reduce the possible load-bearing capability. Generally, in shear-carrying joints, at least one of the adherends will contain angled layers as well as longitudinal plies. The interlaminar shear strength between angled

Fig. 12 Fin spar joint — symmetric multi-step composite/steel

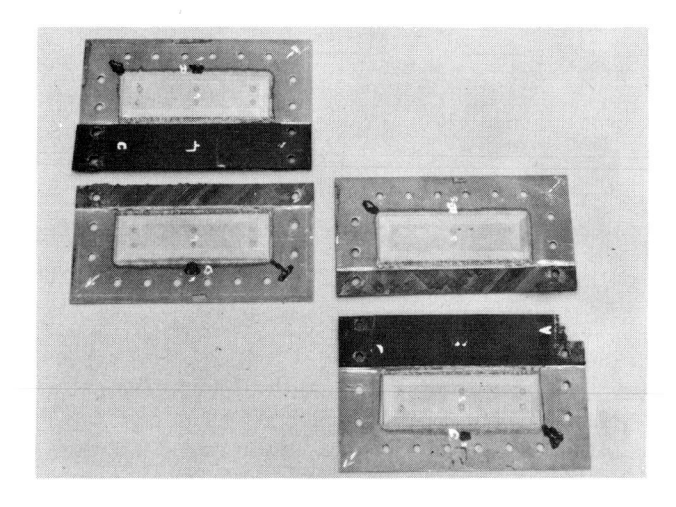

Fig. 13 Failure of composite — metal bonded shear joint

layers may be less than half that of longitudinal material and delamination strength is similarly reduced. The strength limits of bonded shear joints between composite members tend to be determined by the parent material rather than the adhesive as illustrated in the composite-metal joint shown in Fig. 13. Usually it is very difficult, also, to use a sophisticated (eg multi-step) joint geometry in assembly joints because of the difficulty of matching joint surfaces to high tolerance. For these reasons, among others, it is more difficult to design practical bonded shear joints of very high load capacity than it is to design efficient tension joints. In our own work we have not found it possible to make a single face-to-face shear joint capable of carrying more than about 500kN/m and if peeling stresses are present 300kN/m is a more realistic figure. Tapered or stepped joints could, if practical, improve on these figures.

These limitations lead the designer back to more conventional design for highly loaded shear joints and the use of bolted or riveted joints through composite materials is probably increasing. These may either rely on a suitably chosen layup of the composite itself to provide the necessary bearing strength or may use local reinforcement of metal or, for instance, woven glass cloth. Metal reinforcement is particularly effective for riveting and relatively thin facing sheets can provide reaction for setting loads as well as bearing reinforcement. Titanium foil is a particularly suitable material in conjunction with carbon fibre.

Tests carried out at RAE and in industry show that typical $(0°, \pm 45°, 90°)$ laminates in carbon/epoxy can withstand average bolt bearing stresses of the order of $5-600 MN/m^2$ in single shear joints. These are similar to the values obtained with aluminium alloys and hence quite conventional joint designs can be achieved in carbon/epoxy laminates. Again local metal foil reinforcement at or near the laminate surface can be effective in resisting assembly damage.

The beam shown in Fig. 14 provides a good illustration of the integrity of bolted joints in relation to bonded shear joints. Despite excessive flexural strains and consequent additional loads, the bolted joint is undamaged when all surrounding shear joints have failed.

3.2 Fittings

Many workers have found that in designing components to exploit high performance composites the integrity and efficiency depend to a high degree on the design of the fittings for attachment and local loading. Frequently good fitting design is inseparable from good joint design and the previous discussion is relevant.

Generally a good multi-step or shimmed shear joint will provide the most efficient connection to a metal fitting whether this be in a bar, a panel or a tube [as shown, for example in Fig. 3(a)]. Alternative fittings such as those using wedging actions eg Fig. 3(b), necessitate more massive metal parts and in the case of lightweight rods and tubes these have often been found to be heavier than the composite part of the member.

It is particularly important in designing end fittings to bear in mind the disparity in stiffness, ductility, thermal expansion etc between the members being joined. For example if a metal lug is attached to the end of a rod, the strain variation across the metal may cause local yielding and even out the stresses. In a badly designed fitting, if large strain gradients are present at the composite interface, they produce premature failure.

The successful design of fittings for use with composites will demand either the systematic development of classes of

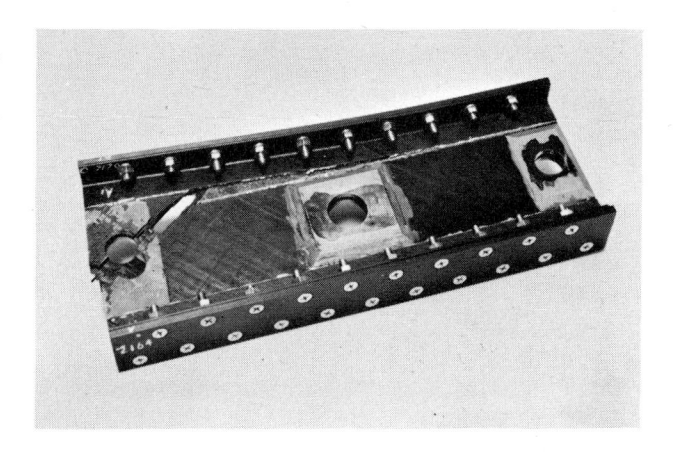

Fig. 14 Beam with bonded and bolted shear joints

fittings for specific load bearing applications or the careful analysis, in detail, of stress and strain distributions and the adjustment of designs to keep stress levels tolerable within the composite and the interface.

4 SOME CONCLUDING REMARKS ON INTEGRITY AND ECONOMICS

4.1 Economic fabrication

The bonded and bolted beam shown in Fig. 14 represents perhaps the ultimate in, apparently, 'wrong design' in a composite member: too many pieces and too many joints. However it was practical fabrication economics and shear joint integrity which led to this design. At the other end of the scale, the one-shot bonded panels in Fig. 15 represent the best use of the peculiar fabrication and handling characteristics of the material, namely the layup and cure in shaped moulds of two composite faces, a sandwich core with its film adhesive and all edge reinforcement in a single processing sequence.

A considerable range of aircraft components exists in which loadings are relatively low but shapes are complex and often stiffness is an important design requirement. Many fairings, doors and hatches come into this category and up to 30% of the external surface (and 10% of the structure weight) may be made up of such items. The composite manufacturing process seems ideal for such components and already grp is being used extensively. The use of higher stiffness materials can only be justified on cost grounds where grp stiffness is inadequate but a useful technology is developing with a wide range of materials which can be tailored to individual requirements. Direct cost savings are sure to emerge and the likely weight saving appears as a bonus.

In most structural applications, however, it is structural efficiency under service conditions which provides the economic justification and it is worth considering a little further just what is implied by using composites in a realistic environment.

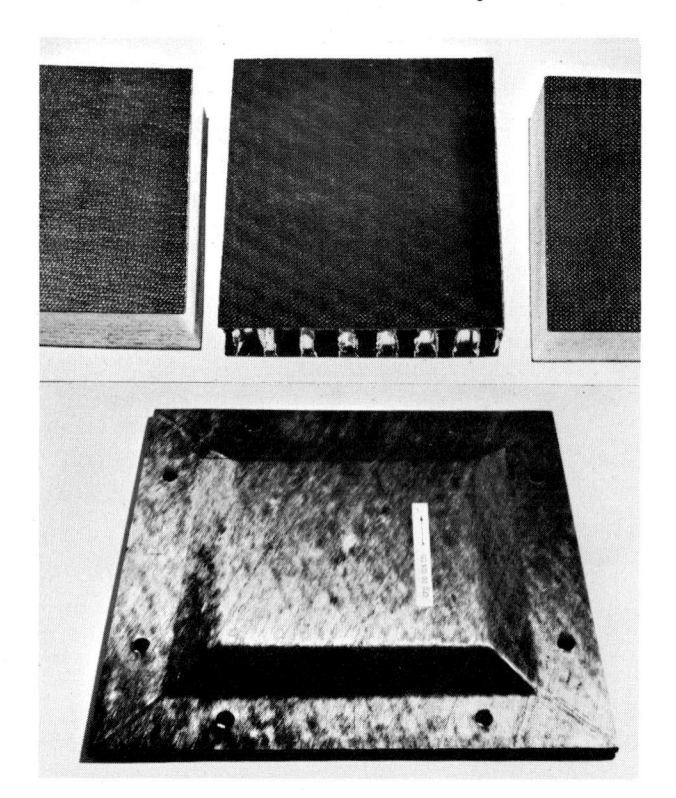

Fig. 15 One-shot bonded and cured panels

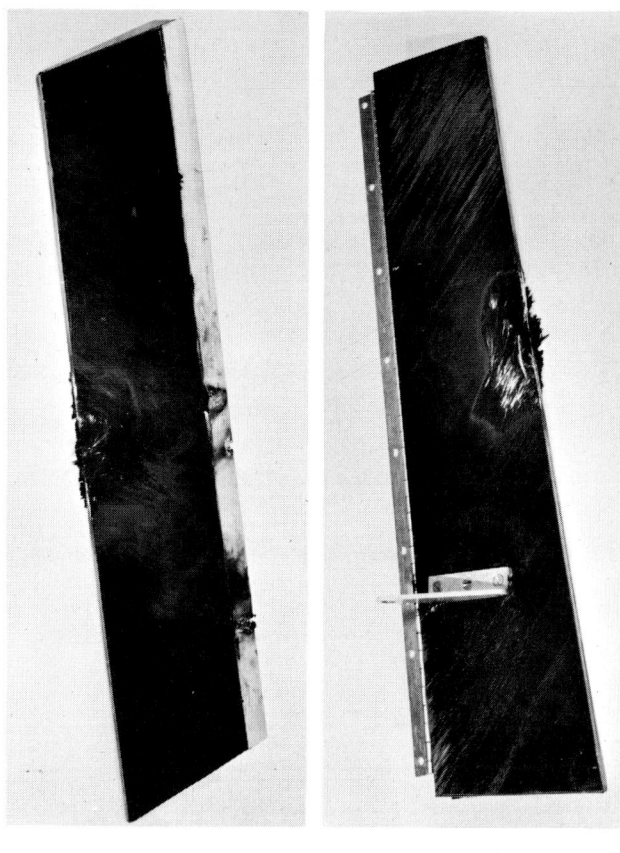

Fig. 16 Damage from simulated lightning

4.2 Integrity in a realistic environment

The high modulus composites are, in many respects, vulnerable materials. They are by their nature liable to impact damage, a situation which can only be improved by providing, either within the constituent materials or by permitting internal relative movement [4] a significant ability to strain without local failure or loss of elastic stiffness. A palliative, but not in any sense a cure, is to introduce reinforcement, by pinning or stitching, in the third dimension (normal to the fibre layers) which may prevent total delamination following impact but not greatly affect the onset of serious fibre and resin damage. At the present time a good rule for the designers is to avoid the use of brittle composites in those areas of primary structure most prone to impact damage. Otherwise they must provide local protection in the form of tough metal sheathing or shields.

In boron and carbon composites the fibre itself and often its surface, are impervious and inert to the various fluids and atmospheres encountered by aircraft. The resin systems currently in use are all affected to some extent by water and moist air but the degree and nature of this effect varies with the resin system and curing conditions. Many are also affected by Skydrol fluid and some by solvent and cleaning fluids in common use. Any resin system suffering catastrophically from contact with these fluids should be ruled out for primary structures. However, water remains as the most aggressive and unavoidable contaminant and composite properties used in design must take account of its effects. This applies even more to glass composites because in this instance the fibre surface is also seriously affected.

Thermal and load cycling may cause resin cracking and local fibre/resin debonding which influence the mechanical properties of the composite directly and also provide paths for the ingress and dispersion of moisture which may do physical (eg expansion during freezing) as well as chemical damage. The effects of cycling are particularly pronounced in multidirectional laminates where transverse and shear strains can be introduced in individual layers such as to precipitate resin cracking. Many composites can survive such cracking — often without apparent loss of static strength — but its occurence is probably the primary cause of 'fatigue' in composites and it can lead to explosive delamination under compression or shear loads.

As a proper background to safe structural design we therefore need data on the strength of representative composite layups after exposure to temperature, load and atmospheric cycling representing normal, or accelerated, airframe environments. A full series of tests on all layups and cycles is obviously impractical and it is necessary to produce such information for a limited range of layups at first and to restrict primary applications to that range.

Testing of multidirectional laminates under variable loads and environments is both difficult and expensive. Two basic approaches are being adopted: one is to make tubular specimens with carefully designed end conditions, in which practically any load combination can be applied but manufacture is usually unrepresentative; the other approach is to make cruciform coupons or beams of representative layup and fabrication and to test them under specific load combinations related to their geometry. Data from such tests are not generally available for the materials of current interest to designers, but the need is becoming increasingly recognised.

The final environmental hazard, regarded by many as perhaps the most important, is lightning damage. The effects of lightning strikes and intense static discharge on grp structures already in service vary from small holes to explosive failure. The higher conductivity of carbon and boron/tungsten fibres compared with glass alters their characteristics and typical damage due to simulated lightning is shown in Fig. 16. It is found that resin burn-out occurs in the immediate vicinity of the strike but that damage is usually contained to within the visibly affected area. There are significant differences in behaviour between different fibre/resin/core material combinations but again the full range of data is not yet available.

Impact (particularly minor erosion), water ingress and lightning stike can all be mitigated by surface protection and a major field exists for development of integrated protection systems to increase the integrity of composite airframe structures.

The overall picture with regard to airframe applications of composites is thus one of marginal but rapidly improving economic justification and a technical problem in ensuring integrity which is now well understood but with many gaps in available design data and great scope for further development.

ACKNOWLEDGEMENTS

The author wishes to acknowledge the support of the Ministry of Defence (Procurement Executive) in carrying out the work on which this paper is based.

REFERENCES

1 Taig, I.C. 'Design concepts for the use of composites in airframes', *AGARD Conference Preprint* No 112 AGARD, France (1972) paper 4

2 Dial, D.D. and Howeth, M.S. 'Advanced composite cost comparison', *SAMPE Conference Proceedings* 16 Anaheim Conference (1971)

3 Lubin, G. and Dastin, S. 'First boron composite production part', *Proceedings of the 26th Annual Technical Conference, SPI Reinforced Plastics/Composites Division* (1971)

4 Morley, J.G. and Millman, R.S. 'The application of self adjusting interfaces to the design of fibrous composites', *J Mat Sci* (to be published); see also Morely, J.G. 'The design of fibrous composites having improved mechanical properties', *Proc Roy Soc A* 319 (1970) pp 117–126

QUESTIONS

1 *On hysteretic heating and design of turbogenerators, by J.S.H. Ross (International Research and Development Co Ltd)*

First of all, I would congratulate Mr. Taig on his most interesting and realistic paper.

Secondly, I want to make a general comment on one aspect of composite materials which has not yet been mentioned at this conference: hysteretic heating. By this I am referring to the production of heat in the body of the composite due to cyclic mechanical stressing and the associated rise in temperature of the composite. This phenomenon has been seen to have a dominant effect on the fatigue behaviour of Permali wood-based laminates and similar effects are evident in grp laminates, particularly in non-unidirectional material.

Before going further, I must mention the application as it is on a much larger scale than has been discussed so far at this conference. I am concerned with the design of large superconducting turbogenerators, with rating of 1000MW and upwards. The ambient temperature stator structure is likely to be a composite, non metallic, material with a mass of perhaps 10 to 30 tonnes, and subjected to mechanical stressing at power frequencies — 50Hz and 100Hz. In such massive sections, thermal runaway conditions are possible due to hysteretic heating and this effect can dominate the design of low modulus composites.

For example, tests at IRD on Permali wood/phenolic composites have indicated that thermal runaway conditions could arise in samples of 100mm to 150mm thickness cycled at 100Hz over a stress range of only 5% of their ultimate strength. Similar results can be expected from CSM grp, and WR grp to a lesser extent, from considerations of the nominal elastic moduli and thermal conductivity values.

I would welcome comments by Mr. Taig or any other delegate with experience of these problems of hysteretic heating.

Reply by Mr Taig

I have not direct experience of this problem and in the case of cfrp the higher stiffness and conductivity of the fibres make it less likely to be troublesome. It has been encountered however, by RAE in their high speed fatigue testing and some carefully measured data should be available from the Materials Dept.

2 On post-buckling strength reserves, failure criterion and fabrication methods, by W.M. Banks (Department of Mechanics of Materials, University of Strathclyde, Glasgow)

The stability problems associated with the use of composite sheets were mentioned by Mr Taig. Were the panels designed to take account of any post-buckling reserve of strength and what was the failure criterion applied?

Also, some details on the fabrication methods used in the production of the carbon fibre reinforced plastic 'I' beams would be of interest.

Reply by Mr Taig

Most composite structures are being designed on the basis that instability constitutes failure and in many tests this has occurred catastrophically. Development specimens have been tested in which substantial amounts of buckling have occurred — see for example Fig. 11b in Reference [1] of my paper (this particular specimen failed at about twice the buckling load). Similar behaviour has been observed in this sandwich panels where core flexibility induced a premature local buckle. We would not reccommend buckled design as good practice.

The 'I' beams shown in the paper are bonded (and bolted where relevant) from previously curved channel, angle and strip elements. All elements are made from hand-laid prepeg tape. We have also made conventional bolted metal 'I' beams and bonded on pre-curved tapered strips of cfrp. The former were laboratory exercises but the latter was developed to production prototype level.

3 On lack of fracture toughness, by Mr Cooper (Institut CERAC SA CH–1024 Ecublens, Switzerland)

Does Mr Taig regard the lack of fracture toughness in cfrp as a disadvantage in aircraft applications?

Reply by Mr Taig

Certainly the lack of impact resistance is a major problem which is touched upon in the written paper. But fracture toughness, whilst it seems amenable to the same type of fracture mechanics analysis as is used for metallic materials, may not be the same problem. I am not sure that all carbon fibre/resin combinations have low fracture toughness and even those which have this property are amenable to design of laminate lay-ups which mitigate the effects. It is known that a judicious mix of strips of different orientation can act as effective crack stoppers within the material itself. I believe in fact that any present-day laminate after several cycles of fluctuating load and temperature is full of local, arrested cracks. Many composite structures can tolerate this condition without complete failure but in my submission one of the most urgent queries regarding composites is how much can be tolerated in a real environment and how incipient failure can be detected.

4 On creep recovery behaviour of cfrp, by K.B. Armstrong (British Airways — Overseas Division)

British Airways have used about 1 tonne of carbon fibre composite in floor panels for Boeing 747 aircraft. Over 500 panels have been produced at British Airways from raw sandwich panels manufactured by Bristol Composite Materials Engineering Ltd. One panel has flown over 14 000 h and continues in service. Several others have reached 10 000 h, many more about 5 000 h and all are continuing. This note is given as an indication of the extent of practical applications of carbon fibre.

Glass fibre panels have now been developed to meet our specification but difficulty has been experienced with creep effects. This appears to have been adequately dealt with but the following points have been of interest to us.

(a) The first type of cfrp panel suffered resin crazing of the bottom skin;

(b) The second type emitted 'ticking' sounds during a deflection test. As a result a creep test was run on a 76.2 mm (3 in) wide beam at $\frac{2}{3}$ of failing load. This resulted in creep which caused after a few days and caused significant permanent set. Some resin crazing was also visible;

(c) The third type which finally met our requirements was tested as above but showed considerable, though not complete recovery after the creep test without any resin crazing visible to the naked eye. Ultimate loads to failure were not affected by prior creep testing.

Does Mr Taig have any comment on the creep recovery behaviour exhibited by our type of cfrp?

Reply by Mr Taig

If 'creep' is associated with progressive fracture of either resin, fibre or the interface then it is most unlikely that a great deal of recovery will occur. If there is little or no fracture, then creep under load will result in permanent relative deformation between fibre and matrix. On removal of load the composite will be self-strained with a balancing load system in the fibres and the resin which may itself cause resin creep. If this occurs the composite will recover because this self balancing load system can only disappear when the fibre (if it is perfectly elastic) has returned to its initial unstrained condition. A combination of an elastic material in parallel with a visco-plastic material is, I believe, the classic example of a mechanical system with 'memory', ie one which will always tend to return to the state of zero internal stress which in this case is the undeformed state. This is an entirely theoretical interpretation of the observations and is not backed up by direct evidence from our own work.

The disproportionate weakening of composites by sub-millimetre defects

L.E. DINGLE, R.G. WILLIAMS and N.J. PARRATT

1 INTRODUCTION

Nearly all the present generation of high modulus composites have only a low elongation to break, which implies that they will crack, at least locally, if significant deformations occur in service, eg at strain concentrations, or by impact, or heavy scoring. It is also difficult to eliminate entirely those flaws which occur infrequently during fabrication, although some types of defects may be located with reasonable certainty using detection equipment, but usually only those exceeding one or two millimetres in size. For design purposes then, it is obviously important to estimate the residual strength of composites in the presence of defects.

Now K_c values (critical stress-intensity) have already been measured or calculated for cracks a few millimetres long, in for example, boron-aluminium [1], SiC whisker-aluminium [2] and cfrp, [3,4]. In all cases, values were in the range 30–50MN/m$^{3/2}$ in the direction of maximum strength, which is perhaps satisfactory, but not outstanding. However, in measuring the fracture toughness of a composite there is no special reason for supposing that plane strain conditions represent the worst case, nor indeed that the work of fracture will be of a constant value locally. It therefore seemed appropriate to introduce into laminates sub-millimetre defects which cut through the fibres, but which could not be detected by inspection. Extrapolation of the existing K_c data suggested that the weakening effect of such defects should be small. In the event, this was not the case.

Weakening was first observed in aluminium reinforced by silicon carbide whiskers, where half-millimetre holes, or quarter-millimetre pairs of edge notches, caused fractures at mean stress levels some 30% lower than for undamaged specimens [2]. A closer examination of the literature then showed some signs of a similar departure from the (crack length)$^{\frac{1}{2}}$ relationship in both boron reinforced aluminium [1,5] and a cross-plied cfrp [3] as the defects were reduced in size. Some cfrp materials in common use were then tested and these were also weakened significantly, as the following results show.

2 EXPERIMENTAL

Holes and notches were introduced in the strip specimens of a standard three-point bend test, normally used for quality control. The holes were drilled, and the edge notches were cut with a scalpel. Holes were 0.5 mm diameter and notches 0.25 mm deep, unless otherwise stated. The type of specimen is shown in Fig. 1, 100 mm long × 10mm wide by 2 mm thick, tested on a span of 80 mm. The artificial defects were offset 4 mm from the centre loading point to ensure that they did in fact cause failure, as they did in every case. The breaking stress was corrected both for this position on the specimen and for the small change in cross-sectional area. The unidirectional laminates used had been compression moulded from prepreg materials containing continuous or discontinuous carbon fibre in both HTS and HMS forms. All fibres were in the 'treated' condition and the resin was Shell DX 210 with BF$_3$400 hardener. Five specimens were tested for each experimental condition, including the 'defect-free' controls, which were taken from the same panel in each case.

3 RESULTS

The initial and residual strengths of the materials are summarised in Fig. 2, while Fig. 3 gives a comparison of the relative weakening effect of different defects of similar size on a single material of HMS discontinuous fibre. It will be

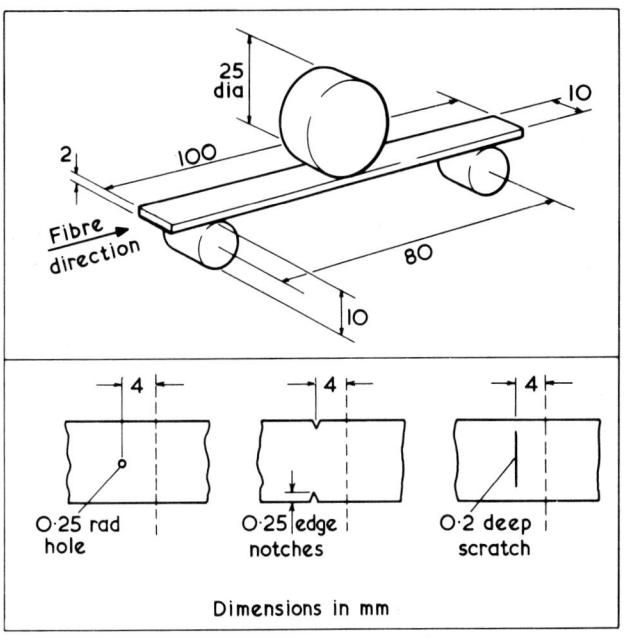

Fig. 1 Flexural strength specimen and type of defect introduced

Ministry of Defence (PE), ERDE Waltham Abbey, Essex, England

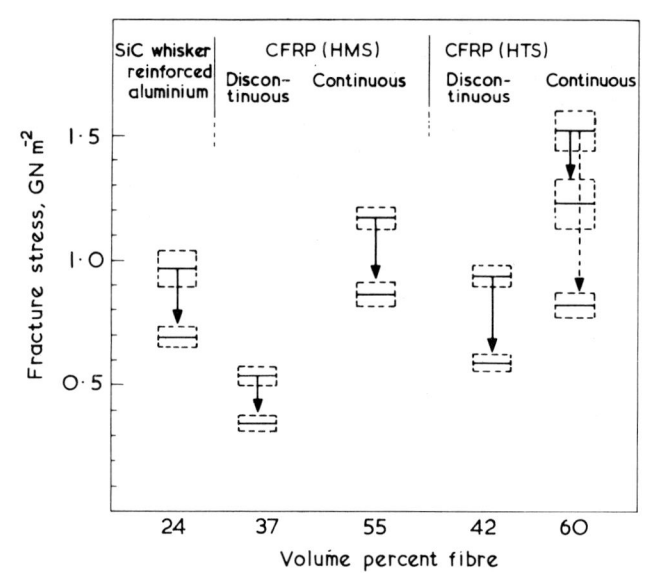

Fig. 2 Effect of 0.25 mm radius holes and 0.25 mm double edge notches on flexural strength

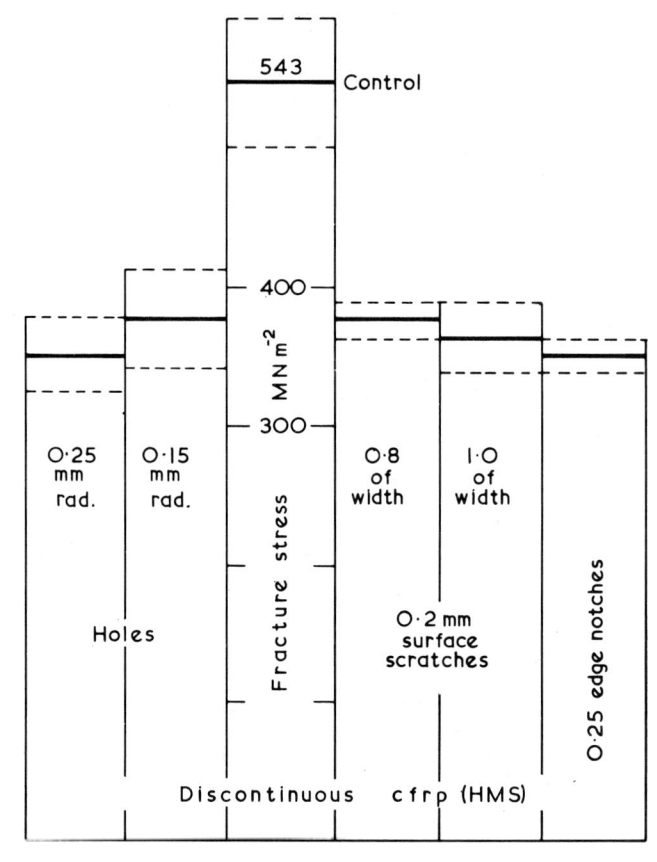

Fig. 3 Effect of type of defect on flexural strength reduction

seen that significant weakening occurs in every case, usually many times the standard deviation, which is also shown. Since the scatter of strength results was, typically, twice that of the modulus measurements made at the same time, one can infer that the scatter is a real material effect, and not a random error of the measurements.

In Fig. 3 it appears that holes and sharp notches of similar width produce similar weakenings, as they do in SiC reinforced aluminium. On theoretical grounds one would argue that the stress concentration around a hole will produce a local material failure at the stress levels applied, thus making the hole virtually equivalent to a notch or crack of slightly larger size than the hole diameter. However, in the one case of continuous HTS material, tiny holes are less damaging than notches, and this distinction is shown in Fig. 2. Nevertheless, it is shown in Fig. 3 that if the edge notches are replaced by surface scratches or slashes of similar depth, the material is still seriously weakened — the probability of attack by boring beetles or flying fragments may be remote, but it seems very likely that minor nicks or scratches will occur in most engineering environments.

In order to make allowances for variations of fibre volume loading between composites, and particularly between discontinuous and continuous, the results for strength and residual strength have been expressed in Table 1 as a fraction of composite modulus. This also approximates to the breaking strain of the composite, and is thus a crude guide to the match of the composite to a metallic structure.

For a given type of composite, one might expect strength, stiffness and the work of fracture all to increase linearly with increasing fibre content. Thus with the axiomatic

Table 1 Strength/modulus (per cent)

	HMS			HTS		
	Discont	Cont	RAE Cont	Discont	Cont	RAE Cont
Control	0.415	0.46	0.65	0.415	1.05	1.13	0.98
Holes 0.15 mm radius	0.31						
0.25 mm radius	0.30	0.50	0.34	0.66	0.95	0.89
Edge Notches 0.25 mm deep		0.33	0.50	0.32		0.63 partial fail	0.85
						0.71 complete	
Surface Scratches 0.2 mm deep		0.32					

assumption of fracture mechanics, the toughness or residual strength will tend to remain a constant fraction of the strength or stiffness, other things being equal.

Results for materials made with ERLA 4617 resin and supplied by RAE Farnborough have also been included in the table, and these emphasise the need for further study. For example it is possible that both the RAE continuous and ERDE discontinuous mouldings of HMS fibre were from early, weaker batches, compared with the recently prepared ERDE continuous. On the other hand, the RAE continuous HTS fibre laminate appears much less sensitive to defects, although its un-notched strength is lower than expected. This result may be related to its very high volume loading (about 75%) compared with the 45–60% loadings in more common use.

4 CONCLUSION

It is evident that fine defects weaken some composites, including cfrp, more than would be expected from K_c measurements based on large cracks. Unless specific tests have already been carried out, for the present it appears unwise when using treated HMS laminates to design for peak stresses in excess of 0.3% of the modulus (less if statistical variation is allowed for) and similarly 0.6% of the modulus for HTS laminates. It is interesting to note that the engineering design figures suggested by Taig [6] are in fact slightly below these values.

It is clear that similar micro-notch tests should first be made in pure tension, but provisionally one might conclude that tests of this type should be made more frequently, especially since there is no evidence that the composites which show the smallest scatter of flexural strength are in fact any tougher or safer to use than those having a larger scatter.

It may be that little can be done to increase work of fracture on a fine scale, when pull-out of fibres has not become fully effective, or a crack front has had little opportunity to become diffused between layers and laminations. It therefore appears best to aim for a material which has a sufficient tolerance of larger cracks, a few millimetres long, to avoid any serious loss of strength.

ACKNOWLEDGEMENT

Crown Copyright is reserved on this paper, which is reproduced by permission of the Controller, Her Majesty's Stationery Office.

REFERENCES

1 Hancock, J.R. and Swanson, G.D. *ASTM STP* 497 (1971) p 299

2 Parratt, N.J. and Cobbin, S. *ERDE Tech Memo* 113 (1973)

3 Zimmer, J.E. *J Comp Mat* 6 (1972) p 312

4 Beaumont, P.W.R. and Harris, B. *J Mat Sci* 7 (1972) p 1266

5 Olster, E.F. and Jones, R.C. *ASTM STP* 497 (1971) p 203

6 Taig, I.C. 'Applications of advanced composites in aircraft structures', *This Conference,* paper 3

Applications of fibre reinforced composites in marine technology

C.S. SMITH

A review is made of existing marine applications of fibre-reinforced composites. Problems arising in the design of marine-type composite structures, including evaluation and selection of materials, structural analysis, buckling behaviour and performance of connections, are discussed with particular reference to large, high-performance grp hulls. Suggestions are made on items of research and development needed in this field.

1 INTRODUCTION

Since the first use of grp for boat construction in the early 1940's [1], a dramatic growth has occurred in marine applications of fibre-reinforced composites. Construction of small boats has in particular been revolutionized, as demonstrated by National Boat Show statistics which indicate an increase from 4% to 80% between 1955 and 1972 in the proportion of boat hulls constructed in grp. Major advances have also been made in the use of fibre-reinforced composites in large, sophisticated hulls culminating in 1972 in the launching of the world's largest grp ship, the 47 m minesweeper HMS Wilton [2].

Despite the success of grp in boat construction and in a wide range of other marine applications, limitations in the properties of presently available materials and in the data and experience needed for efficient design of marine structures have severely restricted the use of fibre-reinforced composites in this area. The purpose of the present paper is to review briefly existing marine applications of fibre-reinforced composites, to discuss problems that have arisen in structural design with particular reference to large, high-performance grp hulls and to suggest items of research needed in order to widen the scope and improve the quality of fibre-reinforced composite applications in the marine field.

2 OUTLINE OF MARINE APPLICATIONS

2.1 Boats of up to 15 m length

GRP is now employed in a very wide range of boat hulls including dinghies, canoes, speed-boats, coastal and ocean-going yachts [3,4], pilot and passenger launches [5], fishing boats [6,7,8], lifeboats [9] and small hovercraft [10]. Its success in this field is attributable to:—

Dr Smith is at the Naval Construction Research Establishment, St Leonard's Hill, Dunfermline, Fife, Scotland

(i) competitive first cost, particularly where many hulls are built to the same design, enhanced by increasing cost and scarcity of wood and skilled shipwright labour;

(ii) trouble-free performance and low maintenance costs resulting from the leak-proof, rot-proof qualities of grp hulls, their resistance to marine boring organisms and their ease of repair;

(iii) the ease with which complex hull shapes, required for hydro dynamic, structural or aesthetic reasons, can be fabricated.

2.2 Larger boats and small ships

Larger scale grp construction has included fishing boats up to 30 m in length [6,11–17], fast patrol boats of up to 25 m for military, fishery protection and other applications [6,18, 19], pilot boats up to 25 m [20], tugs, workboats, lighters [21], military landing craft [20] and minesweepers of up to 47 m [2,22,23,24]. Detailed design studies have also been carried out for minesweepers of up to 60 m [25] and cargo ships of up to 140 m [26].

In ships of over about 35 m steel rather than wood is the main rival of grp. As hull size is reduced, the cost of steel construction per unit weight increases sharply and in boats under 20 m in length, corrosion allowances are likely to lead to uneconomically heavy steel designs. For ships of over 40 m, however, the corrosion margin is a less significant fraction of the required hull thickness and the low cost of welded steel construction is normally a decisive advantage; in ships of over 50 m, grp is at present only likely to be preferable to steel where some special requirement exists, eg for carriage of corrosive bulk cargo. In the case of minesweepers steel and aluminium are both excluded by the need for non-magnetic, non-conducting hulls.

2.3 Deckhouses, masts and funnels

Some use of grp superstructures and masts [27] on steel ships has been made where minimization of topside weight was required for stability reasons. The radar transparancy of grp superstructures is a possible advantage in warships.

2.4 Sonar domes

GRP is now used extensively in warship sonar domes [27, 28], whose purpose is to provide smooth flow round sonar transducers on a ship's bottom, together with protection against damage by hydrodynamic forces. Domes, which are water-filled, must have a high degree of acoustic transparency; in order to avoid signal distortion discrete stiffeners must be eliminated as far as possible and void content minimized. Good results have been obtained from unstiffened single-skin domes formed using polyester resin with alternate plies of woven roving (WR) and chopped-strand mat (CSM) reinforcement.

2.5 Radomes

High specific strength and radar transparency have led to extensive use of grp in warship radomes. Light sandwich construction, based on polyurethane foam or honeycomb cores with epoxy or polyester resin/glass cloth skins, is commonly used in larger domes; single-skin construction is used in small domes. A high standard of fabrication is required with careful matching of dimensions to radar wavelength. Warship radomes must in some cases be able to withstand substantial levels of air-blast loading.

2.6 Submarine casings and appendages

Stiffened single-skin grp laminates based on polyester resin with CSM and WR reinforcement have been used widely [27, 28,29] in free-flooded casings, fins and fairings mounted externally on the steel pressure hulls of submarines. Weight minimization is an important requirement in such structures, which must be able to withstand hydrodynamic forces when submerged and pounding by waves during surface operation. In this application grp has successfully replaced aluminium, which caused severe electrolytic corrosion in earlier designs.

2.7 Sheathing of wood hulls

Effective use has been made of external grp sheathing to protect wood hulls from leakage, rot and attack by marine borers.

2.8 Submersibles

Extensive theoretical and experimental studies have been made of the possible use of filament-wound grp in the pressure hulls of deep-diving submersibles [27,30]. Compressive strengths of up to $1\,250\,MN/m^2$ have been achieved in thick polar-wound S-glass/epoxy resin cylinders with 2/1 distribution of fibres in the circumferential/longitudinal directions and a glass weight fraction of 0.8. If this very high material compressive strength could be used effectively in a submarine pressure hull, diving depths could be achieved far greater than those obtainable from high-strength steel, aluminium or titanium hulls with equivalent weight/displacement ratio (W/D). Major obstacles to this aim remain to be overcome, however; these include susceptibility to interlaminar shear failure and low-cycle fatigue of the fabricated sandwich or stiffened single-skin shells needed to achieve acceptable W/D. Successful use of filament-wound grp has in the meantime been made in low-performance grp hulls, including torpedoes and shallow-water submersibles, and in underwater pipelines.

2.9 Other uses

Other marine applications of grp have included floats and buoys for fishing and minesweeping purposes [27], ship's piping systems and ventilation trunks [29], oil and water storage tanks [27,29] and a wide range of other secondary ship's fittings.

3 MATERIAL CHARACTERISTICS

A comparison of the properties of ship-type grp laminates and other marine construction materials is made in Table 1. Ship and boat hulls are usually of single-skin construction, fabrication being carried out by a cold-cure, contact moulding process [31] in a female mould using polyester resin and E-glass reinforcement. CSM reinforcement is normally used in small boat hulls while WR or mixed WR/CSM reinforcement is used in larger hulls. Polyester resins and E-glass fibres are preferred to higher-strength epoxy resins and S-glass fibres because of lower cost and greater ease of fabrication. Sprayed chopped-strand fabrication and vacuum bag, autoclave and matched-die moulding [3] have also been used to a limited extent in small boat production and sandwich construction of various types is employed in a small proportion of boat hulls.

3.1 Elastic properties of ship-type laminates

The very low stiffness of grp laminates has proved a serious obstacle to development of efficient structural designs. General stress levels in grp hulls have had to be severely restricted, typically to less than 10% of uts, in order to avoid elastic buckling and excessive deformations. Structural design is normally influenced more strongly by laminate elastic properties than by laminate strengths; careful evaluation of elastic moduli should therefore be treated as an important part of the design process.

Approximate theoretical methods [32,33,34] may be used to estimate laminate moduli from the properties and geometry of fibre and resin constituents, providing a useful means of optimizing laminate composition at the initial design stage. A theoretical study of typical ship-type laminates, showing the influence on elastic properties of such variables as resin weight fraction, fibre and resin moduli and Poisson ratios, fibre contiguity and straightness and directional distribution of fibres, is described in Reference [35]. Theoretical methods do not however provide a complete substitute for reliable test data; experimental evaluation of tensile, flexural and shear moduli should be regarded as an essential step in the design of a high-performance hull. A need exists for development of standards governing the numbers and types of specimens and testing and analysis procedures required to establish a sound statistical definition of elastic properties in thick, ship-type WR and CSM laminates.

3.2 Material anisotropy

While CSM laminates can usually be assumed to have membrane and flexural isotropy, orthogonally reinforced materials, including balanced WR laminates, may be strongly anisotropic. Computed directional variations of elastic properties are shown in Fig. 1 for orthogonally reinforced E-glass/polyester laminates having a resin weight fraction of 0.5 with various degrees of fibre imbalance.

Table 1 Comparison of material properties

Nominal material properties	GRP (polyester) laminates		E-glass fibres	High modulus carbon fibres	Unfilled polyester resin	Wood (12% moisture content)		Mild steel	Aluminium (N8 alloy)
	Chopped strand mat	Woven rovings				Oak	Larch		
Resin weight fraction	0.7	0.5	—	—	1.0	—	—	—	—
Tensile strength (MN/m^2)	110	240	3 500	2 000	50	90*	90*	240 (yield)	140 (yield)
Compressive strength (MN/m^2)	130	180	—	—	140	50*	45*	240 (yield)	140 (yield)
Shear strength (MN/m^2)	85	120	—	—	—	13*	12*	140 (yield)	85 (yield)
Interlaminar shear strength (MN/m^2)	20	15	—	—	—	—	—	140 (yield)	85 (yield)
Young's modulus (MN/m$^2 \times 10^3$)	10	14	70	400	3.5	9.5/1.8+	9.0/0.4+	207	70
Shear modulus (MN/m$^2 \times 10^3$)	4	3.5	28	—	1.3	0.75+	0.6+	83	26
Poisson ratio	0.18	0.13	0.22	—	0.33	0.5/0.094+	0.56/0.025+	0.3	0.32
Coeff of expansion (°C)$^{-1} \times 10^{-6}$	30	14	5	−0.6	90	5*	4*	12	22
Specific gravity	1.5	1.7	2.5	1.95	1.2	0.60	0.40	7.8	2.7
Material cost (£/tonne)	560	660	660	40 700 (for 50 ton lot)	460	185	275	90	610
Cost of fabricated structure (£/tonne) (including labour, overheads, material wastage and profit margin)	1 620 (15 m fishing boat)	2 240 (20 m patrol boat) 5 600 (50 m minesweeper)	—	—	—	760 (25 m fishing boat)	760 (25 m fishing boat)	560 (50 m trawler) 275 (200 m tanker) 1 730 (100 m frigate)	1 730 (estimated for 30 m trawler)

* parallel to grain
+ refers to longitudinal/tangential directions relative to grain

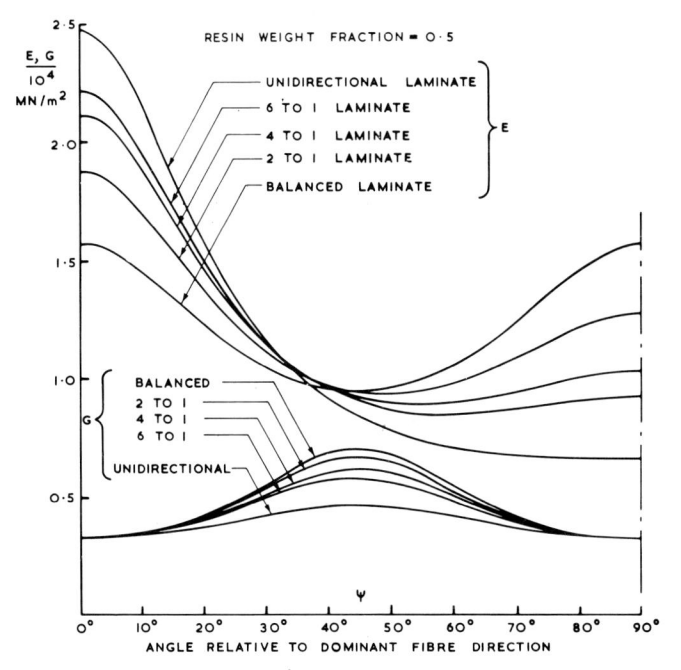

Fig. 1 Directional variation of Young's modulus and shear modulus for orthotropic laminates with various degrees of fibre imbalance

3.3 Static strength

Unlike elastic moduli, the strength of a grp laminate cannot be predicted reliably from the properties of its constituents and must therefore be established by systematic tests. As in the case of elastic moduli, a need exists for development of standards governing statistical assessment of strength in thick ship-type laminates. Properties required for design purposes include tensile, flexural and compressive strengths together with in-plane and interlaminar shear strengths. Although stress levels in a grp hull are normally kept very low in order to avoid elastic buckling and excessive deformation, high stresses may occur locally at structural discontinuities (eg hatch corners in decks) or under concentrated loads; a material failure criterion must then be adopted which accounts for the presence of biaxial and shear stresses in the laminate. Various such criteria, including generalizations for orthotropic materials of isotropic yield conditions [36,37], have been proposed but their validity [38] remains in considerable doubt. A need clearly exists for development of reliable failure criteria for marine-type laminates, based ideally on systematic cylinder tests and accounting for the progressive nature of laminate failure.

3.5 Fatigue strength

Tests on dry, notched and un-notched WR and CSM specimens at temperatures of 15°C to 25°C [39] indicate a fatigue limit of about 25% uts, reducing to about 15% uts in water-immersed specimens. Stress levels in a grp hull are generally kept well below these levels and in existing grp boats serious fatigue failures of the parent laminate have been rare. An improved, quantitative understanding of the progressive development and propagation of fatigue damage in ship-type laminates is however desirable as a basis for specifying maximum stresses at structural discontinuities.

Fatigue failures in a grp hull are most likely to occur at bonded structural connections and thorough fatigue tests on specimens representing fabricated joints are an essential step in the development of a high performance design.

3.5 Impact strength

Drop-weight tests on panel specimens [39], in which water leakage through the laminate may be taken as a failure criterion, probably provide a better comparative indication of impact strength in ship-type laminates than Charpy or Izod tests. As in the case of fatigue, the strength of a complex grp structure under impact loads is likely to depend more critically on the performance of bonded connections than on the strength of the parent laminate.

3.6 Creep effects

Test data for WR and CSM laminates under tensile and flexural loads in wet and dry conditions [39] suggest that creep effects may be ignored where stresses do not exceed 40% uts and temperatures do not exceed 25°C. Under short-duration, wave-induced loads, creep is always likely to be negligible. Significant creep effects might however occur in structural components exposed to high temperatures (eg decks under tropical sunshine) and simultaneously subjected to prolonged load (eg static hull-girder bending or concentrated machinery loads). Further creep test data are needed for marine-type single-skin and sandwich materials at temperatures of up to 70°C.

3.7 Effects of temperature on material properties

Tests on typical marine-type laminates [2,39] indicate losses of up to 35% in modulus and 40% in strength at temperatures of 70°C relative to properties at 20°C. Substantial losses of strength and stiffness at high temperatures also occur in certain sandwich core materials, notably pvc foam [20]. Serious temperature effects are most likely to occur in a ship's decks, sides and superstructure under exposure to tropical sunshine. Scope clearly exists for development of ship-type resins and sandwich core materials with improved retention of mechanical properties at temperatures of up to 70°C. A need also exists for experimental investigation of temperatures in single-skin and sandwich panels exposed to severe insolation, including assessment of the benefits to be obtained from use of light-coloured paint or pigmentation.

A feature of many grp laminates which suggests potentially valuable applications in the containment of low-temperature cargoes such as liquified gas is their retention of strength and stiffness at temperatures as low as −250°C [40]. Design of steel LNG carriers is seriously complicated by the risk of cargo leakage resulting in severe thermal stresses combined with drastic embrittlement of the hull structure. As a material for cryogenic containment grp offers, in addition to strength retention, good insulating properties and an ability associated with its low elastic moduli to withstand sharp temperature gradients without developing high stresses.

3.8 Effects of immersion and weathering

Extensive test data on speciments subjected to water immersion and accelerated weathering under laboratory con-conditions [39] and on specimens cut from boat hulls [41] and submarine casings [42] after prolonged service indicate losses of strength and stiffness in E-glass/polyester laminates of up to about 20%. Further information is needed on the performance of immersed material specimens and structural components under prolonged static and cyclic loads, including particularly sandwich materials in which absorption and migration of water can cause loss of hull buoyancy and debonding between core and skins. A need exists for development of standard test procedures simulating service conditions of simultaneous immersion and loading without unrealistic exposure of the laminate to water penetration.

3.9 Fire resistance

WR laminates based on normal polyester resins have been found, on the basis of flame tests against stiffened single-skin and cellular sandwich panels,[2,29], to provide sufficient strength retention, insulation and fire containment for use in minesweeper construction. In this respect WR laminates, whose fire resistance is attributable to the flame barrier formed by the glass reinforcement after burning off of surface resin, have been found much superior to CSM laminates and equivalent aluminium panels. Flame-retardant resins have been used to a limited extent in boat construction but have been rejected for naval applications [2,29] because of higher cost, lower strength, inferior weathering properties and greater fume toxicity; scope clearly exists for development of improved flame-retardant resins which overcome these difficulties. Intumescent fire-retardant coatings applied to laminate surfaces have however been adopted in minesweeper designs.

3.10 Possible use of carbon fibres in hull construction

The most effective way of increasing laminate stiffness is by incorporation of high modulus reinforcement such as carbon or boron fibres. Bulk use of such materials for ship construction has until recently been out of the question because of extremely high material costs. Substantial reductions [43] in the price of carbon fibre [eg from a previous flat rate of £77 per kg (£35 per lb) for Grafil A to a rate of £18 (£8.2) for lots of 50.8 tonnes (50 tons)] have however revived interest in possible carbon fibre usage. A tentative examination of the cost implications of bulk carbon fibre usage in hull construction [35] has suggested that if high modulus carbon could be obtained at between £11 to £22 per kg (£5 and £10 per lb), increases in a ship's hull stiffness and reductions in weight could be achieved without cost penalty in a labour-intensive design.

The influence of carbon fibre on the elastic properties of possible ship-type materials is illustrated in Fig. 2, which shows calculated principal moduli, Poisson ratio and specific gravity for balanced laminates having various mixtures of carbon and glass fibres, each having a total fibre volume fraction of 0.324 (corresponding in the case of zero carbon content to a weight fraction of 0.5); fibres and resin were assumed to have the properties indicated in Table 1. More efficient use of carbon fibres may be achieved by their selective application in highly stressed components, eg stiffener tables which in the top-hat stiffeners used in ship construction are commonly formed by alternate plies of WR and unidirectional reinforcement. The increase in Young's modulus obtained by introduction of varying proportions of carbon to the unidirectional reinforcement in such a laminate is shown as a dotted line in Fig. 2.

It should be noted that if insufficient carbon fibre is included in a mixed glass/carbon composite excessive load may be carried by the carbon fibres resulting in premature failure of these fibres; load will subsequently be carried by glass fibres, so that laminate strength may not be drastically affected, but the stiffening influence of the carbon will be lost. This effect has been demonstrated experimentally [44]. A simple indication of minimum permissible carbon content for avoidance of premature failure may be obtained from 'rule of mixtures' considerations [35].

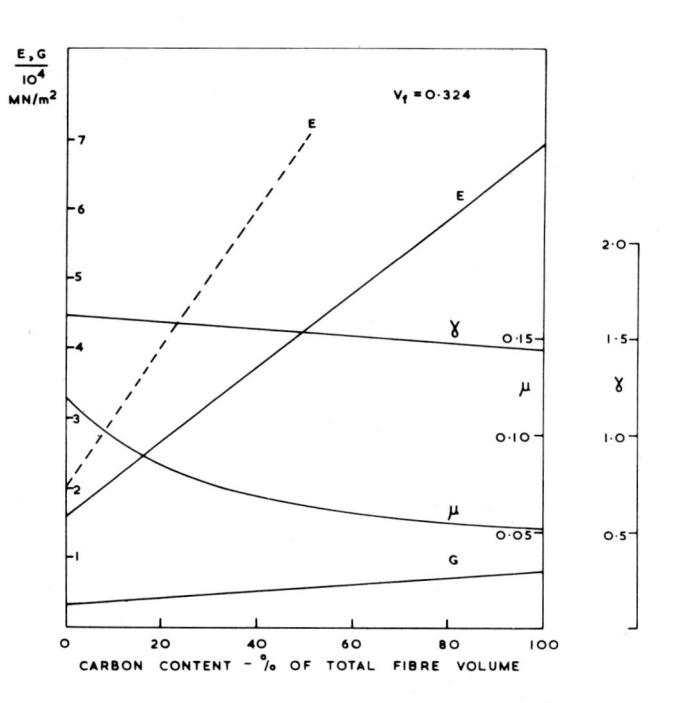

Fig. 2 Influence of carbon fibre content on properties of balanced laminate

4 STRUCTURAL DESIGN

Structural design of a small boat is usually aimed at provision of a robust, durable, low-cost hull capable of withstanding bumping, abrasion and concentrated forces arising from berthing, minor collisions and such conditions as beaching on stony ground; the hull thickness needed to meet these requirements normally ensures that stresses and deformations associated with overall bending, shear and torsion of the hull, including wave-induced loads, are small. Design of small boats up to 15 m in length does not normally include explicit prediction of hull loads and analysis of structural response; scantlings are instead based on boat-builders' knowledge of previous successful designs, incorporated to some extent in Classification Society Rules [45,46].

Increase in hull size is usually associated with an increase in structural slenderness resulting in higher general stress levels. In designing a large grp hull, careful consideration must be given to bending of the hull caused by wave action, account being taken of the cyclic nature of such loading and consequent danger of fatigue damage. Very little design experience and no service data exist for grp ships over 40 m in length; as demonstrated in the case of HMS Wilton [2,47], such ships must be designed from first principles with explicit predictions of loads, evaluation and selection of materials and thorough structural analysis and structural materials testing with reference to buckling, fatigue and the performance of connections.

4.1 Load evaluation

Theoretical methods and computer programs are now available, eg [48], for evaluation of ship motions and loads caused by regular and irregular waves. These methods involve:—

(i) prediction of the wave conditions, defined in terms of directionally distributed energy spectra, which a ship will encounter during its life;

(ii) calculation of ship response at various speeds and headings in regular, long-crested waves using linear strip theory [49, 50];

(iii) analysis of ship response to irregular, short-crested seas by linear superposition of regular wave responses [51, 52].

Analysis of this type has been used in the design of HMS Wilton and other grp minesweepers to obtain qualitative estimates of wave-induced motions and loads. Some typical results are indicated nondimensionally in Fig. 3, which shows an extreme-value probability distribution and histogram of midship bending moment (BM) computed for a 58 m minesweeper design. Also shown in Fig. 3 are design bending moments obtained by extrapolation from strain data recorded in larger steel ships. As discussed in [47], however, the accuracy of such theoretical or extrapolated estimates is open to serious doubt; uncertainty about wave loads contributed substantially to the need for high factors of safety in grp minesweeper designs. The lack of reliable wave load data for ships in the 30 to 100 m range is a serious obstacle to development of efficient grp designs. The best prospect of obtaining such data probably lies in systematic measurement of service stresses in existing steel coasters and trawlers, both for direct empirical application in the design of similarly proportioned grp hulls and for evaluation and empirical correction of theoretical predictions.

Other forms of load which must be considered in design include static hull bending, static and wave-induced pressures on the hull surface, slam-induced pressures, wave-excited vibrations, dead and inertial forces caused by machinery and other large masses, docking loads and berthing loads. Temperature-induced stresses and deformations must be considered and in minesweeper designs the effects of underwater explosions must be examined carefully. Some of the special problems associated with these forms of loading in grp ships are discussed in [47].

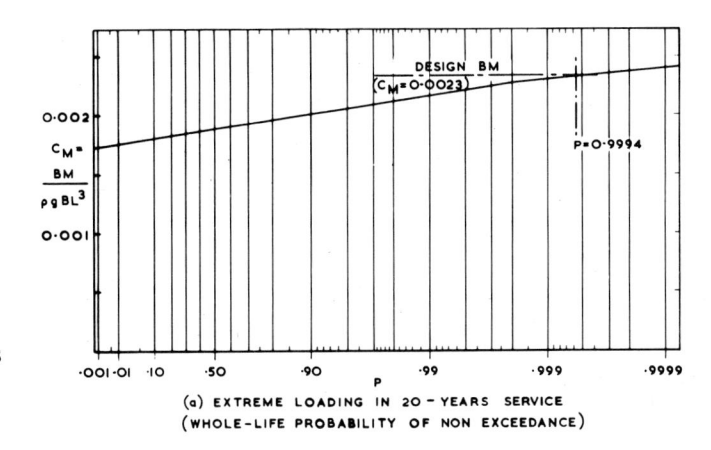

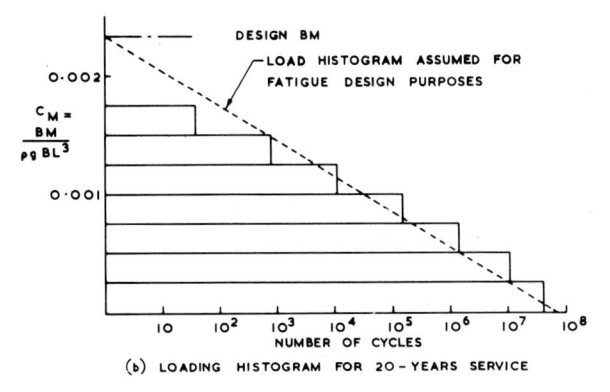

Fig. 3 Computed midship vertical bending moments for a 58 m grp minesweeper

4.2 Structural analysis

While design of small grp boats is normally carried out with little or no structural analysis by basing scantlings directly or indirectly on previous successful designs, development of a new high-performance grp hull requires careful structural analysis with the objects of:—

(i) ensuring that excessive deflections do not occur and that direct and shear stresses at any point do not exceed permissible levels relative to specified static and fatigue failure criteria;

(ii) establishing stress distributions for use in buckling calculations;

(iii) evaluating bending moments and direct and shear forces transmitted at structural connections.

Computer programs for analysis of plane-stress systems, grillages and three-dimensional shells should therefore be regarded as an essential part of the designer's equipment.

In analysing a grp structure it is important to represent material anisotropy correctly: if for example a balanced ship-type WR laminate is assumed to be isotropic, normal design values of E and G ($E/G = 4$) imply an isotropic Poisson ratio of 1.0 with disastrous results in the computation of laminate stiffness and stresses. Most existing finite element programs make provision for material orthotropy. In examining stress concentrations at structural discontinuities, allowances may have to be made for the permanent reduction of modulus which occurs in ship-type laminates at tensile stress levels of about 20% uts as a result of partial fibre debonding [53]; an incremental or iterative analysis procedure is then strictly necessary.

Single-skin grp hulls are normally stiffened by top-hat frames and girders which have the virtues of being highly stable, of minimizing unsupported spans of shell between stiffeners and of being easy to fabricate. Accurate folded plate and finite element calculations have shown [47] that bending, torsion and shear of such frames can be represented satisfactorily by beam theory. The high torsional rigidity of top-hat stiffeners, which can be approximated satisfactorily by the Bredt formula for closed sections, has a beneficial stiffening effect which should always be represented in grillage and shell analyses. In analysing sandwich structures care must be taken to allow for shear deformation in low-modulus core materials [54].

4.3 Longitudinal versus transverse framing

In small, stocky grp hulls transverse strength requirements usually predominate and stiffening is normally provided by closely spaced transverse frames with few, if any, longitudinal girders, shell curvature contributing substantially to hull rigidity and buckling strength. In larger, more slender grp hulls, longitudinal hull bending strength is likely to be the dominant design requirement; longitudinal framing (ie closely spaced longitudinal stiffeners with comparatively widely spaced transverse frames) should then normally be employed in bottom and deck structures unless an overriding requirement for transverse framing is imposed by special loading or fabrication considerations. The superiority of longitudinal framing in such designs lies primarily in the provision of buckling strength under longitudinal compression, augmented by the contribution of longitudinal stiffeners to the hull section modulus. Substantial savings of cost and/or weight may be achieved by adopting this form of framing in a large, slender hull.

Fig. 4 (a) shows a typical structural section in HMS Wilton [2]; transverse framing was adopted in this design partly in order to provide high strength under explosive loads and partly for fabrication cost reasons. Fig. 4 (b) shows a trial alternative design for Wilton in which the shell is longitudinally stiffened by moulded corrugations allowing three out of every four transverse frames to be eliminated, transverse flexural and shear rigidities being maintained by increasing frame dimensions. Designs (a) and (b) have similar weights; the main objective in design (b) is a saving in hull cost, which in the case of Wilton was strongly influenced by the cost of fabricating frames and of reinforcing the frame to shell connection against explosion effects by insertion of closely spaced metal bolts.

Figure 4(c) shows a trial design for a 25 m single-skin grp fishing boat, again incorporating moulded longitudinal corrugations in the hull. This form of stiffening, which has been adopted in at least three large fishing boat designs [11, 16,58], has a beneficial stabilizing effect on hull rolling motions and appears to have negligible influence on hull resistance at normal speeds.

4.4 Single-skin versus sandwich construction

Sandwich materials employing grp skins with cores formed by polyurethane foam [19,55], rigid pvc foam [17,20,24, 55,56], solid balsa wood [55,57] and grp boxes [29] or corrugations [19,59] have been investigated actively for use in large fishing boats, patrol boats and minesweepers and have been used with some success in small boat construction. Sandwich construction offers some important potential advantages over single-skin construction, including:—

(i) greater specific flexural rigidity of the shell and hence a reduced need for framing;

(ii) an ability in some cases to sustain puncture of the outer skin without loss of hull water-tightness, possibly combined with a reserve of buoyancy built into the shell;

(iii) improved thermal and acoustic insulation together with increased vibration damping.

These virtues are however undermined by a number of serious disadvantages including:—

(i) risk of debonding between core and skins originating from small manufacturing imperfections, from berthing, beaching, docking or slamming loads and above all (in a minesweeper) from explosive loads; such debonding, which is likely to propagate by a peeling action and is difficult to detect by present inspection methods, may cause drastic loss of buckling strength;

(ii) difficulty of inspection and repair;

(iii) possible degradation of core materials, including rot of balsa wood and softening of pvc foam at elevated temperature;

(iv) risk of absorption and migration of water within the sandwich;

(v) difficulties associated with fabrication of structural connections and attachment of fittings.

Cellular sandwich construction was originally favoured for Wilton [29] but was subsequently rejected because of poor performance under explosive loads [2]. The disadvantages outlined above have led also to rejection of sandwich materials in feasibility studies for other large grp hulls [11, 12,15,22,26].

The potential benefits of sandwich construction combined with present uncertainties indicate a need for further active development of sandwich materials with emphasis on long-term durability in a marine environment and for development of reliable and economical inspection techniques and repair procedures.

Although single-skin construction is at present likely to be preferable in the main hull of a large grp ship, use of sandwich materials may be advantageous in parts of the hull less directly exposed to impact loads and water penetration. Fig. 4(b) shows a trial alternative design for Wilton in which decks, bulkheads and superstructure are of unstiffened sandwich construction with grp skins and cores formed by a combination of pvc foam and corrugated grp.

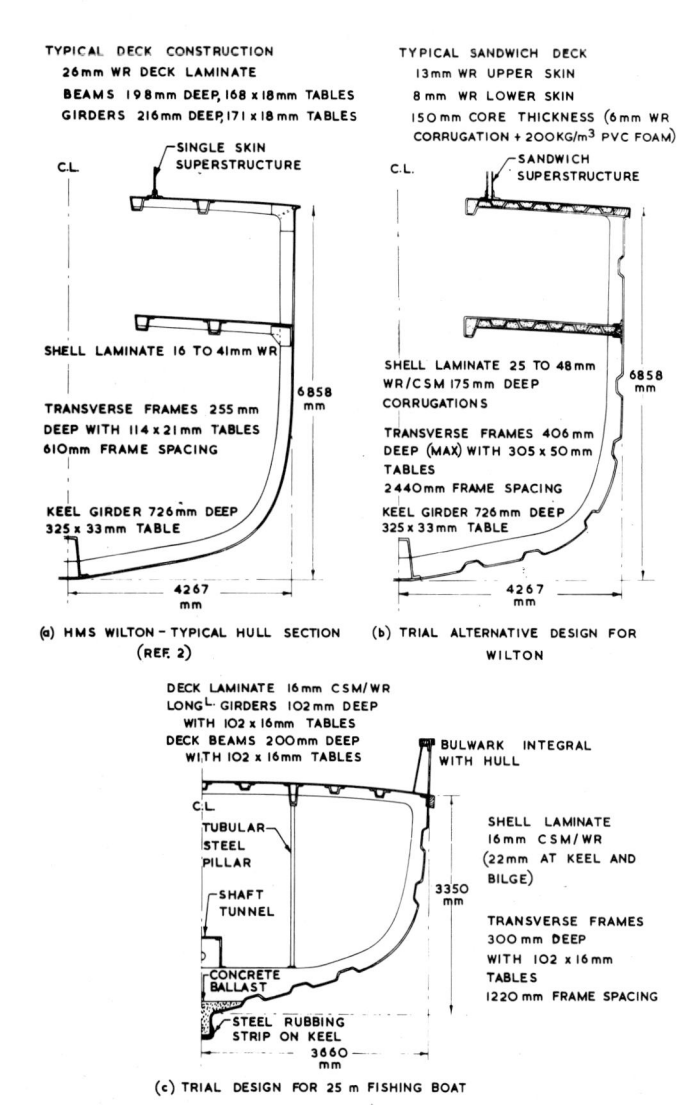

Fig. 4 grp hull sections

4.5 Local instability

Because of the low elastic moduli of grp laminates, design of grp structures must include careful consideration of elastic instability. Such instability is particularly likely to occur in bottom and deck structures under longitudinal compression caused by wave-induced bending of the hull, combined in some cases with transverse compression caused by hydrostatic pressure on the ship's sides. Instability of bulkhead and side shell structures under combined compression and shear must also be examined, together with local buckling of stiffener webs and tables under direct and shear stresses. Buckling behaviour is complicated by material anisotropy and by the top-hat form of frames normally used in grp hull construction.

In a transversely framed hull, failure of bottom or deck panels under uniaxial or biaxial compression is likely to occur by local instability having one of the forms shown in Fig. 5(a). Buckling stresses corresponding to these modes can be estimated approximately from data curves [60] or may be computed accurately using folded-plate or finite element analysis [60–62]. Experiments on large and small-scale models [47,63] have demonstrated these forms of instability and established that transversely framed grp panels are likely to fail catastrophically at loads close to theoretical critical levels with little or no post-buckling

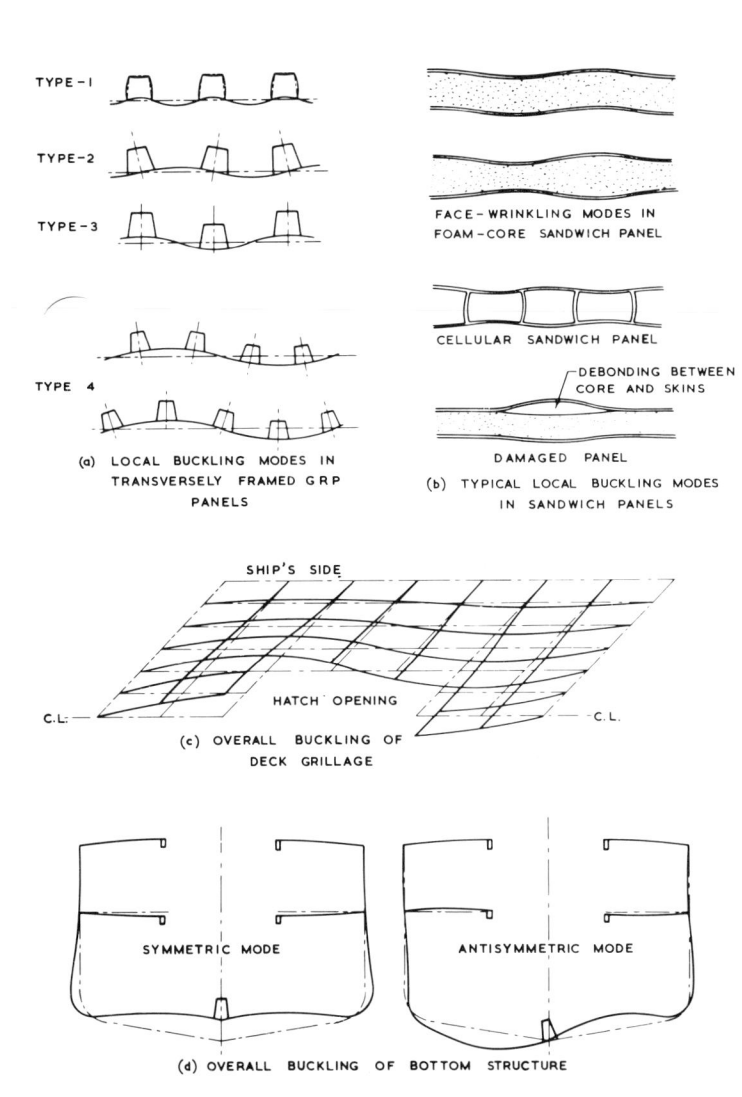

Fig. 5 Buckling modes for various materials and structures

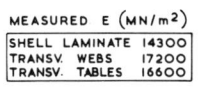

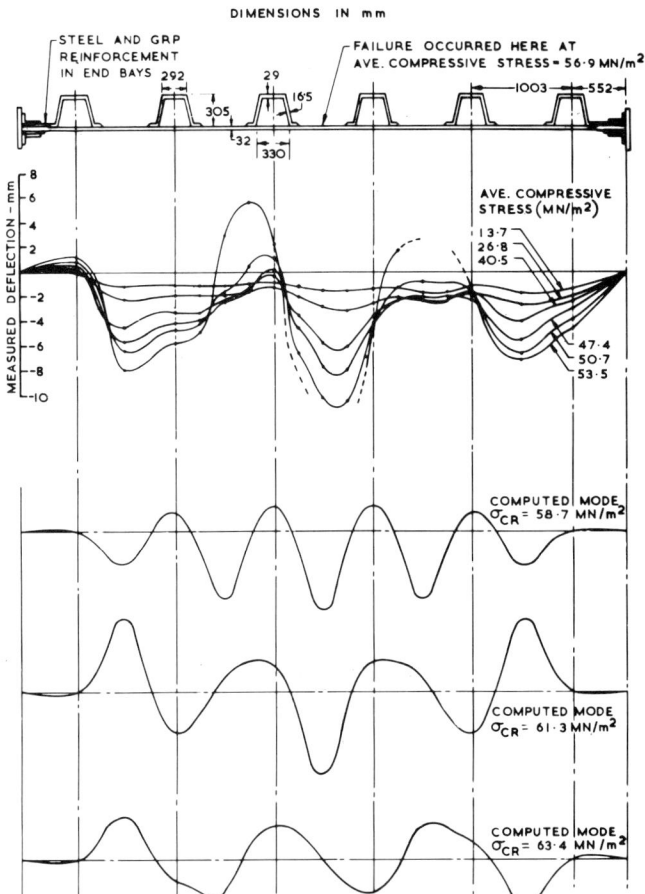

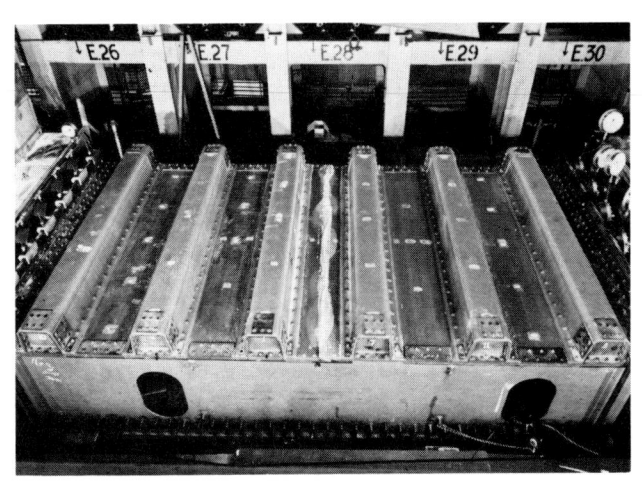

Fig. 6 Transversely framed grp test panel: comparison of experimental deflections and computed buckling modes

reserve of strength. Figs. 6a and 6b shows details of one of a series of large-scale compression tests carried out at NCRE.

Local instability of a longitudinally stiffened bottom or deck structure is likely to have one of two forms:—

(i) 'Strut-type' buckling, in which longitudinal stiffeners together with the attached shell buckle into a single half-wave between transverse frames; ignoring rotational restraint provided by transverse frames, buckling stresses may be found accurately [60] from the Euler strut formula (corrected for shear effects) in which the flexural rigidity is taken as that of the stiffener acting with an assumed effective breadth of shell;

(ii) 'Panel' buckling, in which strips of laminate forming the shell and the walls of top-hat stiffeners buckle into several half-waves over their lengths; upper or lower-bound estimates of buckling stress may be obtained by applying the formula for long clamped or simply supported orthotropic strips to the most slender element of the cross-section, normally the strip of shell between stiffeners; comparison with accurate folded-plate analysis [60] indicates that panel buckling usually occurs at a stress substantially higher than the lower-bound value.

Strut-type instability is likely to result in a catastrophic failure at a stress somewhat lower than the theoretical critical value. Compressive stress in a longitudinally stiffened panel may theoretically exceed panel buckling stresses without causing failure, but post-buckling behaviour is likely to involve large panel deformations imposing severe loads on bonded connections; experimental justification should therefore be sought before placing reliance on post-buckling strength in a fabricated grp structure. Fig. 7 shows a large-scale, longitudinally stiffened panel tested at NCRE, in which collapse was precipitated be debonding of longitudinal stiffeners at a compressive stress only slightly higher than the initial panel buckling stress.

Local instability under shear or combined direct and shear stresses can usually be examined satisfactorily using formulae for long orthotropic strips [64]. More accurate analysis using folded-plate [62] or finite element [25] methods is also possible.

Local instability of cellular sandwich panels can be examined approximately by applying plate formulae to elements forming the cross-section or again by use of folded-plate or finite element analysis. Local buckling of foam or solid-core sandwiches, including the forms shown in Fig. 5(b), may be evaluated using standard formulae [54]. The most serious buckling problem is likely to be that arising in a damaged or initially imperfect sandwich with incomplete bonding between skin and cores, as illustrated in Fig. 5.

4.6 Overall instability

While buckling of a transversely framed hull usually occurs in an interframe mode, the lowest buckling stress for a longitudinally stiffened bottom or deck structure may correspond to an 'overall' mode in which stiffeners and shell deflect together into one or more half-waves over the span between transverse bulkheads.

Overall buckling of a uniform deck structure can often be examined satisfactorily using the formula for a simply [64] or elastically [65] supported orthotropic plate in which flexural and torsional rigidities represent the properties of stiffeners acting with assumed effective breadths of deck laminate. Assessment of non-uniform or geometrically irregular decks requires application of a finite element grillage buckling analysis. A design case requiring particular care is that of a deck with large hatch openings, as shown in Fig. 5(c), in which the free edge at the hatch side, unless suitably reinforced, may have a serious destabilizing influence; treatment of this problem may require an initial plane-stress analysis to establish the distribution of de-

Fig. 7 Longitudinally framed grp test panel: collapse under compressive load

stabilizing stresses [60]. Overall deck buckling stresses may be raised or suppressed either by increasing stiffener rigidities, by introducing pillars, or by relying on the stiffening influence of secondary structural components such as minor bulkheads, engine casings and deckhouses; if the latter remedy is adopted, as often proves necessary in design, it is important that the secondary structure concerned should be designed with adequate strength and stiffness and sufficiently strong connections to provide the required support.

Overall buckling of a ship's bottom structure may occur in either a symmetric or an antisymmetric mode, as shown in Fig. 5(d). It has been shown [66] that vee-bottom structures with deadrise angles of less than $10°C$ will usually buckle symmetrically while those with angles over $20°$ will usually buckle antisymmetrically; in the range $10°$ to $20°$ either mode may occur. Antisymmetric buckling may be examined conservatively by applying the formula for a simply supported orthotropic plate to the flat portion of structure between the keel and turn of bilge. Symmetric buckling, which depends on the stiffness of the keel girder and involves the coupled membrane and flexural stiffnesses of the bottom structure, may be examined approximately using recently developed data curves [66].

It should be borne in mind that calculated buckling stresses refer to initial elastic instability which need not correspond closely or even approximately to collapse loads. In all cases where intended working stresses approach theoretical critical levels, an experimental assessment of buckling strength should be made. Much scope remains for experimental investigation of unstable collapse in fabricated grp structures and for optimization of ship designs against such failure.

4.7 Design of structural connections

Fabricated grp structures are particularly susceptible to failure at bonded connections, in which weakness arises from the absence of load-bearing fibres across bonded interfaces, from the low tensile and shear strengths of the thin layer of resin forming the bond and from the inevitable occurrence of stress concentrations caused by geometric irregularities and imperfections (eg regions of incomplete bonding). Under ideal laboratory conditions the tensile and shear strengths of secondary polyester bonds may be as high as $15MN/m^2$; under practical fabrication and service conditions, however, in which bond imperfections are likely to exist and in which tensile and shear stresses usually occur simultaneously, effective tensile and shear strengths may be as low as 1.5 and 3 MN/m² respectively [67]. Lack of ductility in the bond material and the absence of crack-arresting action by fibres may result in rapid propagation of small cracks by a 'peeling' action until catastrophic failure of the connection occurs.

The main types of structural connection occurring in a grp ship are shown in Fig. 8. These include:—

(i) Butt connections, normally of butt-strap or scarf type, which must be able to transmit tensile, compressive and in-plane shear forces, together with bending and twisting moments and interlaminar shear;

(ii) Frame/shell connections, whose primary purpose is to transmit shear between the shell and the frame flanges under transverse bending of the hull caused by lateral pressure or concentrated loads, but which must also be able to withstand tensile stresses: in minesweepers hulls severe transient tensile stresses across the frame/shell bond occur under explosive load: incipient failure caused in this way is shown in Fig. 9;

(iii) Intersections between longitudinal and transverse stiffeners, at which it is necessary to allow for transmission across the intersection of direct and shear stresses carried by the webs and table of each stiffener, together with exchange of shear forces between intersecting members: compensation must be provided for material cut out of the deeper stiffener in order to allow continuous passage of the smaller member; exchange of shear force between the webs of intersecting members is normally provided by grp angles;

(iv) Bulkhead/shell connections, which must be designed to transmit direct and shear forces and bending moments caused by normal hydrostatic and wave loads together with hydrostatic pressure on the bulkhead under damage conditions in which compartments on either side of the bulkhead are flooded; because of the difficulty of attaching vertical bulkhead stiffeners rigidly to the shell, these stiffeners may be snapped off at their ends as shown in Fig. 8(b), leaving a short unstiffened strip of laminate at the base of the bulkhead which acts as a virtual hinge;

(v) Deck-edge knee and tee connections, as shown in Fig. 8(e) and (f), which must be able to transmit forces and moments associated with transverse bending of the decks and shell, together with membrane shear stresses associated with longitudinal hull bending.

Safe and efficient design of structural connections requires a knowledge of the forces and moments to be transmitted, from which estimates may be made of the average direct and shear stresses acting across bonded interfaces and hence of required bond areas. Reliable estimates of such moments and forces are unlikely to be available unless thorough structural analysis of the hull has been carried out for a full range of design load conditions. Detailed analysis of internal stress distributions within a structural joint, using approximate analytical [68] or finite element [47] methods, may

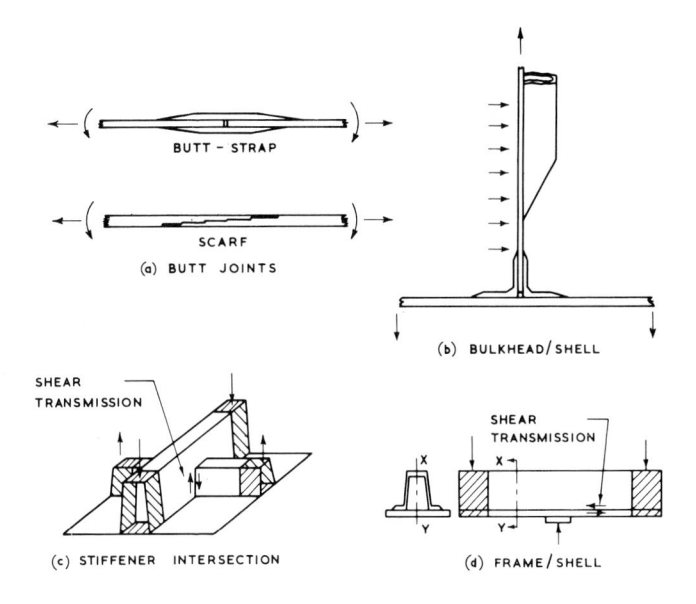

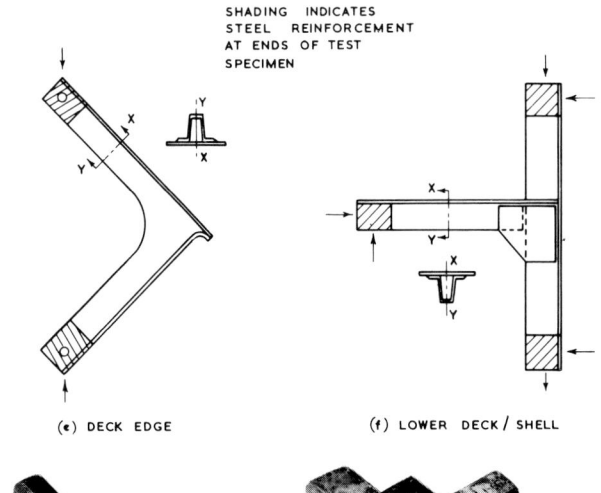

Fig. 8 Types of grp structural connection

be worthwhile in some cases as a basis for rough optimization of joint geometry.

Fig. 10(a) shows, for example, a finite element idealization of a bulkhead/shell connection. Stress distributions near the tip of the bulkhead stiffener, in particular tensile and shear stresses transmitted between the bulkhead laminate and the stiffener webs, were computed for various geometries, variations being made in the angle of taper at the end of the stiffener and in the depth of the unsupported strip of laminate at the edge of the bulkhead. Fig. 10(b) shows failure of an initial, unsatisfactory design under hydrostatic load applied to the bulkhead; Fig. 10(c) shows an improved design in which, by modification of the stiffener geometry in accordance with theoretical results and by improvement of the arrangement of metal fasteners at the tips of stiffeners, the bulkhead failure load was more than doubled.

While evaluation of forces and moments transmitted at structural connections may substantially reduce margins of uncertainty referring to joint performance, purely theoretical estimates of joint strength are unacceptable as a basis for design because of uncertainty about joint imperfections. Development of a high-performance grp design should therefore include a thorough programme of static, fatigue and where necessary impact tests on all important connections. Care must be taken to subject specimens to realistic combinations of direct, bending and/or shear loads and care must also be taken, by local thickening of the laminate and use of steel reinforcement, to avoid premature failure at points of load application.

Fig. 8 shows the types of test specimen and manner of loading used in static and fatigue tests carried out at NCRE in support of minesweeper designs. Fatigue tests were based on load histograms having the form shown in Fig. 3. Because of uncertainty about the accuracy of predicted wave loads (which did not include vibratory effects) and because of the need to restrict test programmes in the face of high manufacturing and testing costs, a conservative procedure was followed involving severe proof tests on each type of

Fig. 9 Incipient debonding at frame/shell connection

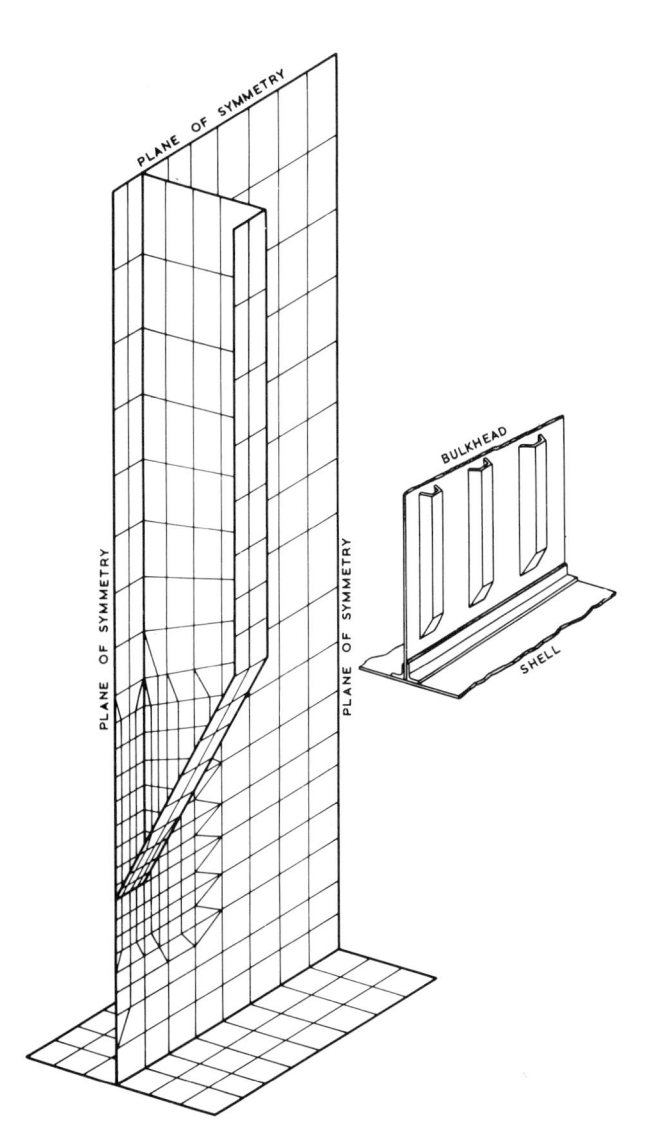

Fig. 10 (a) Finite element idealisation of bulkhead/shell connection

Fig. 10 (b) Unsatisfactory design (bulkhead/deck connection is shown: failure at shell connection was similar)

Fig. 10 (c) Improved design (bulkhead/deck connection)

specimen; specimens were required to withstand 10^6 cycles at 25% and 10^4 cycles at 50% of their static collapse loads, a factor of safety of 4 (failure load/estimated maximum working load) being required against static failure. This severe requirement was largely met in the HMS Wilton design [2].

At connections in which high tensile stresses act across bonded interfaces, reinforcement by metal fasteners is likely to be necessary. Bolts, rivets and screws of many types may be used for this purpose. In grp minesweeper designs [2,23,47] use of metal fasteners below and to a lesser extent above the waterline has proved essential to counteract the effects of underwater explosions, particularly in frame/shell connections and at seatings for massive items of machinery and equipment.

In counteracting bond tension, metal fasteners act by carrying a proportion of tensile load and in some cases by imposing an initial compression across the bond, thus raising its apparent tensile strength. Fasteners also act as 'peel-arrestors', inhibiting the propagation of local debonding. Metal fasteners introduced with the primary object of preserving the resin bond should where possible be designed with the secondary, fail-safe capability of transmitting all or most of the design loads in the event of bond failure. Care must be taken to avoid premature damage by fastener 'pull-through', in which failure of the laminate occurs by local bending and interlaminar shear as shown in Fig. 11(a) and (b), together with damage under shear loads in which fasteners may tear or gouge the laminate as shown in Fig. 11. (c) and (d). Some information on permissible fastener spacings and minimum distances from free edges in ship-type laminates may be obtained from published data [36,67]; however, development of a new design is likely to require proof tests on specimens of the type shown in Fig. 11 to check the adequacy of selected fastener dimensions, spacings and head or washer sizes.

Much scope exists for research aimed at improved design of structural connections in grp ships. It is suggested that such work should include:—

(i) further theoretical studies of stress distribution in bonded connections aimed at optimization of joint geometry;

(ii) micromechanical studies of joint behaviour referring to the initiation and spread of debonding under static, fatigue and impact loads;

(iii) development of improved bonding techniques, possibly including methods of introducing fibres across bonds and use of fibrous and other fillers to provide a crack-arresting action;

(iv) development of provisional standard designs, backed by thorough test data, for joints of the type shown in Fig. 11.

5 CONCLUSIONS

GRP is now firmly established as the most effective and widely used material for small boat construction. Reliable methods of design and construction, developed on a trial and error basis over the last 30 years, exist for a wide range of boat types; room for improvement remains, however, in boatbuilding resins and reinforcement and in some areas of design and fabrication, notably provision of trouble-free gel coats and durable structural connections.

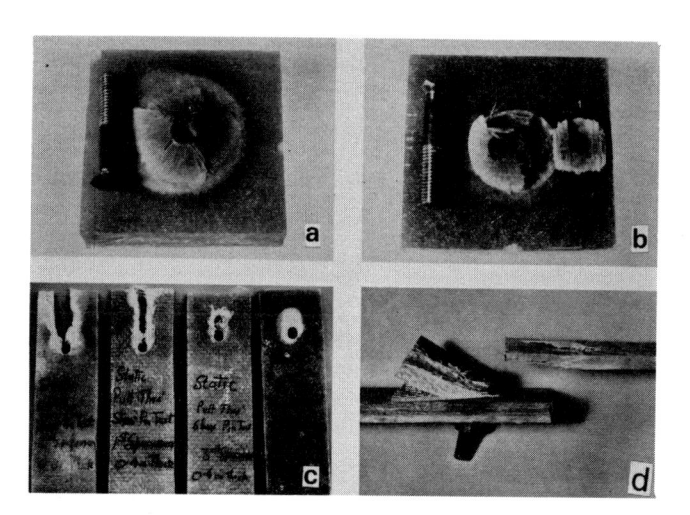

Fig. 11 Forms of laminate failure at metal fasteners

The greatest scope for increased application of fibre-reinforced composites in the marine field appears to lie in construction of medium-size hulls in the 20 to 40m range, including particularly fishing boats, together possibly with large patrol boats and service vessels for the offshore gas/oil industry. As hull size is increased beyond 30m, existing semi-empirical design procedures are likely to prove unreliable and an increasing need is likely to arise for design from first principles, involving analysis and tests of the type described in the present paper.

While design studies for larger ships [69] have indicated that grp tankers, trawlers and ferries of up to 80m might prove economically viable, present cost figures suggest that grp is unlikely to compete with welded steel construction in ships of over 50m unless a special requirement arises, eg for carriage of corrosive cargo or, in the case of minesweepers, for a non-magnetic hull. Construction of grp ships substantially over 100m in length seems unlikely to become feasible unless

(i) major reductions are achieved in costs of material and fabrication relative to those for steel;

(ii) a substantial increase in laminate stiffness is achieved allowing more effective use to be made of high specific strength;

(iii) construction techniques are developed allowing large hull components to be prefabricated and subsequently connected, thus avoiding the need for very large air-conditioned laminated sheds.

ACKNOWLEDGEMENTS

The author is indebted to colleagues who have contributed to grp investigations at NCRE, in particular Messrs P. Christopher, D. Crabbe, W. Kirkwood, L. Somerville and G. Wallace, and to colleagues in the Design and Materials Sections at DGS, Bath, with whom grp design problems have been frequently and profitably discussed. Acknowledgements are also due to the White Fish Authority and BSRA for provision of cost data referring to fishing boats and steel merchant ships and to Forest Products Laboratory for information on the mechanical properties of wood.

REFERENCES

1. **Rosato, D.V.** 'History of composites' in *Handbook of fibreglass and advanced composites*, ed G. Lubin, Van Nostrand Reinhold Co, NY (1969)
2. **Dixon, R.H., Ramsay, B.W.** and **Usher, P.J.** 'Design and build of HMS Wilton', *Proc of symposium on grp ship construction*, RINA, London (October 1972)
3. **Buermann, T.M.** and **Della Rocca, R.J.** 'Fibreglass reinforced plastics for marine structures', *Trans Soc Naval Arch and Mar Engrs (SNAME)*, **68** (1960)
4. **Wildman, D.** 'Reinforced plastics for small craft', *Proc of symposium on small craft*, RINA, Southampton (September 1971)
5. **Sharples, A.K.** and **McLeod, J.D.** 'Pilot launches – design and operation', *Proc of symposium on small craft*, RINA, Southampton (September)1971)
6. *International ship structures congress*, Hamburg (September 1973). (Report of Committee 12 on Materials other than steel, ed W. Muckle)
7. **Takehana, M.** 'All-plastic fishing vessels', *Fishing boats of the world*, Vol 3, Fishing News (Books) Ltd, London (1967)
8. **Verweij, D.** 'Comparison between plastic and conventional boatbuilding materials', *Fishing boats of the world*, Vol 3, Fishing News (Books) Ltd, London (1967)
9. **Wilks, T.M.** 'Boat hulls' in *Glass reinforced plastics*, ed B. Parkin, Iliffe, London (1970)
10. 'GRP in sealand hovercraft', *Reinforced plastics* **17** No 4 (April 1973)
11. **Pike, D.S.** and **Yeatman, M.** 'Commercial fishing vessels in glass fibre reinforced plastics (construction techniques and future trends)', *Proc of conference on fishing vessel construction materials*, Montreal (October 1968)
12. **Cobb, B.** 'Design, construction and economic considerations in fibreglass trawler construction', *Proc of conference on fishing vessel construction materials*, Montreal (October 1968)
13. **Fraser, D.J.** 'Estimated hull work and material content for 100 ft combination fishing vessel in different materials', *Proc of conference on fishing vessel construction materials*, Montreal (October 1968)
14. **Della Rocca, R.J.** 'A 110 ft Fibreglass reinforced plastic trawler', *Fishing boats of the world*, Vol 3, Fishing News (Books) Ltd, London (1967)
15. **Hallett, H.R.** and **Simpson, J.H.** 'Practical reinforced plastics shipbuilding', *Plastics and Polymers* (February 1968)
16. **Hallett, H.R.** and **Simpson, J.H.** 'Fabrication of large RP trawlers', *Reinforced plastics*, **12** No 6 (June 1968)
17. **Takehana, M.** and **Kimpara, I.** 'On the design of fibreglass reinforced plastic ship hull structures', *J Soc Naval Arch, Japan*, **123** (June 1968); **125** (June 1969); **126** (December 1969)
18. 'Vosper Thornycroft 75 ft patrol craft', *Reinforced plastics*, **17** No 3 (March 1973)
19. **Flacken, H.** et al. 'Prufsektion für ein schnelles Marinefahrzeug aus glasfaserverstarktem Kunststoff' *Schiff und Hafen*, Heft 11/1971, 23 Jahr gang
20. **Wimmers, H.W.** 'Consideration of the design and construction of larger glass fibre reinforced polyester ships', *Schip en Werf* (March 1966)
21. 'Line production of GRP lash barges', *Shipbuilding and Shipping Record* (September 1972)
22. **Spaulding, K.B.** and **Della Rocca, R.J.** 'Fiberglass reinforced plastic minesweepers', *Trans SNAME* **73** (1965)
23. **Lankford, B.W.** and **Angerer, J.F.** 'Glass-reinforced plastic developments for application to minesweeper construction', *Naval Engineer's J* **83** No 5 (October 1971)
24. **Gardin, A.** 'Notes on the Swedish GRP minesweeper design: contribution to discussion', *Proc of symposium on grp ship construction*, RINA, London (October 1972)
25. **Harrhy, J.** 'Structural design of single skin glass reinforced plastic ships', *Symposium on grp ship construction*, RINA, London (October 1972)
26. **Scott, R.J.** and **Sommella, J.H.** 'Feasibility study of glass reinforced plastic cargo ship', *US Ship Structures Committee*, Report SSC–224 (1971)
27. **Fried, N.** 'Marine applications' in *Handbook of fibreglass and advanced plastics composites*, ed G. Lubin, Van Nostrand Reinhold Co, NY (1969)
28. **Cheetham, M.A.** 'Naval applications of reinforced plastics', *Plastics and Polymers* (February 1968)
29. **Henton, D.** 'Glass reinforced plastics in the Royal Navy', *Trans RINA* **109** (October 1967)
30. **Fried, N.** 'The potential of filament wound materials for the construction of deep submergent pressure hulls', *Conference on filament winding*, The Plastics Institute, London (October 1967)
31. *Crystic Monograph No 2: Polyester Handbook*, Scott Bader Co Ltd (1971)
32. **Hashin, Z.** and **Rosen, B.W.** 'The elastic moduli of fiber-reinforced materials', *J Appl Mech* (June 1964)
33. **Whitney, J.M.** and **Riley, M.B.** 'Elastic properties of fiber reinforced composite materials', *AIAA J*, (September 1966)
34. **Tsai, S.W.** 'Structural behaviour of composite materials', *NASA CR–71* (July 1964)
35. **Smith, C.S.** 'Calculation of elastic properties of grp laminates for use in ship design', *Proc of symposium on grp ship construction*, RINA London (October 1972)
36. **Hill, R.** 'The mathematical theory of plasticity', Oxford Univ Press (1950)
37. **Azzi, V.D.** and **Tsai, S.W.** 'Anisotropic strength of composites', *Proc Soc for experimental mechanics*, **22** No 2 (September 1965)
38. **Owen, M.J.** and **Bishop, P.T.** 'The effect of stress concentration on the failure of fibre reinforced plastics', *MOD(N) Contract Report* (April 1971)
39. Gibbs and Cox Inc. 'Marine design manual for fiberglass reinforced plastics', McGraw-Hill (1960)
40. **Hertz, J.** 'The effect of cryogenic temperatures on the mechanical properties of reinforced plastic laminates', *SPE J* (February 1965)
41. **Cobb, B.** 'Long-term durability of resin glass boats', *Ship and Boat Builder* (February 1963)
42. **Fried, N.** and **Graner, W.R.** 'Durability of reinforced plastic materials in marine service', *Marine Technology* (July 1966)
43. *The Engineer*, **234** No 6052 (March 1972)
44. **Dukes, R.** 'Fibre reinforced plastics', *Eng Mat and Des* (November 1970)
45. Provisional rules for the application of glass reinforced plastics to fishing craft, *Lloyds Register of Shipping* London
46. Rules for the building and certification of glass fibre reinforced plastics pleasure boats, *Det Norske Veritas*, Oslo (1969)
47. **Smith, C.S.** 'Structural problems in the design of grp ships', *Proc of symposium on grp ship construction*, RINA, London (October 1972)
48. 'Simulation by computer of motions and sea loads for the design of ships and offshore structures', *Lloyds Register, Research and Technical Advisory Service Report* No 5105 (1972)
49. **Korvin-Krukovsky, B.** 'Investigation of ship motions in regular waves', *Trans SNAME* (1955)
50. **Salveson, N., Tuck, E.O.** and **Faltinsen, O.** 'Ship motions and sea loads', *Trans SNAME*, (1970)
51. **St Denis, M.** and **Peirson, W.J.** 'On the motions of ships in confused seas', *Trans SNAME*, (1953)
52. **Conolly, J.E.** 'Ship motions in irregular waves', *Seminar on*

53 Owen, M.J. and Dukes, R. 'Failure of glass reinforced plastics under single and repeated loading', *J Strain Analysis* **2** No 4 (1966)

54 Allen, H.G. 'Analysis and design of structural sandwich panels', Pergamon Press (1969)

55 Spaulding, K.B. 'Cored fiberglass reinforced hull construction', *Proc of conference on fishing vessel construction materials*, Montreal (October 1968)

56 Brandl, K. 'Cellular plastics of pure PVC as a sandwich core for large FRP boat hulls', *Proc of conference on fishing vessel construction materials*, Montreal (October 1968)

57 Lippay, A. and Levine, R.S. 'End grain balsa cored reinforced plastic as a fishing vessel construction material', *Proc of conference on fishing vessel construction materials*, Montreal (October 1968)

58 Article on Halmatic 50 ft Trawler, *Reinforced plastics*, **17** No 3 (March 1973)

59 Guiton, J. 'Glass reinforced plastic ship construction', *MSc Thesis*, University of Newcastle (1968)

60 Smith, C.S. 'Buckling problems in the design of fiberglass reinforced plastic ships', *J of Ship Research*, **16** No 3 (September 1972)

61 Smith, C.S. 'Bending, buckling and vibration of orthotropic plate-beam structures', *J of Ship Research*, **12** No 4 (December 1968)

62 Wittrick, W.H. 'General sinusoidal stiffness matrices for buckling and vibration analyses of thin flat-walled structures', *Int J Mech Sci*, **10** (1968)

63 Smith, C.S. 'Investigation of ship buckling problems using small-scale plastic models', *5th Int Conf on Experimental stress analysis*, Udine (May 1974)

64 Lekhnitskii, S.G. 'Anisotropic plates', Gordon and Breach (1968)

65 Smith, C.S. and Faulkner, D. 'Dynamic behaviour of partially constrained ship grillages', *Shock and Vibration Bulletin*, No 40 (December 1969)

66 Smith, C.S. 'Buckling and vibration of a ship's vee bottom structure' *Trans RINA* (1974)

67 Rufolo, A. 'A design manual for joining of glass reinforced structural plastics' *US Naval material laboratory Report* Navship 250-624-1 (August 1961)

68 Perry, H.A. 'Adhesive bonding of reinforced plastics', McGraw-Hill (1959)

69 Guiton, J. 'Production and design aspects of GRP craft over 90 ft in length', *Plastics and Polymers* (August 1973)

QUESTIONS

1 On wood and cfrp duplex compositions, by M. Bedwell (Lanchester Polytechnic, Rugby CV21 3TG)

Dr Smith referred to the 'skinning' of wooden boat structures with grp. From the structural point of view, is there not a strong case for further work on wood + cfrp duplex composition? My question is prompted by:

1. Dr Kelly's reminder, earlier in the conference, that by many criteria, notably the 'index of merit' $E^{1/3}/\rho$, wood stands closer to cfrp than most of conventional structural materials, and much closer than grp.

2. The experimental racing oars made by GKN in 1971 (see *Composites* **2** No 4 (1971) p 208).

Reply by Dr Smith

Where grp sheathing is employed on a wooden boat hull the object is usually to protect the hull from leakage or attack by marine boring organisms. Such sheathing is normally very thin, typically one or two plies of chopped-strand mat which contribute little effective structural stiffness or strength. This application should not be confused with the use of wood, eg balsa [57] or plywood, as a sandwich core material between grp skins.

I agree with Mr Bedwell that wood should be borne actively in mind by designers as a material for possible use in combination with both grp and cfrp. The merits of wood, notably high specific stiffness (or $E^{1/3}/\rho$ where high flexural stiffness is required), must however be weighed against a number of serious disadvantages including susceptibility to splitting, warping and rot, marked variations in strength, stiffness and density under conditions of varying moisture content [72] and a tendency to swell or shrink which is likely to cause problems at interfaces with grp or cfrp.

REFERENCE

70 Hearmon, R.F.S. 'The elasticity of wood and plywood', *Forest Products Research*, Special Report No 7 (HMSO 1948)

2 On theoretical failure stress/buckling stress ratio for panel (Fig. 7) and post buckling strength reserves, by W.M. Banks (Lecturer, University of Strathclyde, Department of Mechanics of Materials, Montrose Street, Glasgow G1 IXJ)

Dr Smith indicates that in longitudinally stiffened panels the compressive stress may theoretically exceed the buckling stress without causing failure. I wonder if some indication could be given of the ratio of the theoretical failure stress to the buckling stress for the panel shown in Fig. 7.

In tests completed in our laboratories at Strathclyde, we have found large post buckling reserves of strength with flat orthotropic plates. I would be interested to have the author's comments on the possibilities of increasing the effectiveness of the bonded connections in his structures to utilise some of this reserve of strength.

Reply by Dr Smith

As Mr Banks is no doubt well aware, calculation of the post-buckling load-deformation relationship (let alone the failure load) for a longitudinally stiffened grp panel is a formidable task; a difficult non-linear analysis is involved, complicated by the progressive nature of laminate failure under combined membrane and bending stresses with consequent changes in stress/strain and moment/curvature relationships. In my opinion it is not possible, even assuming a perfect stiffener to panel connection, to make reliable theoretical predictions of collapse load for structures of this type. Any reliance on a post-buckling reserve of strength should therefore be supported by sound experimental data.

The structure shown in Fig. 7 collapsed at a compressive stress about 10% higher than the theoretical panel buckling stress. Collapse was clearly precipitated by debonding of the longitudinal stiffeners associated with panel buckling deformations. I have no doubt that an increase in compressive strength, possibly a substantial increase, could have been achieved by reinforcing the bonded connections, eg by insertion of closely spaced metal fasteners. The increase of strength likely to be obtained in this way must however be

3 On water induced degradation of grp and use of 'Marine type' resins, by E.S.G. Elkin (Brighton Polytechnic)

What is the mechanism of water induced degradation of grp and in what way are special 'Marine type' resins involved?

Reply by Dr Smith

Water-induced degradation of grp is associated with penetration of the laminate by water, both by diffusion through the resin and by capillary flow through cracks and voids and along debonded fibre-resin interfaces. Diffusion of water through the resin involves an osmotic process, influenced by impurities in the resin and by solutes in the water: it is of interest that water absorption and loss of strength and stiffness occur more markedly in fresh than in salt water. Degradation of mechanical properties is probably attributable mainly to chemical attack by the water on the fibre/resin bond [70], much reduced but not eliminated by treatment of fibres with a silane coupling agent, together with mechanical effects associated with water-induced volume changes in the resin and build-up of osmotic pressure between fibres and resin [71]. Resins for marine application have been selected in some cases partly on the basis of water absorption and wet-strength test data [2]; it appears however that the factors most strongly influencing retention of strength and stiffness in a water-immersed grp laminate are correct choice and application of size and coupling agent to fibres, careful handling of fibres to avoid contamination of fibre surfaces and careful control of the laminating process to ensure an intimate chemical bond between fibres and resin throughout the laminate.

REFERENCES

71 **Norman, R.H., Stone, M.H.** and **Wake, W.C.** 'Resin-glass interface', in *Glass-reinforced plastics* edited by B. Parkyn (Illiffe 1970)

72 **Ashbee, K.H.G.** and **Wyatt, R.C.** 'Water damage in glass fibre/resin composites', *Proc Roy Soc* A 312 (1969)

COMMENTS

1 On competitiveness of grp to steel, by J. Guiton (Consultant Naval Architect)

I much enjoyed Dr. Smith's thorough yet lucid paper; he has taken great pains in his research and analysis of the subject.

However I would disagree with him on his final paragraph. In this he suggests that grp is unlikely to be competitive with steel for vessels over 50m, unless special requirements exist. I would like to try and demonstrate how grp can in fact be competitive with mild steel construction today, at present material and labour prices.

If you look at the two full curves (A and B) in Fig. 1, you will see the present possition of grp and steel merchant ship construction prices: grp is competitive with steel below a length of about 18m, but thereafter the curves diverge until at 150m the grp price is about double the steel price. The divergence is largely due to the reduction in plate curvature and hence the increased ease of construction in steel as vessel size increases. The dotted curve (C) shows the effect of an improved laminate design and structural configuration on the cost of grp construction. The reduction in price is achieved solely by a reduction in the structure weight without loss of strength, in this case a 75% weight reduction over steel construction (compared to the 50% generally achieved by conventional grp design*). Weight reduction is the most effective method of reducing cost in grp construction since it results not only in lower material costs but also, by virtue of the layer on layer thickness build-up of the laminating process, in lower labour and overhead costs per hull.

The greater weight reduction is achieved:

(1) by the use of higher strength reinforcements — woven and uni-directional rovings rather than mat and woven roving,

(2) by employing these in a structural arrangement which allows their higher strength to be utilized without restrictions due to buckling and deflection with the low modulus material, ie stiffness where necessary is achieved by section shape and panel size and not by material properties and

(3) by employing an improved design of structural connection.

The increase in complexity of the structure is negligible, but an increase in construction price/ton of £500 (30–40%) over conventional construction is adopted to cover the more stringent quality control requirements for the more sophisticated structure. (See Notes and assumptions to Fig. 1)

If, now, suitable ship types are selected which can benefit from the weight saving afforded by grp over steel, then building can be competitive up to some 106m. Consider for example a small oil/chemicals tanker. In this case the amount of relatively high density cargo carried is restricted more by deadweight (buoyancy) capacity than available cubic capacity, and consequently any structural weight saving which can be utilized to the full in the carriage of additional cargo. Alternatively, and more relevant here, a smaller vessel may be used to carry the same cargo deadweight, and this reduction in dimensions provides an additional building cost saving which enables grp to compete with steel.

In Fig. 1 the lengths of two tankers of the same deadweight capacity (3400T), one in steel and one in grp, are plotted, and it will be seen that the smaller grp ship is cheaper than the steel vessel (by £22 000). Thus by a combination of improved structural design and judicious ship type selection, grp can be competitive with steel in small merchant ship construction today.

I have not described the proposed structural design in detail because it is still being evaluated, but I hope to have positive results in the near future, and then details can be given.

One simple method of achieving competitive cost has been described, but there will be many others. For example improving the rate of laminating and/or the consistency of laminate quality will achieve economies, and this can be done by improved construction methods or efficient mechanization of the lay-up process.

The effect of such improvements on construction costs is illustrated in curve D (applied here to the improved design), while curves E and F show further savings which could result from improved materials and methods, all possible with materials and technology available today.

Reply by Dr Smith

Mr Guiton has presented an interesting and persuasive case in support of his view that grp would be economically competitive with steel for construction of a 3400 T deadweight tanker. I should be delighted to see this argument proved correct, but suspect that Mr Guiton has been over-optimistic in some of his assumptions. In particular:

(i) the estimates of cost per ton for fabricated grp structures indicated in Mr Guiton's diagram seem too low; I suggest for example that in the case of the 80.7m tanker, 1973 costs of £2 000 and £3 000 per ton for 'conventional' and 'improved' designs would be more realistic than the figures of £1 300 and £1 800 tabulated;

(ii) I believe that Mr Guiton may have underestimated the technical problems involved in designing a 100m hull having a weight of only 25% that of an equivalent steel hull; the resulting slender and relatively highly stressed hull structure would be particularly susceptible to buckling failures; large local and overall structural deformations under wave forces and other loads would tend to cause problems in shaft alignment and attachment of pipes and other fittings; low hull-girder frequencies would be likely to fall in a range of wave-encounter frequency containing significant energy, resulting in serious wave-excited vibrations; a thorough and expensive programme of structural analysis and testing would probably be needed to establish the reliability of the design;

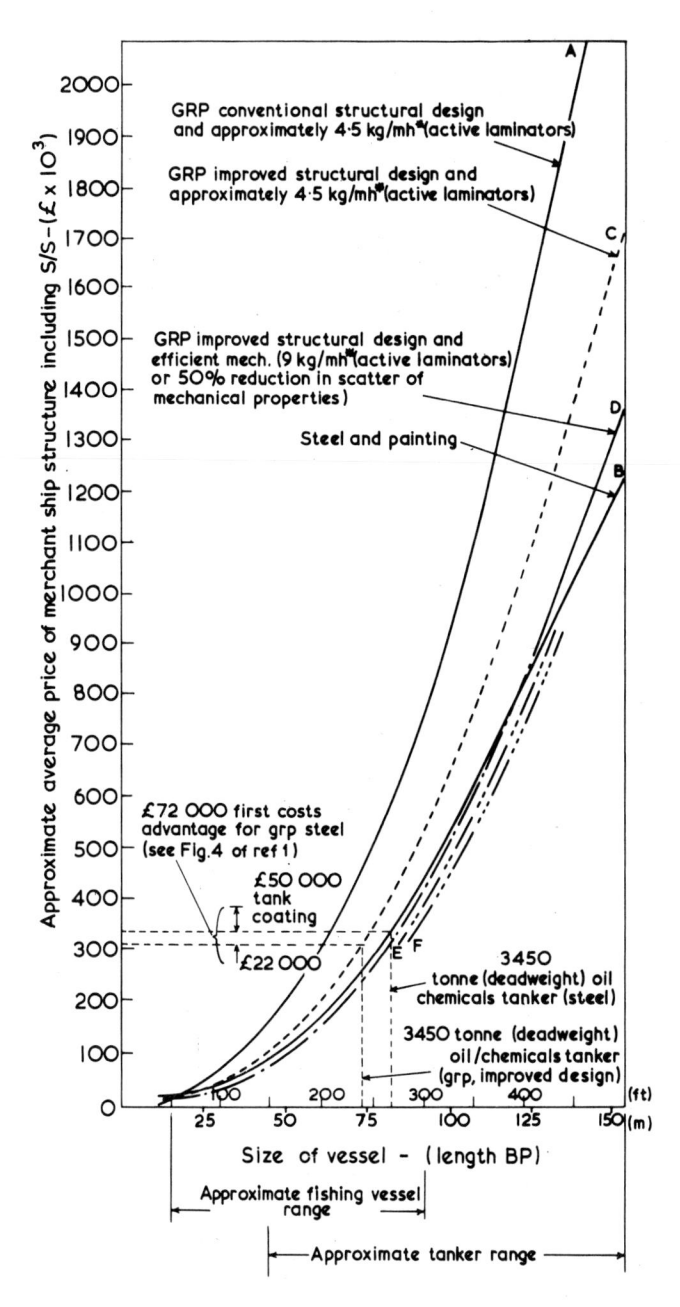

(iii) considerable difficulties and substantial first costs seem likely to be involved in the construction of a 100m hull, including provision of a large air-conditioned fabrication shed or alternatively development of techniques which would permit pre-fabrication and subsequent joining-up of hull components, as in steel construction.

I hope that Mr Guiton will publish a full account of his analysis, perhaps validating his cost figures and indicating how the various design and construction problems might be solved; until he does so I think his case must be judged 'not proven'.

Fig. 1 Average price versus size for grp and steel vessels

* mh = man-hour

1 Prices shown are for a single vessel in a production run of (such runs are now becoming common in the fields of steel tankers, bulk carriers, liberty ship replacements, oil rig supply vessels, etc).
2 Prices are for the UK (1973). The relative positions of the curves are considered to remain valid for mid-1974.
3 Prices include material, labour, general overhead costs, mould (grp only), workshop and equipment amortization allowances plus profit.
4 The higher quality requirement for the improved grp design accounts for its higher price rate/ton over the conventional grp design.
5 The improved grp design achieves a weight saving over heavy steel construction of 75%, while the conventional grp design achieves a 50% weight saving (in accordance with UK and US commercial experience).
6 The steel construction price/ton reduces with increase in vessel size due to reduction in curvature and hence increased ease of construction grp costs do not reduce so markedly as the shape becomes simpler — hence the divergence of the building cost curves.
7 Curves E and F: further savings resulting from improvements in productivity, quality, materials, methods etc, or by efficient carbon/glass composite construction etc (all possible with materials and technology currently available).
8 Average price rates used in the construction of basic curves A, B and C.

Vessel size and type	Price – £/net tonne of structure		
	Steel	Conv. grp	Impr. grp
24.4m trawler	508	1524	2032
73.2m fast ferry	406	1321	1829
79.4m tanker	356	1321	1829
143.3m general cargo	285	1219	1727

Pyrolytic surface treatment of graphite fibres

D.J. PINCHIN and R.T. WOODHAMS

A coherent coating of pyrolytic graphite was deposited in a continuous process on Thornel 50 fibres to improve the fibre matrix bond strength. The pyrolytic coating was produced from acetylene at low temperatures (900°C–1200°C), atmospheric pressure, and with short deposition times (5–45 s). The interlaminar shear strength and flexural strength of unidirectional graphite fibre/epoxy composites were increased significantly by the pyrolytic graphite coating. The fibre strength and resistance to abrasion were also increased.

1 INTRODUCTION

One of the early problems encountered in the use of carbon fibre reinforced plastics was the poor fibre/matrix bond strength compared to other fibrous composites which resulted in composite failure in shear [1]. Methods of fibre oxidation have been developed to improve the composite shear strength from the values of 30 MN/m² typical of untreated high modulus rayon-based fibres, to 60 MN/m² after oxidation [1,2]. Composite shear strength of rayon-based fibres used in this work are consistently lower than for polyacrylonitrile based fibres due likely to the filament shape or the twisted yarn of the rayon fibres. The oxidation process is not entirely satisfactory because the method is quite lengthy and the fibre strength is decreased to some extent [3]. No other method of fibre treatment has proven entirely satisfactory.

The fibre/matrix bond strength can be related to the properties of the fibres surface although the nature of the bond is not fully understood. The composite shear strength has been found to increase with decreasing crystallite size at the fibre surface [4]. An increase in the 'amount of crystal boundary' at the fibre surface, measured by Raman Spectroscopy is accompanied by an increase in composite shear strength [3,4].

In the present work a coating of pyrolytic graphite was deposited on the high modulus carbon fibres in order to reduce the surface order and hence increase the bond with an epoxy matrix without degrading the fibre strength.

2 EXPERIMENTAL

The fibre used in the work was rayon based Thornel 50 supplied by Union Carbide Co Limited. The fibre was drawn under slight tension into a closed Pyrex reaction tube where it was heated by passing an electric current through it. Electrical contact was made by passing the fibre over polished brass contacts. The acetylene source gas was diluted with nitrogen to reduce gas phase reactions. X-ray diffraction of the fibres was carried out on a Siemens Goniometer Diffractometer using copper $K\alpha$ radiation and a nickel filter.

Unidirectional bar specimens were prepared with a standard epoxy resin (Shell Epon 828 cured with Nadic methylanhydride/benzyldimethylamine) by pressing in a leaky mould [5]. The specimens were tested for flexural strength and modulus (50:1 span:depth ratio) and interlaminar shear strength (ILSS) in the short beam test (5:1 span:depth ratio). The fractured surfaces were examined by scanning electron microscopy.

3 RESULTS AND DISCUSSION

Rapid rates of graphite deposition were achieved at temperatures from 900°C to 1200°C. These low temperatures should not result in a degradation of the fibre strength observed on heating above 1200°C even in inert atmospheres [3]. Typical treatment conditions are shown in Table 1. The pyrographite was deposited on each filament in a coherent sheath which reproduced the filament shape as seen in Fig. 2.

Table 1 Experimental conditions for the pyrolytic vapour treatment of Thornel 50 with acetylene

Fibre Temperature, °C	1 130 ± 40
Treatment duration, s	10.25 ± 0.25
Lineal speed of graphite yarn, m/min	3.0
Acetylene flow rate, cm³/s	1.0
Nitrogen flow rate, cm³/s	4.0
Percentage weight increase of treated fibre, %	7.37
Power supplied to fibre, watts/cm	5.24

Mr Pinchin is at the Cavendish Laboratory, Madingley Road, University of Cambridge, Cambridge, England.

Dr Woodhams is at the Department of Chemical Engineering and Applied Chemistry, Wallberg Building, University of Toronto, Toronto, Ontario, Canada.

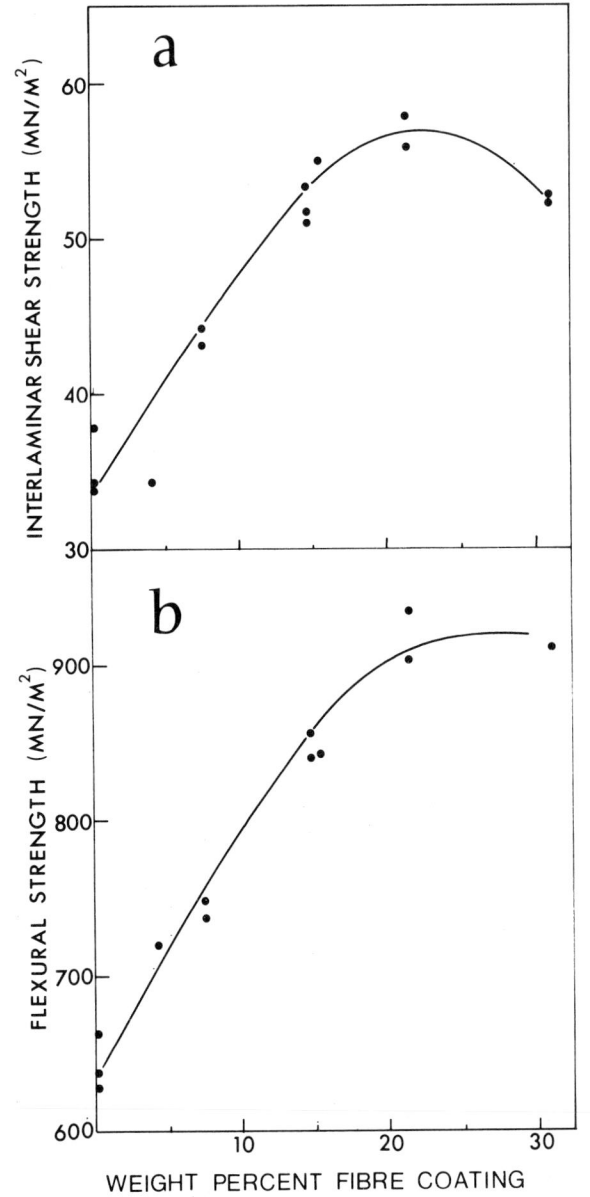

Fig. 1 Interlaminar shear strength (a) and flexural strength (b) versus weight percent pyrographite coating on the fibres for Series I specimens: volume fraction of Thornel 50 (without volume of coating) was 0.50 in all specimens: weight percent coating based on original fibre weight

Fig. 2 Tensile face of flexural specimen prepared from fibres coated to a 14.7% weight increase. Note the short pullout lengths and material adhering to the filaments

Fig. 3 Shear face of specimen shown in Fig. 2. Note the fractured coating and the considerable matrix deformation

X-ray examination of coated fibres showed the pyrographite to consist of smaller crystallites, less highly aligned than the Thornel 50 fibres. This was expected since pyrographite produced at atmospheric pressure and low temperatures has been shown to consist of small, disordered crystallites [6] which should bond well to the epoxy matrix.

Two series of specimens were prepared and tested. The first series contained a constant volume fraction ($V_f = 0.50$) of Thornel 50, not including the volume of pyrolytic coating in the volume of fibre. The second series contained a constant volume fraction ($V_f = 0.50$) of coated fibred including the volume of pyrographite.

The ILSS and flexural strength of the first series of specimens (Series I) are shown in Figs. 1(a) and 1(b). These contained a decreasing amount of resin as the amount of pyrographite coating increased. The decrease in ILSS above 20–25% weight of coating based on original fibre weight was believed due to insufficient resin and fibre–fibre contact. When the total volume fraction of coated fibre was reduced to 0.50 (Series II) the ILSS continued to increase to over 60 MN/m^2 at the maximum coating level used. The increase in composite flexural strength was shown by tests on Series II specimens to be mainly due to improved fibre/matrix adhesion rather than an increase in fibre strength or fibre volume fraction.

The flexural modulus was unchanged with the first series of specimens and decreased somewhat with the second series of specimens since the pyrographite coating would have a lower modulus than the original fibre.

The tensile fracture face of a flexural specimen (Fig. 2) shows short pullout lengths compared to untreated fibres and material adhering to the fibre surface. The adhesion between the pyrographite coating and the matrix appears better than between the fibre and coating since the fibre appears to pull out of the sheath. The shear face of a flexural specimen (Fig. 3) shows a great deal of matrix deformation not seen in untreated composites and areas where the coating has fractured, some adhering to the fibre and some to the matrix which has fractured away from the fibre.

The presence of a weak fibre/matrix bond has been suggested by Morley [7] as a possible energy-absorbing mechanism in a 'duplex' fibre composite. Notched impact tests were carried out on specimens containing coated fibres. The impact energy absorbed was found to decrease as the ILSS increased but not to the same extent as found with surface oxidized fibres, although more work is necessary before any definite conclusion can be drawn. Details of the impact testing have been reported by Pinchin [8]. The pyrographite imparted greater resistance to abrasion of the fibres and should provide greater resistance to fibre oxidation [6], although the yarn flexibility was decreased.

Recent work [3, 9] has shown that pyrolytic graphite fibre coatings produced from polymer coatings possesses a similar surface to coatings produced by vapour decomposition. The higher shear strength achieved using pyrolysed polymer coatings is believed due to improved adhesion between the coating and fibre.

4   CONCLUSIONS

The deposition of a pyrographite coating on Thornel 50 fibres has been found to increase the composite interlaminar shear strength and flexural strength while increasing the fibre strength and resistance to abrasion. The method is rapid and should provide improved impact resistance compared to fibre oxidation methods.

ACKNOWLEDGEMENTS

The authors are indebted to R.E. Steele for many helpful discussions and suggestions. The research was supported by the Defence Research Board of Canada (Grant 7501–88).

REFERENCES

1   Goan, J.C. and Prosen, S.P. 'Interfacial bonding in graphite fiber-resin composites', *Interfaces in Composites, ASTM STP 452*, American Society for Testing and Materials (1969) pp 3–26

2   Goan, J.C., Joo', L.A. and Sharpe, F.E. 'Surface treatment for graphite fibers', *Proceedings of the 27th Annual Technical Conference, Reinforced Plastics/Composites Division, Section 21E*, The Society of the Plastics Industry (1972) pp 1–6

3   Larsen, J.V., Smith, T.S. and Erickson, P.W. 'Carbon fiber surface treatments', *Naval Ordinance Laboratory Technical Report* 71–165

4   Tuinstra, J., and Koenig, J.L. 'Characterization of graphite fiber surfaces with Raman spectroscopy', *J Comp Mat* **4** (October 1970) pp 492–499

5   *Grafil Test Methods*, Carbon Fibres Unit, Courtaulds Limited, PO Box 16, Coventry CV6 5AE, England

6   Fialkov, A.S. et al. 'Pyrographite (Preparation, Structure, Properties)', *Russian Chemical Reviews* **34** No 1 (January 1965) pp 46–58

7   Morely, J.S. 'The design of fibrous composites having improved mechanical properties', *Proc Roy Soc A* **319** (1970) pp 117–126

8   Pinchin, D.J. 'Pyrolytic surface treatment of graphite fibres for improved composite shear strength', *MA Sc Thesis*, Department of Chemical Engineering and Applied Chemistry, University of Toronto, Toronto, Canada (1972)

9   Duffy, J.V. 'Carbonized polymer coatings as surface treatments for carbon fibers', *Naval Ordinance Laboratory Technical Report* pp 73–153

QUESTIONS

1   On effect of deposition on tensile strength, by C.A. Cross (TAC Construction Materials)

Has Mr Pinchin measured the effect of the deposition upon the tensile strength of the fibre? His first photomicrograph shows a considerable increase in cross sectional area, and it seems to me that some part of the improvement in laminate properties he has observed might be due to this factor instead of to improved fibre/matrix bonding.

Reply by Mr Pinchin

Fibre tensile strength measurements were performed on epoxy impregnated fibres over a 25.4 mm (1 in) gauge length. Fibres treated to a weight increase of 14.7% showed the breaking load to be increased approximately 20% while the tensile strength based on the new fibre cross section (including the area of pyrographite) was increased only 5%. Flexural strength of Series II specimens containing 0.50 volume fraction fibre coated to a 20% weight increase (ie total volume fraction coated fibre = 0.50) were increased by 30% over composites containing 0.50 volume fraction untreated Thornel 50. It is therefore believed that the marked improvement in flexural performance is due mainly to improved fibre/matrix adhesion.

Fatigue effects in carbon fibre reinforced glass

K.R. LINGER

Two carbon fibre reinforced glass materials were used as model systems to investigate the stability of multiple matrix cracking in a brittle matrix composite. Specimens were subjected to a cyclic strain in a flexural mode at 50 Hz and the load-deflection behaviour, after fatigue, correlated with changes in microstructure. It has been clearly established that the composite structure is degraded by cyclic strain, which may result in significantly reduced fracture toughness.

1 INTRODUCTION

Recent studies [1,2,3] have shown that the introduction of fibres into a brittle matrix can produce materials of increased strength and toughness. One fundamental difference between carbon fibre reinforced ceramics, and carbon fibre reinforced plastics is that the ceramic matrix will fail at a significantly lower strain than the fibre. If the full potential of such materials are to be available to designers and engineers it is of prime importance that information be available on the mechanical stability of brittle matrix materials at strains in excess of those necessary to produce failure of the matrix.

2 MATERIALS

The development, at Harwell, of a hot-pressing route to carbon fibre reinforced glass materials (crg) provided a readily available model brittle matrix material of closely controlled properties.

Two Pyrex materials were prepared containing continuous unidirectional type 1 carbon fibres, surface treated, and the fibre volume fraction and composite porosity were determined by a combined density and burn-out technique (Table 1). Flexural tests were carried out on long and short beams to establish the mechanical properties of the non-fatigued material (Table 2).

Materials Development Division, Atomic Energy Research Establishment, Harwell, Didcot, Oxfordshire, OX11 0RA, England

3 THE OCCURRENCE OF MATRIX CRACKING

The load deflection curves for crg materials tested in three-point bending show a noticeable change in slope at some stress σ_B below the fracture stress (Fig. 1). Critical examination of the tensile surface at this point reveals multiple failure of the glass matrix (Fig. 2).

Assuming the properties of the material can be predicted by the 'rule of mixtures' and further that equal strain exists in fibre, and matrix, the stress in the matrix at the 'bendover' point may be related to the composite stress σ_B thus:

$$\sigma_B = \sigma_m^* (1 + V_f(E_f/E_m - 1)) \qquad (1)$$

where E_f and E_m are the Young's modulus of fibre and matrix respectively and V_f the fibre volume fraction. The value of σ_B and hence the matrix stress σ_m^* at which cracking was observed was higher in the material containing the greater fibre content demonstrating the suppression of cracking by the fibre (Table 2).

Table 2 The physical properties of crg materials

Material Reference	Bend strength MN/m²	Standard deviation MN/m²	Coefficient of Variation %	'Bendover' stress σ_B MN/m²	Interlaminar shear stress MN/m²
1	387	61	16	202	31
2	482	58	12	426	69

Table 1 The compositions of crg materials

Material Reference	Fibre			Matrix			Volume closed porosity %	Volume closed porosity %
	σ Strength MN/m²	Modulus GN/m²	Volume fraction %	σ_m^* Strength MN/m²	Modulus GN/m²	Volume fraction %		
1	2 805	395	26.66	116	65	73.02	0.32	–
2	1 210	349	45.96	138	63	52.40	0.62	1.02

4 FLEXURAL TESTS

As a beam of crg material is bent, tensile and compressive forces are built up until the critical strain of the matrix is reached on the tensile face, and multiple matrix failure occurs. The material in this cracked region is now of reduced modulus and the stress distribution becomes asymmetrical with the neutral axis displaced towards the compressive face (Fig. 3).

Assuming the modulus of the material in the uncracked and cracked regions, E_1 and E_2 respectively, is given by

$$E_1 = E_f V_f + E_m V_m \qquad (2)$$

$$E_2 = E_f V_f \qquad (3)$$

and further that in both regions the materials are linearly elastic such that

$$E_1 \propto \mathrm{Tan}\,\theta$$

and

$$E_2 \propto \mathrm{Tan}\,\phi$$

Because there is no applied force along the beam the tensile and compressive forces must balance and in the nomenclature of Fig. 3.

$$\mathrm{Area\ ACDF} = \mathrm{Area\ FGH}$$

and the displacement of the neutral axis is related to the

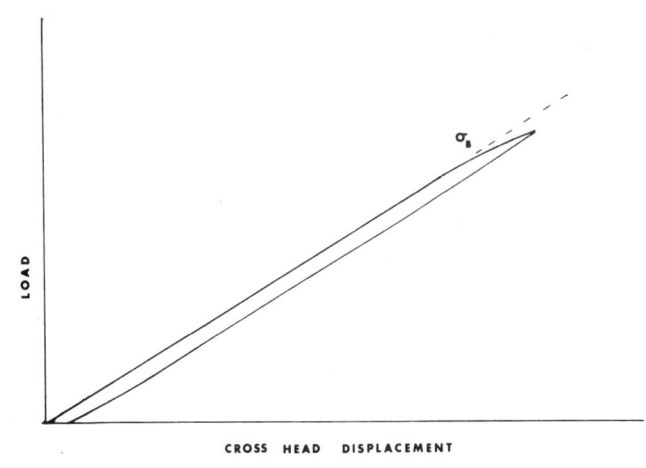

Fig. 1 The onset of matrix cracking in a crg beam during flexure

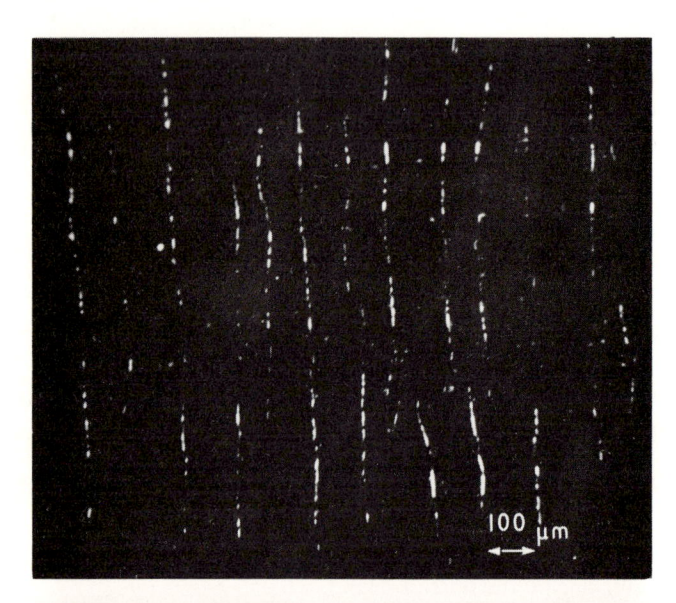

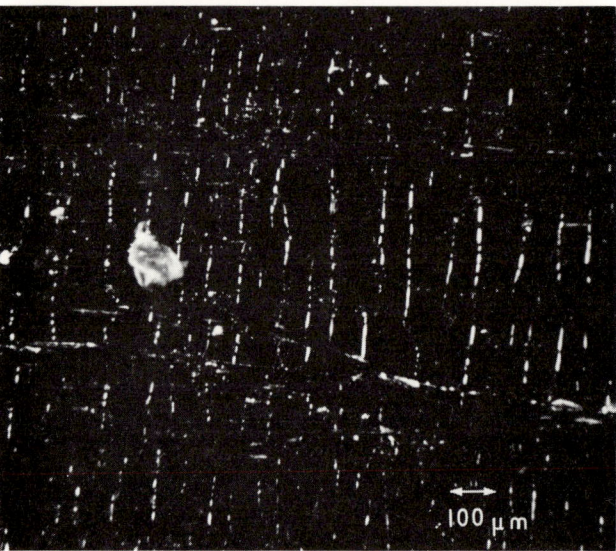

Fig. 2 Multiple matrix cracking after treatment with dye-penetrant: (top) material 1; (bottom) material 2.

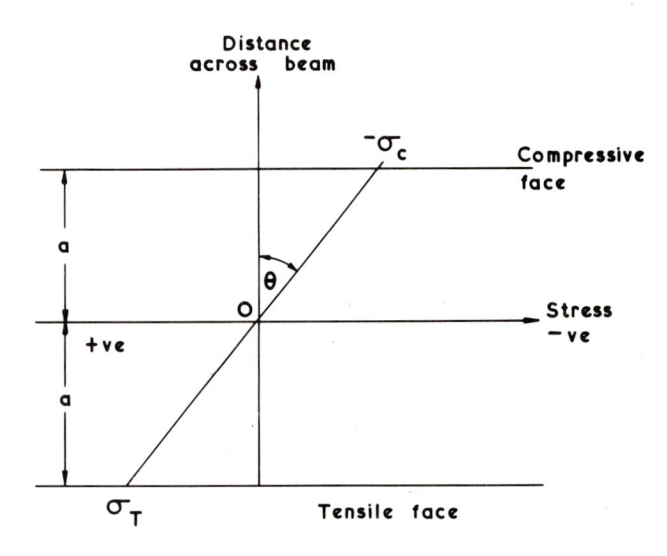

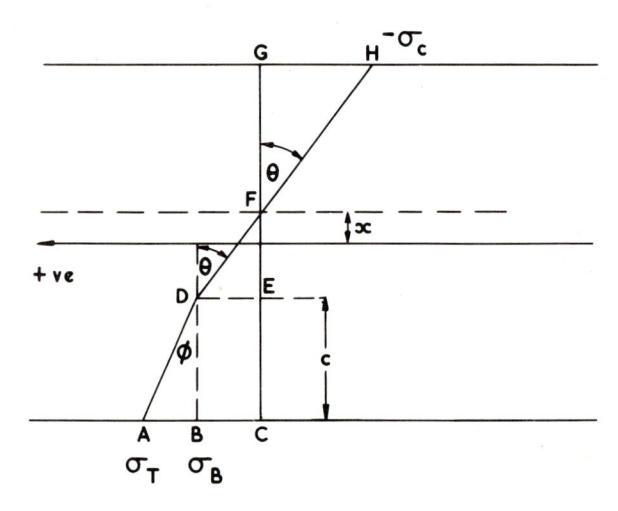

Fig. 3 The change in stress distribution at the onset of matrix cracking

extent of the cracking c by the expression:

$$c^2 \left(1 - \frac{E_2}{E_1}\right) = 4ax \qquad (4)$$

where $2a$ is the depth of the beam.

In the three-point bending test once matrix cracking has occurred further loading extends the cracks until the surface stress reaches the critical fibre stress and fracture results. The actual surface tensile stress at fracture in the cracked beam is dependent on crack penetration. Assuming that at the limit of the cracking the composite stress is σ_B, and that linear changes in stress occur across the specimen (Fig. 3) it is possible to estimate σ_T from the measured values of σ_B and c using the geometric relationship

$$\sigma_T = \sigma_B + \sigma_B \frac{c}{(a + x - c)} \cdot \frac{E_2}{E_1} \qquad (5)$$

5 DYNAMIC FATIGUE TESTS

All tests were carried out on accurately ground rectangular specimens tested in flexure using a three-point loading configuration. Generally specimens were $50 \times 1.5 \times 1.5$ mm with one face and side polished with $2 \mu m$ alumina to facilitate the observation of the matrix.

Tests were carried out using small bench fatigue machines designed and manufactured at AERE Harwell (Fig. 4). These machines utilise a motor driven centre anvil, the displacement of which is controlled by the position of two eccentric cams and they operate at frequencies in the range 0–100 Hz. The deflection of the specimen was measured by a sensitive displacement transducer positioned under the anvil against the tensile face of the specimen. The meter output from the transducer fed into an oscilloscope allowed the displacement cycle to be monitored continuously.

6 FATIGUE STRENGTH

In a first experiment nine specimens of both materials were cycled between a maximum deflection which corresponded to a stress on the surface of approximately 80% of the flexural strength, and a lower limit. The deflections were measured statistically and monitored continuously during fatigue. After 3×10^7 cycles at 50 Hz the specimens were loaded to failure on an Instron and the fracture stress calculated from simple beam theory

$$\sigma = \frac{1.5 P l g}{b (2a)^2} \qquad (6)$$

where l is the span, P the applied load and b and $2a$ the beam width and depth respectively. The results are summarised in Table 3 and include the values of σ_T calculated from Equation 5. The mean fracture stress for both materials fell within the scatter of the values obtained for non-fatigued specimens.

7 MICROSTRUCTURE

A technique was developed for the decoration of the fine matrix cracks using a dye-penetrant visible under uv light. A number of experimental observations were recorded:

(i) The crack spacing appears regular and is different for materials 1 and 2 (Fig. 2);

(ii) The penetration of the matrix cracking increases with increasing load;

(iii) The density and penetration of matrix cracking is increased by cyclic loading between fixed strain limits (Fig. 5).

8 FRACTURE TOUGHNESS

Two series of four specimens each of the higher fibre volume fraction material 2, were strained individually on the Instron to a maximum stress from 57–93% of the mean flexural strength, and then unloaded. They were then cycled between a deflection corresponding to the maximum and a lower limit. After 3×10^6 cycles at 50 Hz the specimens were loaded to failure, (Table 4). The strengths were again similar to the non-fatigued specimens but the failure modes were noticeably different.

With non-fatigued specimens of material 2 the normal failure was that of controlled fracture with high work of fracture. The fracture behaviour of both series were almost identical with the more highly strained materials showing a more catastrophic failure (Fig. 6).

Observations of the microstructure at each stage of the experiment revealed that crack extension had occurred during the strain cycling. The more catastrophic failure in these tests could generally be associated with less branching in the fracture as it progressed through the crg beam.

Fig. 4 The AERE flexural fatigue machine

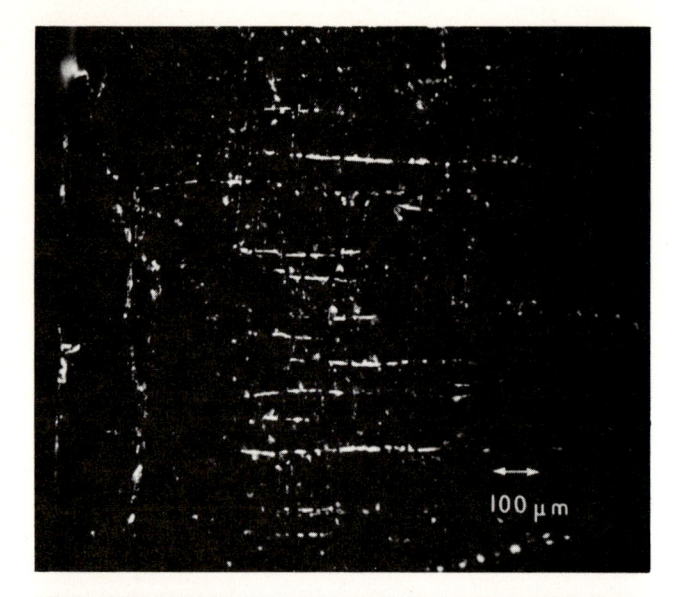

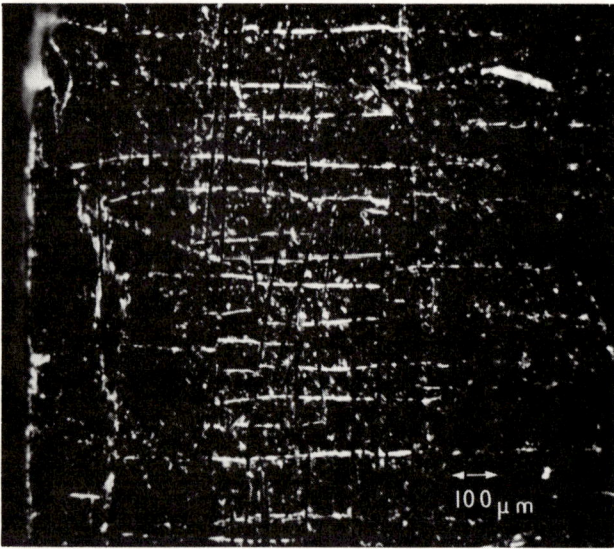

Fig. 5 The effect of cyclic loading on the density of matrix cracking (top) before fatigue; (bottom) after 3 × 10⁶ cycles

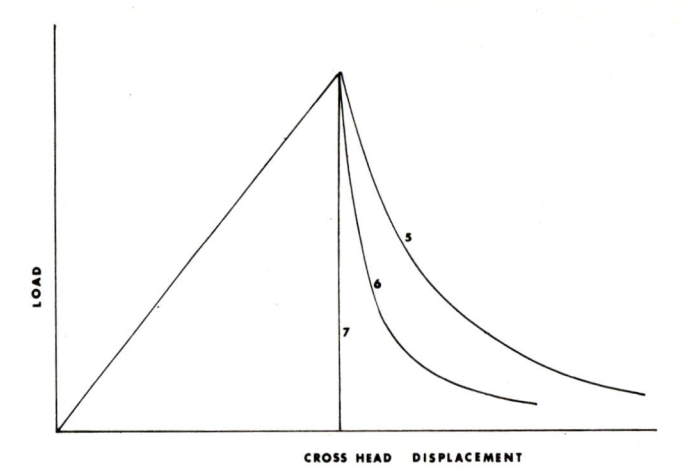

Fig. 6 The fracture behaviour of material 2 after fatigue

Table 3 Flexural strengths of fatigued materials after 10^7 cycles at 50 Hz

No	Material 1			Material 2		
	σ MN/m²	c mm	σ_T MN/m²	σ MN/m²	c mm	σ_T MN/m²
1	465	0.53	309	592	0.30	643
2	346	0.70	436	581	0.25	607
3	427	0.57	388	623	0.33	684
4	438	0.70	433	505	0.21	554
5	358	0.75	448	582	0.23	566
6	360	0.70	470	504	0.25	527
7	313	0.65	296	529	0.25	592
8	366	0.65	332	Failed in		
9	386	0.47	245	loading for fatigue		

Table 4 The flexural strength of material 2 after 10^6 cycles for increased maximum strain limits

Specimen No	σ_{max} (% flexural) strength	σ MN/m⁻²	σ_T MN/m⁻²
1	57	438	505
2	62	525	462
3	67	477	502
4	85	515	667
5	67	458	440
6	85	488	697
7	88	442	540
8	94	535	655

9 SUMMARY

Flexural fatigue tests offer a simple and rapid means of examining the mechanical stability of btittle matrix composite materials. The results indicate that even under the mild fatigue conditions imposed by controlled strain cycling the composite structure is degraded. Although the load bearing capability of the fatigued material appears unchanged the failure may be catastrophic.

It is possible to explain the crack extension in the glass matrix in terms of fatigue, but more difficult to explain the consequential reduction in fracture toughness.

In conclusion it would seem that a safe workable limit for brittle matrix composite materials, in environments where the material is subjected to a cyclic stress, may well be below the critical stress for the matrix severely limiting their use in engineering applications.

10 ACKNOWLEDGEMENTS

The author would like to thank the Ministry of Defence for permission to publish this paper and colleagues at Harwell for contributions.

REFERENCES

1 **Sambell, R.A.J., Bowen, D.H.** and **Phillips, D.C.** *J Mat Sci* **7** (1972) p 663

2 **Sambell, R.A.J.** et al. *J Mat Sci* **7** (1972) p 676

3 **Majumdar, A.J.** 'Glass fibre reinforced cement and gypsum products', *Proc Roy Soc A* **319** (1970) p 69

Acoustic emission and fatigue of reinforced plastics

M. FUWA, A.R. BUNSELL and B. HARRIS

Stress waves emitted in the course of cyclic loading experiments on unidirectional glass and carbon fibre reinforced epoxy resins have been studied in an attempt to explore the different ways in which these two materials respond to fatigue conditions. It is shown that whereas acoustic emission accompanying debonding can occur continuously during cycling of grp, even at low stress levels, the noise emitted by fibres breaking in cfrp will usually cease as the stress distribution between fibres and resin reaches a state of equilibrium through visco-elastic or plastic flow of the resin. True fatigue does not appear to occur in cfrp.

1 INTRODUCTION

We are familiar, from the work of Owen [1] for example, with the response of glass reinforced plastics to cyclic loading. It appears that because of the low moduli of the glass and resin, cycling to quite small fractions of the static failure stress leads to irreversible damage in the form of fibre/resin debonding which results first in a reduction of the composite rigidity and finally in failure by splitting. In carbon fibre reinforced plastics, on the other hand, the high fibre modulus prevents the development of critical strains in the resin and along the interface, and failure under cycling tensile loading usually occurs only within the static fracture stress scatter band, [2]. Indeed doubt has been expressed as to whether true fatigue failures occur at all in unidirectionally reinforced cfrp loaded along the fibres. We have sought to clarify the differences in behaviour of the two types of composite by cycling similar, hot-pressed samples of aligned grp (60 volume % E-glass/epoxy resin supplied by IMI Engineering Composites) and cfrp (40 volume % type I surface treated fibre/epoxy resin, supplied by Ciba-Geigy Ltd), and recording the stress waves emitted by the samples during deformation in repeated tensile loading.

2 EXPERIMENTAL WORK

Thin plates were hot pressed at 180°C and 15 Kg/cm² pressure for 60 minutes. Parallel-sides samples, some 0.35 mm thick and 120 mm long were cut from these plates and aluminium end tabs were glued on for testing. The sample cross-section and the tab glued area were deliberately chosen to ensure that no noise was heard from shearing failure in the grips. Acoustic emission was recorded by means of a Dunegan system with PZT resonant transducer (120kHz) at gains of 80 dB for crfp and 60 dB for grp which are much noisier. Cycling was carried out at a nominal rate of 1.25%/min.

3 RESULTS AND DISCUSSION

3.1 Glass reinforced plastics

Fig. 1 shows how the acoustic emission from grp cycled at very low loads (eg up to 25% of the static failure load) appears to die out after a few tens of cycles, although we feel that damage does continue to occur, albeit at a very low rate. If the load is raised to only 40% σ_f, however, small bursts of emission continue to occur intermittently and it is clear that even at such low levels the safety of a component

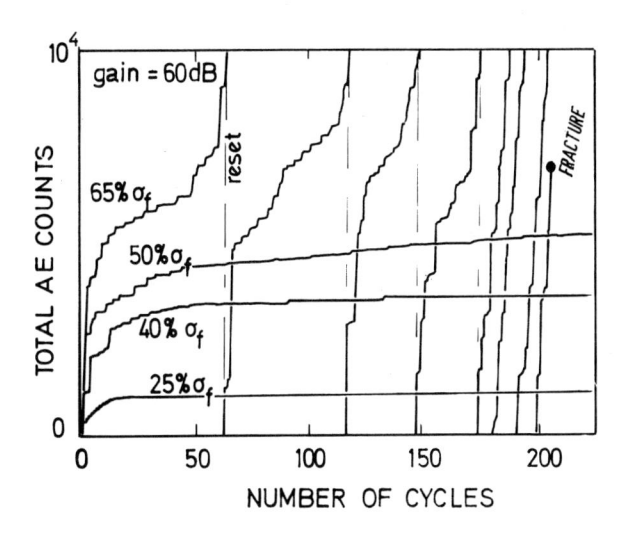

Fig. 1 Acoustic emission recorded during zero-tension cycling of a grp sample to increasingly greater fractions of its static tensile failure stress

---

Mr. Fuwa is a research engineer with Mitsubishi Petrochemical Co., Japan. Dr. Bunsell and Dr. Harris are, respectively, Research Fellow and Reader in Materials Science, School of Applied Sciences, University of Sussex, Brighton, England

cannot be guaranteed. As the stress level is increased the AE curve no longer levels out, and at only 65%σ_f cycling results in noisy splitting failures after some hundreds of cycles only. It should be emphasised that the noise heard during cycling at stresses up to the 50% level is not accompanied by any *visible* sign of damage other than a measured reduction in elastic modulus.

3.2 Carbon fibre reinforced plastics

The emission recorded on first loading of a cfrp sample (Fig. 2) is not repeated on reloading. Instead, a small amount of noise occurs just prior to peak *load* and during subsequent cycling the amount of noise recorded falls until it can no longer be detected. If, subsequently, the sample is reloaded beyond the previous maximum further emissions are recorded as soon as that maximum is exceeded. The accumulated AE as a function of the number of cycles is shown for a number of samples loaded to various levels in Fig. 3. Samples which fail invariably do so before the AE level has stabilised: in fact it appears that if stabilisation occurs the sample is unlikely to fail. We interpret Figs. 2 and 3 in terms of the occurrence, during cycling, of small amounts of viscoelastic or plastic flow of the matrix which permit redistribution of stress so that the weakest fibres, or those most perfectly aligned at the start of the test, are broken during the first few hundred cycles and emit stress waves. As more fibres take their share of the load, resin flow can no longer occur and noise dies out because fibres cease to break. This suggestion is supported by our observation that small permanent strains accompany the early acoustic emissions. Cycling is thus equivalent to holding a sample under load and permitting it to creep.

If a sample is deformed monotonically to a given load, cycled between that level and zero load until AE ceases (Fig. 4), and then reloaded monotonically to failure, the AE curve after cycling naturally starts at a higher point than that at which it would have been if the sample had been loaded only monotonically, but it quickly rejoins the original curve and its accumulated AE output is soon indistinguishable from that of a sample loaded non-cyclically. This observation seems to confirm our suggestion that 'tensile' and 'cyclic' damage are identical.

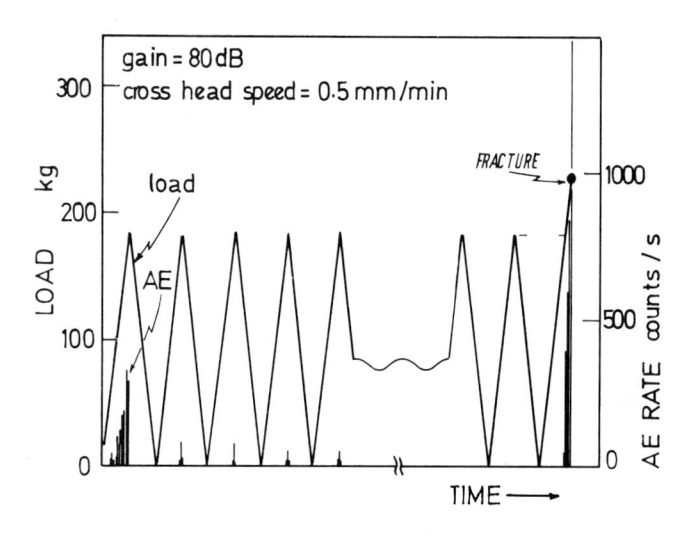

Fig. 2 Acoustic emission recorded during zero-tension cycling of crfp samples

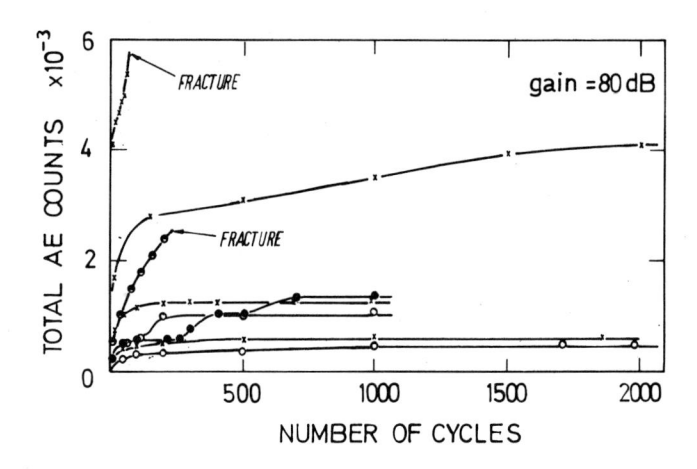

Fig. 3 Accumulated acoustic emission counts as a function of number of cycles during zero-tension cycling of cfrp samples

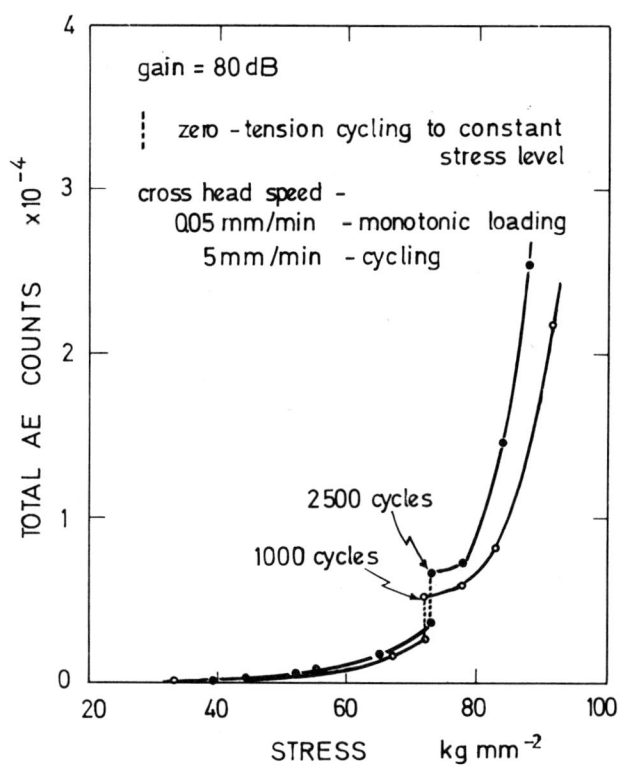

Fig. 4 Acoustic emission recorded when a sample of cfrp is monotonically loaded to a given stress, cycled between that stress and zero load, and subsequently re-loaded monotonically

Unlike grp, cfrp samples fail not by splitting but by homogeneous fracturing of fibres throughout the sample until sufficient numbers of breaks have occured in one cross-section to permit the joining up of microcracks leading to final separation. We have observed a reproducible relationship between the overall levels of acoustic energy emitted during fracture and the failure stress, as shown in Fig. 5. The differences between the samples obtained from different batches of pre-impregnated sheet shown in this figure are apparently due to variability in raw material quality, but it is interesting to note that had the data in Fig. 5 been plotted in terms of failure strains, the two sets of points would have been superposed.

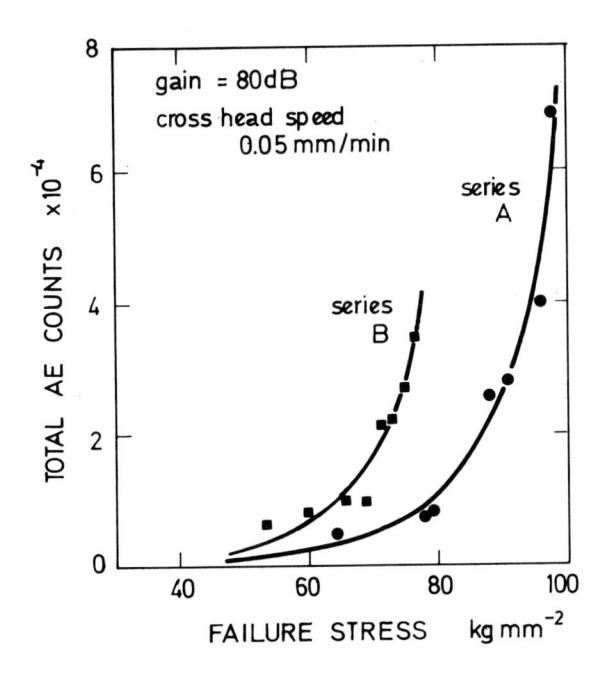

Fig. 5 Accumulated acoustic emission recorded during fracture of a number of different samples manufactured from two different batches of pre-impregnated carbon fibres

4 CONCLUSIONS

Acoustic emission analysis of composite behaviour during service provides an important means of non-destructive testing. It has been used to study fatigue in grp and cfrp, and the results suggest that whereas grp may be unsafe under cyclic loading at almost any stress level, cfrp are probably not susceptible to fatigue unless the stress system is such that large repeated strains occur in the matrix or at the interface. Failure in aligned samples stressed along the fibres occurs only within the first few hundred cycles and only in samples in which the cumulative AE record does not become stabilised.

ACKNOWLEDGEMENTS

The authors are grateful to the Science Research Council and the Mitsubishi Petrochemical Company for financial support, and to IMI Engineering Composites Group for samples of grp.

REFERENCES

1 **Owen, M.J.** in 'Glass reinforced plastics', edited by B. Parkyn, Illiffe (1971) p 251

2 **Beaumont, P.W.R.** and **Harris, B.** *International Conference on carbon fibres, their composites and applications* Plastics Institute, London (1971)

Composite materials in civil engineering: their current use, performance requirements and future potential

L.H. McCURRICH* and M.A.J. ADAMS†

The paper describes the general philosophy required for civil engineering materials and illustrates a number of applications for the main composites discussed, concrete, fibre reinforced resin and fibre reinforced cement. The methods used in design for determining safe allowable stresses in the materials are described and this demonstrates where information is required on the new composites. The paper concludes with suggestions for immediate work to provide the necessary information to enable the prediction of safe design stresses and elastic modulus for building applications where the design life is generally at least 30 years.

1 INTRODUCTION

It has been said that when civil engineers encounter an obstacle in their path they simply find a means of going round it whereas scientists prefer to analyse the obstacle in detail and possibly try to change it. This comment is to some extent relevant to the field of composite materials: for example, the engineer simply adds reinforcing rods to overcome the problem of poor tensile strength in the cement matrix, whereas the scientist is rather more interested in the fracture mechanics of the cement and ways in which micro-cracks can be arrested to improve the matrix strength.

The most notable civil engineering composite is concrete. The addition of sand and aggregate to the cement paste binder reduces the cost and reduces thermal, drying shrinkage, elastic and creep movements. The resulting composite is a remarkably durable low cost material having high compressive strength (up to $90 N/mm^2$) but poor tensile strength (up to $7 N/mm^2$). To circumnavigate the obstacle of poor tensile strength the engineer adds steel bars or precompresses the concrete using stressed steel cables, and the resulting material is in some ways the ultimate in fibre reinforcement, the steel rods being placed exactly where required, in the correct orientation, and at the optimum volume fraction. We even incorporate end hooks or bar deformations to improve bond!

A second composite used in large volumes is timber, a material which also makes use of oriented fibres, and these provide a high strength and stiffness to weight ratio. This paper however will concentrate on man-made composites and timber will not be discussed in any detail.

The difficulty of obtaining long term durability and of providing moulded shapes in timber, together with limitations on the supply and hence increase in cost, have resulted in civil engineers looking towards other fibre reinforced materials such as glass fibre reinforced polyester resin (grp) and fibre reinforced cement.

* Now with Chemical Building Products, Hemel Hempstead, Herts.
† Taylor Woodrow Construction, Southall, Middlesex, England.

An important fact which must be considered in relation to the use of materials in civil engineering is that, in general, construction products do not have to move once constructed, and therefore weight is not critical to the performance as it would be in the case of cars or aircraft. For this reason civil engineers are accustomed to the use of low cost, heavy materials such as concrete or bricks, and this is illustrated by comparing the cost of products from other branches of engineering with those of the construction industry, see Table 1.

The second fact about civil engineering materials is that for most applications they must have a long service life, often at least 30 years, and must withstand severe exposure conditions.

2 CURRENT USE

Trends in the use of composite materials are shown in Fig. 1 [1,2]. This shows that a remarkable volume of concrete is used each year and that the newer materials account for a very small fraction of the total consumption. It can also be

Table 1 Comparative costs of products from various industries

Product	Approximate cost £/tonne
Lockheed Tristar aircraft	62 500
Boeing 747 Jumbo jet aircraft	32 500
Rolls Royce Silver Shadow car	5 600
Ford Escort 1300 car	1 300
Bulk carrier cargo ship	800
Detached 3-bedroomed house	200
Concrete oil platform for North Sea	160
Prestressed concrete bridge	150

seen from Fig. 1 however that the growth rate of grp as an example of the new materials has been strong in recent years and it seems reasonable to expect a steady growth in the future both of grp and of fibre cement materials. Before considering the problems of performance assessment and requirements for these composite materials it is worth briefly reviewing some of their more interesting applications in the construction industry.

2.1 Concrete

Everyone is familiar with the variety of uses to which concrete is put and we shall therefore only illustrate three examples, but each of these requires a design understanding of the materials' performance which is beyond that normally called for.

The use of prestressed concrete pressure vessels for nuclear power stations [3,4] is illustrated in Figs. 2 and 3. The scale of these structures is impressive with a height of 31 m, a diameter of 27 m and a maximum wall thickness of 6.4 m. They have to contain gas at a pressure of 4.0N/mm^2 and the concrete operates at continuous temperatures of up to 70°C. The materials engineer has to provide the designer with knowledge on strength, on the creep of concrete at elevated temperatures, on thermal movements, on the effect of moisture migration

Fig. 2 Model of one of the prestressed concrete pressure vessels being constructed for the Hartlepool nuclear power station. An idea of the scale can be obtained from the man standing at the bottom of the vessel. The concrete is pre-compressed by wires wound in channels round the outside of the vessel and by prestressed tendons passing through vertical ducts. A very thorough understanding of concrete performance is required to enable safe prediction of stresses and deformations in such a structure. (Designed and constructed for the CEGW by Taylor Woodrow Construction)

Fig. 3 A view of the Hartlepool pressure vessel during construction, illustrating the point that concrete is a macro-fibre reinforced composite

through the structure, and on the effect of radiation on concrete.

Concrete is being increasingly used for long span bridges [5], see Fig. 4, where apart from mechanical and physical properties one may have to consider the durability of its surface appearance. There is now a strong trend towards the use of reinforced and prestressed concrete for the construction of North Sea Oil production platforms [6] and a model of a proposed rig is shown in Fig. 5; in this case concrete is used in water depth not previously encountered and under severe exposure conditions including dynamic fatigue condition not normally experienced in civil engineering structures. The engineer must have data available on the performance of concrete under these particular conditions [7]

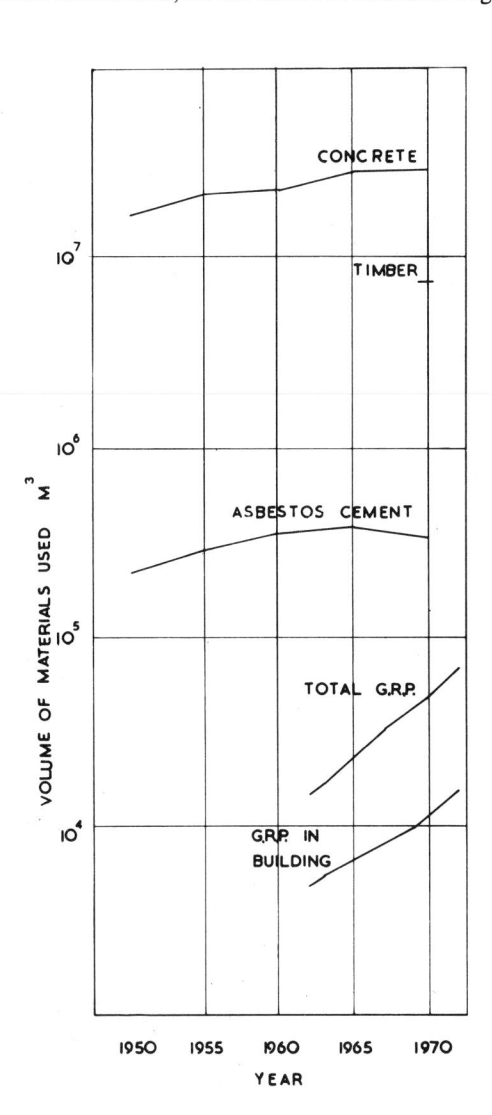

Fig. 1 Consumption of composites in UK building and civil engineering. A log scale has been used to enable the new materials to be plotted; see also [1,2]

to ensure that satisfactory durability can be achieved with absolute confidence.

2.2 GRP

Glass fibre reinforced polyester has been steadily gaining acceptance as a construction material. Being one of the few translucent structural materials it has been widely used for translucent roof sheeting, and on a limited scale for shell roofs such as those used to enclose swimming pools. Grp has recently been extensively used as a semi-structural material

Fig. 4 Example of an advanced use of concrete in bridge construction. This shows a model of the Staples Corner bridge being tested in the Taylor Woodrow structures research laboratory

Fig. 5 Concrete has recently been adopted as the construction material for North Sea offshore oil production platforms. For the safe design of these structures detailed materials performance knowledge is required on the effects of deep sea immersion, fatigue loading and severe splash zone corrosion environments

Fig. 6 An exciting use of grp for the construction of a training centre at Haselmere, Surrey, England. Parts of the inside of this building are lined with grc for fibre protection. (Client: British Olivetti; architect: James Stirling and Partners; contractors: Myton Ltd; consulting engineers (plastics): Polyplan Ltd)

Fig. 7 GRP was used extensively in the construction of the Dubai airport terminal. This shows the umbrella units constructed from chopped strand glass fibre and polyester resin. (Client: Government of Dubai; architects: Page and Broughton; main contractor: Costain Civil Engineering; grp manufacture: Mickelover Ltd); see also [8]

for wall cladding and Fig. 6 illustrates an exciting use in the construction of a company training centre in Surrey. A further example of the use of grp to provide a decorative yet functional building is shown in Fig. 7, [8].

Major structural uses of grp have been limited by problems of fire behaviour, deflection and high cost. Nevertheless a number of interesting grp structural applications have been investigated [9] and it may well be that in the future some of these will gain acceptance, and the material will be adopted for structural use as well as the semi-structural applications found at present. Grp has also been effectively used for parts of buildings not normally seen such as formwork for the casting of concrete, machinery covers, and pipes.

2.3 Resin concretes and polymer modified cements

These materials are made by a combination of polymers with inert mineral fillers and have a limited use for specialized applications such as industrial flooring, bearing pads, bridge joint nosings and chemically resistant pipes, [10].

2.4 Fibre reinforced cement

Materials under this heading broadly fall into two distinct categories [11]:

(1) Fibre reinforced concrete in which chopped lengths of fibre are incorporated in the concrete mix and give random 3-D fibre orientation with a volume fraction limited to about 2% before severe problems of poor workability and compaction are encountered.

(2) Thin sheet fibre reinforced cement or mortar materials which are made by laminating fibre and cement in a similar way to grp to give 2-D planar fibre orientation with fibre contents of 3–15% by volume. Ferro cement, [12,20] in which layers of fine wire mesh are used for reinforcement, should also be included in this category.

Reinforcement that could be moulded into shape during the casting of concrete or even included in the original mix has long been the aim of much research and development, for example attempts to include nails, steel swarf and other lengths of reinforcement. In 1964 however Romualdi [13] in the USA published results for cement reinforced with short lengths of chopped steel wire, the concept being not that the fibre acted as conventional reinforcement obeying the rule of mixtures, but that it arrested crack growth in the cement matrix and by virtue of this increased the tensile strength, ie the strength of the material is proportional to the spacing of the fibres for a constant volume fraction. Since then, there has been considerable controversy as to whether this effect can be achieved in practice, but arising from this there has been a growing use of chopped wire in concrete. There have been several misconceptions as to what can be usefully achieved, but Fig. 8 illustrates the typical tensile stress/strain curve for fibre reinforced concrete. This shows that fibres should be considered as a useful 'ductility' admixture in concrete and whilst the strength is not significantly increased, the flexural resistance and more particularly the impact resistance are improved and the fibres can also impart crack control properties.

In general steel fibres are used for this application because of their high modulus though a wide range of both natural and man-made fibres are available and these have been reviewed in detail in the Concrete Society report [14] 'Fibre reinforced cementitious materials'. Current applications for fibre concrete [15] include road and runway slabs such as that shown in Fig. 9, concrete overlays, industrial flooring, shell piles (West's piles which include polypropylene fibre reinforcement) removable car park decks, explosion resistant structures and refractory linings. Work is also progressing on the use of fibres in conjunction with conventional reinforcement to enable the conventional reinforcement to be taken to high stresses without unacceptable cracking in the concrete, [16]. Fibres have also been used to provide impact resistance to mesh reinforced concrete used for marina pontoons.

Thin sheet fibre reinforced cement materials have long been available in the form of asbestos cement. This material is in large scale production for corrugated sheet, cladding, water tanks, pipes and fire resistant components, [17]. Asbestos fibres however have limitations due to their short length and the high cost of sophisticated separation processes. There are also the problems of health hazards associated with handling some of the asbestos fibre materials, [18]. Asbestos cement sheets tend to be rather brittle and they require sophisticated machinery for their manufacture. Steel fibres are difficult to mould into shape and suffer from rust spotting where exposed on the surface, and the most promising fibre for new types of thin sheet fibre cement material appears to be glass fibre in particular the alkali resistant glass Cem-FIL [19] (developed by Building Research Establishment and Pilkingtons).

Current applications for glass fibre cement sheet materials include cladding panels such as those shown in Figures 10 and 11, permanent shutters, marina pontoons, chimneys,

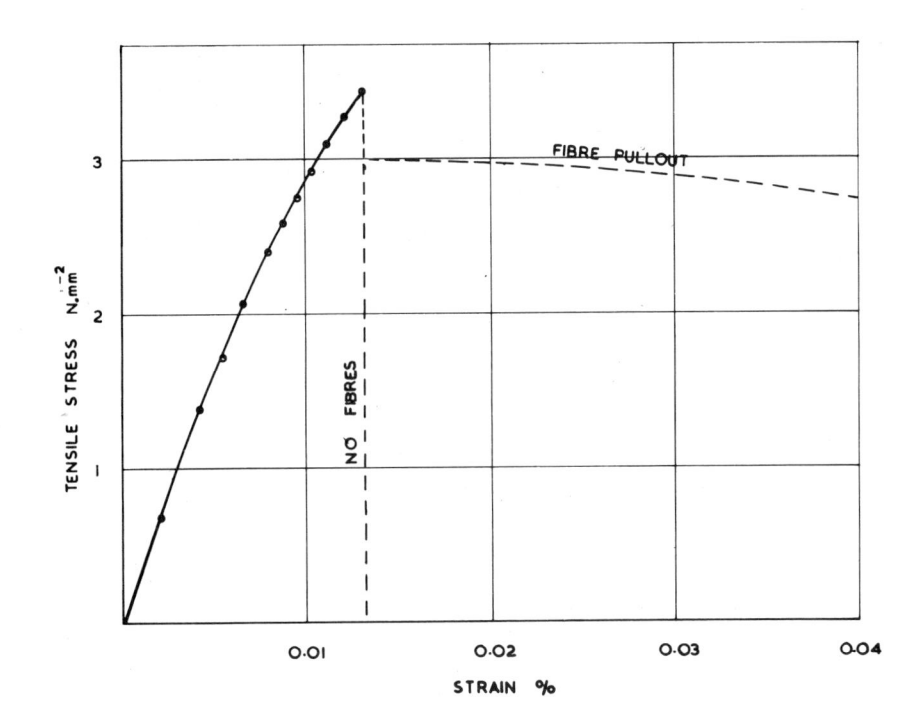

Fig. 8 Fibre reinforced concrete tensile stress/strain curve obtained from direct tension test shown in Fig. 19. This mix contained 2% by volume of 37 mm long 0.5 mm indented steel wire fibre and the curve illustrates how fibre addition to concrete mixes acts as an effective ducility admixture. The direct tensile strength is not increased but there is a considerable degree of post cracking strength which results in improvements in impact and flexural strength. The age of the concrete was 1000 days.

L. H. McCurrich* and M. A. J. Adams†

Fig. 9 Road slab being cast using wire reinforced concrete, in Ohio, USA. In this application fibres are used to give crack control and tests indicate that thinner slabs can be successfully used. (Photograph by courtesy of National Standard Co Ltd)

flues and refuse shutes. These products are in many cases fabricated by techniques somewhat similar to those used for grp. They have properties and costs generally falling in the gap between asbestos cement and grp and have the distinct advantage of being incombustable.

Ferro cement [20] has been successfully used for shell roofs and has been used as a boat building material for over a hundred years.

Having briefly described some uses and applications for these composite materials in building and civil engineering this paper will now attempt to outline the data which is available and used by design engineers, to identify shortcomings in this data and to assess the materials with regard to their future potential.

3 PERFORMANCE REQUIREMENTS AND ASSESSMENT

The engineer is interested in a quantitative value for the mechanical and physical properties of composites which he can safely use in design. These values are in general based on short term test data, modified to take account of likely variability in quality and time dependant properties in particular environments. It is interesting to look at the three main composite materials previously described (concrete, grp and fibre cement) to see how safe design values are arrived at, to indicate where knowledge is lacking, and to see how the information is communicated as design data.

3.1 Concrete

Basically two levels of design data are required. The first is simple data relevant to the immediate requirements of the design engineer and which can be readily incorporated in Codes of Practice or British Standard Specifications. These data must have fairly universal application and not be too limited to specific conditions. The second level of design data is where the performance of the material is more critical, where further properties need to be considered and where more specific values are required; these data will then be relevant to the specific conditions involved.

For simplified engineering design the performance of concrete is defined by the compressive strength of a cube from the material tested at 28 days. The required strength is obtained by careful selection of the concrete mix proportions; a low water cement ratio is required for high strength, but the water content is also controlled by the need to obtain adequate workability, [21]. The reinforcing steel performance is based on a tensile strength test to determine its yield point or proof stress at 0.2% strain. The way in which these figures are reduced to allowable design stresses in the code of practice for the structural use of concrete CP110 Part 1, 1972 [22] is illustrated in Fig. 12.

To illustrate the effect of the above allowance for the strength of concrete, a mix with a target cube strength of about 58 N/mm^2 would have a characteristic cube strength of 50 N/mm^2 which would be reduced by the factors shown to give an allowable compressive stress in bending of 22 N/mm^2. This gives an allowable design stress about 38% of the main cube strength, and this value is further reduced for members in direct compression where redistribution of stress through local over-loading would be more critical.

For simplified design the thermal expansion coefficient would generally be taken as similar to reinforcing steel (ie 12×10^{-6} per $^{\circ}$C) and where required, for example in prestressed concrete, Young's modulus is taken as a function of the cube strength; shrinkage and creep are allowed for by simplified formulae which for shrinkage is typically a strain of 200×10^{-6} and for creep a strain of 36×10^{-6} per N/mm^2. For steel a Young's modulus of 200 KN/mm^2 is assumed

Fig. 10 Shop front in Bognor, England using Elkalite glass fibre reinforced cement thin sheet cladding panels.

Fig. 11 Glass fibre reinforced cement thin sheet cladding panels used on a Pilkington building in St Helens, Lancs, England. (Panels manufactured by Taylor Woodrow Anglian. Photograph by courtesy of Fibreglass Ltd)

for reinforcing rods and small diameter strand, and in the case of large diameter prestressing strand a slightly lower value of 175 KN/mm².

In addition to the materials allowances described above, load factors are applied to the anticipated loading conditions to give the safe design loads. The values described above are used for calculating the ultimate failure condition of the structure, and separate factors are applied to check certain servicibility conditions for example deflection and cracking.

For more sophisticated structural applications of concrete, such as the prestressed concrete pressure vessels for nuclear power stations described previously, the properties of concrete must be more closely defined and this is a particularly involved process since the material is sensitive to the following:

(1) Variation in the properties of the basic components which are naturally won materials and therefore vary depending upon their source. For example Thames Valley gravel would have a thermal expansion coefficient of 10×10^{-6} per °C whereas limestone might be as low as 5×10^{-6} per °C, and this will set up greater interfacial stresses between aggregate and cement paste (18×10^{-6} per °C) under conditions of thermal cycling.

(2) Variation due to the wide range of mix proportions which are required for workability and economic considerations and which also depend on the particle shape of the given source of aggregate. Some of these effects are illustrated in Fig. 13, [3].

(3) Maturity in that concrete gains strength progressively with age, unlike plastics or resin concrete for example which may tend to lose strength; this is illustrated in Fig. 14, [3];

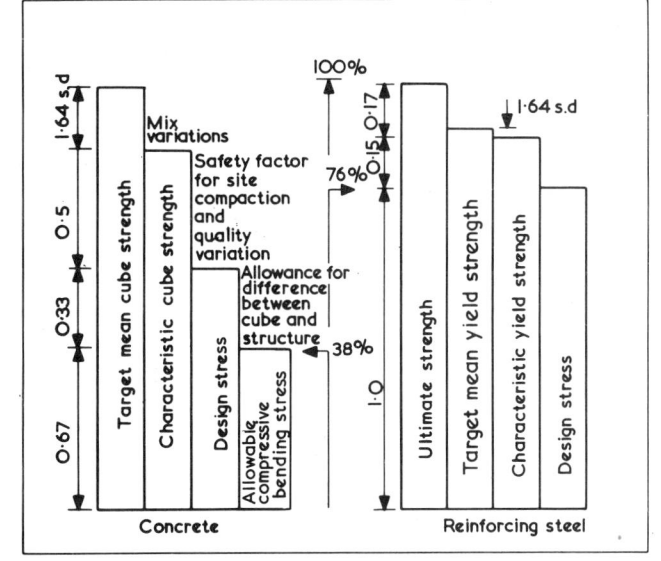

Fig. 12 Determination of allowable design stresses for reinforced concrete structures, from CP 110 Part 1, 1972 [22]

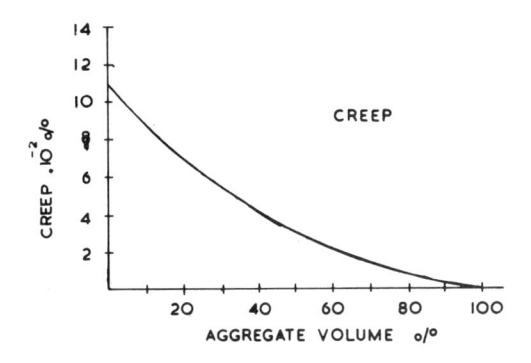

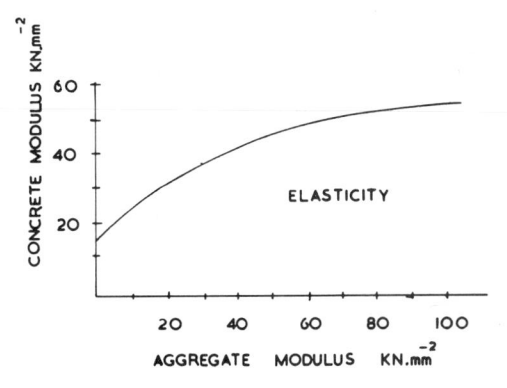

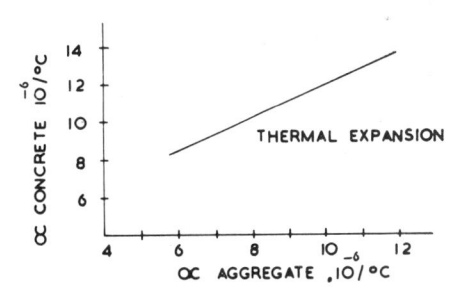

Fig. 13 Effect of various aggregate proportions and properties on the properties of the resultant concrete. Where the performance of concrete has to be accurately determined, for example in pressure vessel work, the effect of the local source of available aggregate must therefore be taken into account. See also [3]

(4) Moisture state in that concrete properties vary with the moisture content, or the moisture migration conditions existing within the concrete; this will for example effect shrinkage movements, thermal expansion and creep characteristics as illustrated in Fig. 15 [3].

Data have been built up over decades of research work, and over the last 10 years they have been analysed with new experimental work to provide the necessary design data on concrete performance for pressure vessel applications. The process by which this is done is too detailed to discuss in this paper, but the approach adopted by Taylor Woodrow to predict long term (30 year) deformations, stress distribution and ultimate strengths for the Wylfa, Hartlepool and Heysham pressure vessels has been described in a comprehensive paper by Browne and Blundell, [3]. The basic principle involved is to see how particular properties vary in given environments, and know how the property curves can be built up to predict the effect of combinations of properties. The principle generally used for example, in the prediction of concrete creep under varying stress conditions, is that of superposition.

Another example of this sophisticated property prediction technique are the methods which have been developed to predict durability in severe operating environments, such as those existing for North Sea oil production platforms [7]. In this case a thorough assessment is made of corrosion mechanisms in both the concrete and the reinforcing steel so that test programmes can be designed or data sought from the literature, which are strictly relevant to the problem. Examples of this include tests to measure current flows in macro cells which may form due to differential salt concentrations, or fatigue tests on bars both in sea water and in sea water plus cement environments similar to those that reinforcing bars would experience at 100 m depth in the North Sea.

The comments above illustrate how data has been made available to enable engineers to design novel concrete structures with confidence in the material.

3.2 GRP

Unfortunately for the use of grp in building and civil engingeering, no standards or codes of practice for safe design stresses, deformations or durability requirements at present exist. There are however large volumes of miscellaneous data available, much of which can be used, though rarely are, to provide a basis for safe design. Most of these data have been obtained in other industries particularly aeronautical and marine engineering. Since many applications of grp are non-structural, the thickness and type of laminate used is based on intuition and previous experience rather than calculated strength design.

In some respects grp presents a more difficult material for standardized design data than concrete, in that a vast number of different resins are available covering a spectrum from rigid brittle materials to soft flexible materials all of which have temperature sensitive properties. On the other hand one is saved from some of the moisture migration problems within the material such as those that occur within hardened cement paste.

To illustrate the state of the art in some of the design aspects which are relevant to civil engineering use, the strength, deformation and fire aspects will be briefly considered.

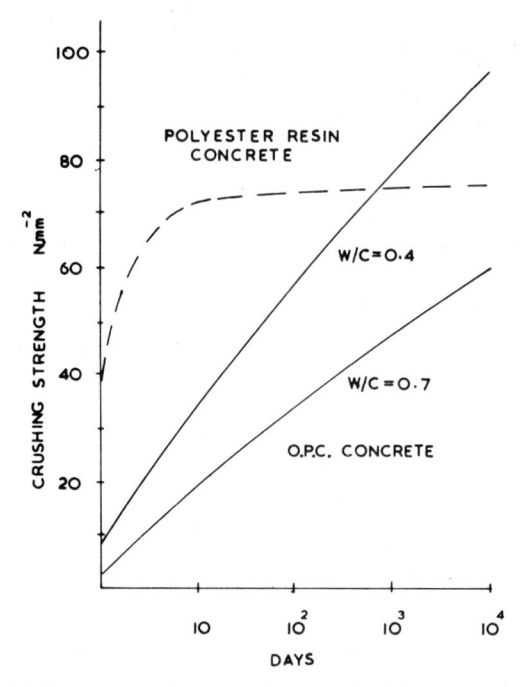

Fig. 14 Concrete continues to gain strength with age, which is an added safety factor in the use of the material, whereas other composites, for example resin concrete, reach a high strength early in their life but then may progressively lose strength with time. The strength of concrete is largely controlled by the water/cement ratio in the original mix proportions and the effect of this is also shown. See also [3]

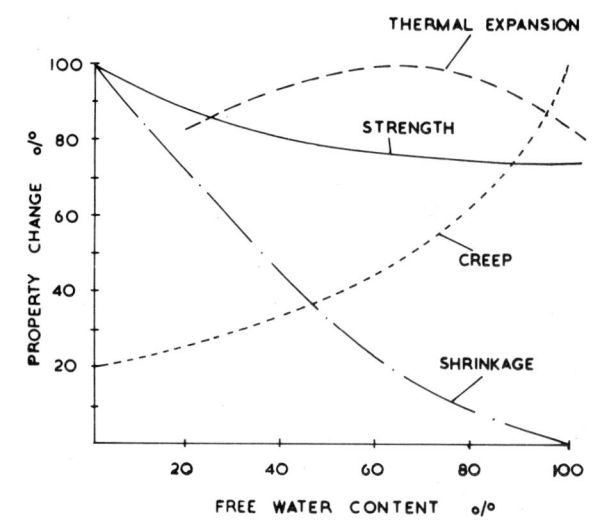

Fig. 15 The moisture state within concrete plays a significant role in modifying the properties of concrete as illustrated by these curves. A changing moisture condition will also effect properties and for this reason moisture diffusion calculations have to be carried out where long term performance needs to be accurately predicted. See also [3]

Fibre content and orientation are critical to the strength performance. The effect of the resin matrix properties will be more significant in chopped fibre reinforcement where it transfers stress from fibre to fibre than it will be with unidirectional continuous fibres.

The effect of fibre content is usually considered by the simple rule of mixtures, ie

$$\sigma_c = \sigma_m V_m + \eta \sigma_f V_f$$

where σ = stress at failure, V = volume fraction, η = orientation effectiveness and the suffixes c, m and f are for

Table 2 Effect of fibre orientation on effectiveness (η)

Orientation	Effectiveness, %
Unidirectional	100
Orthogonal	40–50
2-D array random plane	30–37
3-D array random solid	0–20

composite, matrix and fibre respectively. Fibre orientation is considered by a number of techniques [14,23] and may broadly be summarised as in Table 2.

To illustrate this simple method of calculating strength for a typical chopped strand mat polyester laminate with a 3–1 resin-glass ratio by weight, used for a cladding panel the expected strength is as follows:

volume fraction of fibre	=	0.12
fibre strength	=	2 000 N/mm^2
effectiveness of fibres	=	33%
matrix strength	=	28 N/mm^2
expected tensile strength	=	104 N/mm^2
measured tensile strength	=	84 N/mm^2

It can be seen this simple technique gives a reasonable prediction of short term ultimate strength, but for structural applications of grp it is the long term strength which must be assessed by the design engineer and for building applications this may well be at least 20 years. Not only must the effect of the environment such as water soak or outdoor weathering be considered, but also the effect of continuous loading combined with the environment. At this point the designer starts to run into difficulties because whilst adequate data are usually available for short term ultimate strengths for the particular composite composition to be used, it is likely that little extra data exist on long term performance — for example creep rupture. Examples do exist and creep rupture results for a chopped strand laminate at 20°C are given in Fig. 16 [24,25] from which it can be seen that substantial reductions in strength occur under long term loading especially in wet conditions.

Fatigue may also need to be considered for example to allow for wind loading, and the effect of fatigue on a chopped strand laminate is shown in Fig. 17, [24].

There are as yet no authoritative documents to state what safety factors should be applied to the short term test data to allow for some of these effects but the Gibbs and Cox Marine Design Manual [26] (now 14 years old) makes tentative suggestions for design values to be applied to the test results, see Table 3.

Recently a British Standard, BS 4994 1973, has been issued for the design of vessels and tanks in reinforced plastics and this may also be used as a guide for safe design values in building. This standard uses a design factor which is made up from five partial safety factors which cover the following aspects; manufacture, long term loading, temperature of operation related to the resin heat distortion temperature, cyclic loading and curing conditions. These factors result in a minimum factor of safety of six and this may increase depending on the partial safety factor.

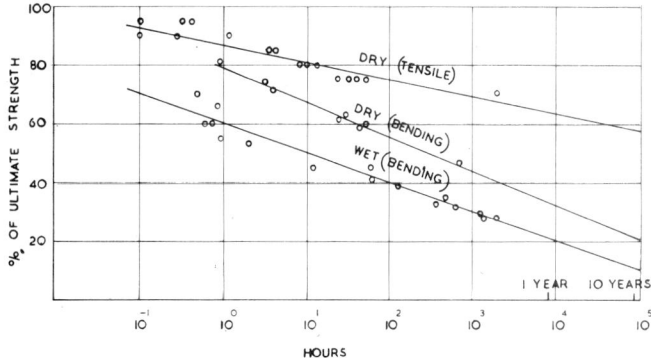

Fig. 16 Creep rupture curves for grp. This curve illustrates the significant loss in strength which occurs for a chopped strand mat glass fibre reinforced polyester laminate held in a wet condition under continuous load. For long term loading the strength may well reduce to 50% of the original strength and when a further factor of safety is applied to this, the allowable design stress for grp may be as low as 10% of the short term ultimate strength. To predict the safe design stresses for grp, it is essential to have these types of data available. See also [24,25]

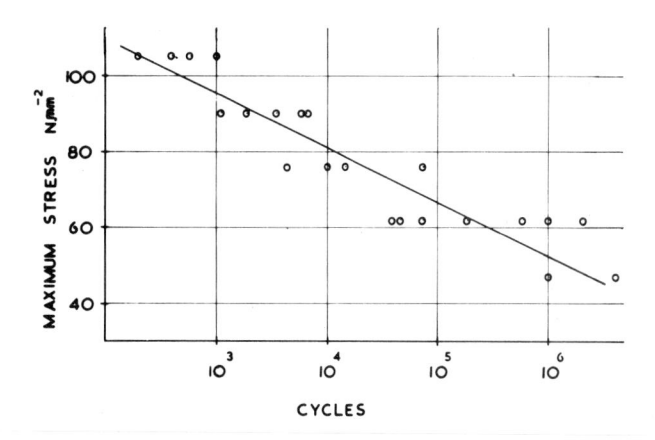

Fig. 17 Cyclic loading results in a loss of strength for grp. This curve illustrates the loss which occurs for a chopped strand mat glass fibre reinforced polyester subject to fatigue testing in the dry condition. There is a significant loss in strength and this must be taken into account for safe allowable design stresses under conditions of cyclic loading. See also [24]

Table 3 Safety factors for various loading conditions (after Gibbs and Cox Marine Design Manual [26])

Loading condition	Minimum factor of safety on test results
Static short term	2
Static long term	4 on reduced values from creep rupture tests. This is likely to increase the required factor to 10 minimum.
Variable or chaning	4
Repeated	6
Fatigue or reversed	6
Impact, repeated	10

For long term deformation calculations, creep under the relevant environmental conditions must be considered and only limited data are available on this. Because of the low Young's modulus of grp it may become a critical factor in the design where deflection becomes the design criteria rather than strength. This is particularly true for members in compression where design is invariably controlled by buckling criteria, [27]. Typical creep data is shown in Fig. 18 [28] and the effect of temperature is particularly noticable. Cladding panels used on buildings in the UK can be expected to go through the following temperature extremes, [29]: dark colour +65 to −20°C; light colour +50 to −20°C.

The problem of fire in relation to grp can be divided into three categories: fire resistance, surface spread of flame and toxic fumes. The creep curves illustrate that grp is unlikely to have adequate load carrying capability under fire conditions where for example in the BS 476 fire resistance test the temperature rises to 840°C after 30 min and 1000°C after 2 h. In the case of surface spread of flame characteristics, laminates can now be made with very low flame spread characteristics down to Class 1 in BS 476 but this often is at the expense of outdoor durability, and once burning the laminate will contribute to a fire in a similar way to a non-fire retarded resin, [30]. There is also the further and more potentially serious problem of smoke and fumes and whereas no legislation exists to control this at present, it seems likely that at some stage in the future legislation may be introduced and grp will be at a severe disadvantage in this respect.

3.3 Fibre concrete

This is the first group of fibre cement materials as defined above and, apart from continuous fibre reinforced ferrocement, cannot yet be considered as a structural material. Performance data required by the design engineer are therefore data which are relevant to its crack control characteristics. These can then be related to modifications in flexural, shear and impact strength. A test which has been found useful in establishing this performance is the direct tensile strength for concrete developed by Johnson and Sidwell [31]

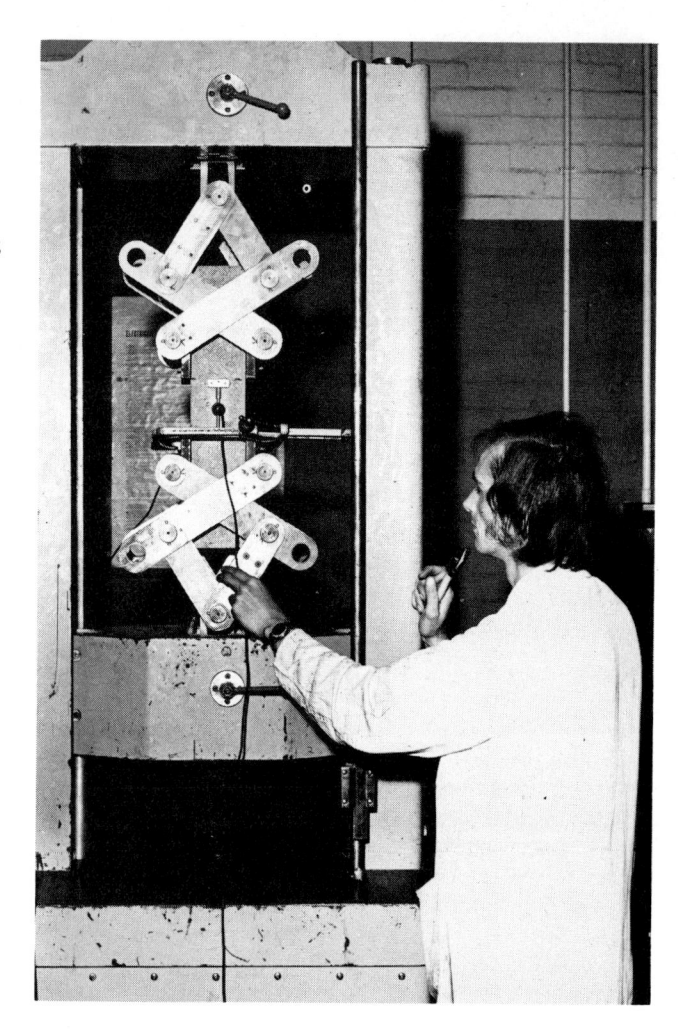

Fig. 19 Direct tensile test using scissor grips for measuring the tensile stress/strain characteristics of fibre reinforced concrete. The shape of the curve obtained from this test (see Fig. 8) is essential for understanding the performance of chopped wire fibre addition in concrete

and illustrated in Fig. 19. This test enables the stress/strain characteristics of the fibre concrete materials to be analysed so that impact performance can be assessed by the area under the curve, and crack distribution performance from the shape of the curve.

Although considerable research is being done into the performance of fibre concrete [32–38] very little useful performance data of this nature are at present available for design purposes and it would then be necessary to extend this test to enable assessments of durability and therefore suitable factors for safety. For example what degrees of temporary loading are required before the long term durability of the composite is affected.

3.4 Fibre cement sheet materials

These materials are in many ways similar to grp and the methods of assessment described for grp should also be applied to fibre cement sheets. A point of significant difference however is the incompatability which exists in fibre cement between the strain capacity of the matrix and that of the fibres. For example glass fibres have an elongation at break approaching 2% whereas the cement matrix will fail at 0.03%. This results in a stress/strain curve shown typically in Fig. 20 [39] in which quite early on, a point is reached where the matrix starts cracking but the load carrying

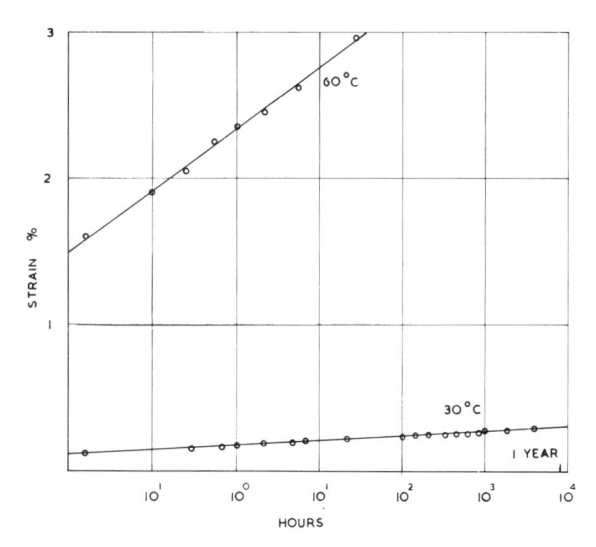

Fig. 18 Creep of chopped strand mat reinforced polyester loaded in compression at 20N/mm² in the dry condition. This figure illustrates the significant increase in creep with increasing temperature. This type of data must be available to enable prediction of long term deflections and the risk of buckling in stability. See also [28]

capability continues to increase. During this subsequent increase the matrix is progressively cracking as illustrated in Fig. 21 and the question to which design engineers require an answer is how the durability of the fibre cement composite is affected either by intermittent loading or long term loading in this matrix cracking area. Methods are available [40] for predicting likely crack spacing and widths.

It has been shown that fibre cement composites may lose strength with time under no load conditions [41], but what are really required before the composites can be used for semi-structural or structural applications with confidence are creep rupture data for laminates held under load in outdoor or wet environments. Despite the considerable volume of research on grc [39–48] no design data exist.

Until these types of data become available it is unwise to use fibre cement for load carrying applications. It is also not possible to assess the effect of additives such as polymer modifications to cement paste to extend its strain capability in terms of what effect these would have on the long term load carrying capability.

Tests to establish long term performance of fibre cement panels would also have to include the significance of changes in the cement paste with time such as carbonation from CO_2 in the atmosphere, which will have a greater significance to the thin members in question compared with the mass of cement used in conventional concrete.

Data must also be obtained on local overloading and crushing such as that which might occur round bolt holes or other methods of fixing, and comprehensive data on this should be made available such as the recommendations which already exist [26] for bolt spacing and edge distances in grp.

In the case of ferrocement, (continuous fine wires in cement), much practical experience has now been gained but the re-

Fig. 21 Excellent crack distribution occurring in the cement matrix of an Elkalite glass fibre reinforced cement sheet under a tensile load of $10 N/mm^2$. This was a 6 mm thick laminate, and the crack spacing was approximately 3 mm

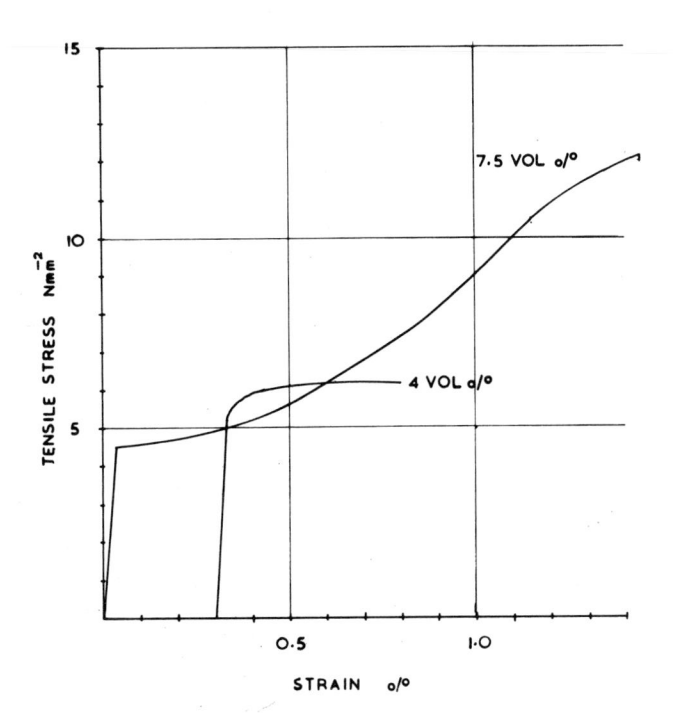

Fig. 20 Tensile stress/strain results for thin sheet glass fibre reinforced cement material; the initial steep part of the curve is where the cement matrix is intact. The matrix then starts to crack at a low strain (about 0.03%) and progressively cracks until the laminate ultimately fails, generally by pullout of the glass fibre reinforcement from the cement matrix. See also [39]

marks above for design values on glass fibre cement are also applicable to ferrocement.

4 FUTURE POTENTIAL

The future potential of particular composites in building and civil engineering is controlled by their performance and cost. Performance may include strength and stiffness to weight characteristics, durability, fire safety and ease of use. Assessment of future use must in general be based on quantitative figures which are the safe allowable figures for design purposes.

The newer composites, especially fibre reinforced resin and cement, are hindered in their use for semi-structural or structural applications by large volumes of research which give results of little value to design engineers. The results which are available to determine safe long term working stresses and deformations need to be put together in a simple design manual so that they can be readily used.

Cost comparisons are difficult to make especially as in recent years the rapid increases in labour costs have meant that it is often economic to use more expensive materials if an overall saving in labour can be achieved. Table 4 and Table 5 attempt to relate some of the performance and cost factors, for the basic materials and for a typical wall cladding application.

Table 4 Comparison of materials costs and performance

	Reinforced concrete	Plywood	GRP	Glass reinforced high alumina cement (grc)
Ultimate bending strength N/mm^2	58	90	140	30
Allowable bending stress N/mm^2	22	14	14	3
Materials cost $£/m^3$	25	200	500	200
Fabricated cost $£/m^3$ (very simple units, eg flat sheet)	65	200	900	400
Fabricated cost $£/m^3$ divided by allowable stress N/mm^2	3	14	65	130

Note: This table is intended to illustrate how materials should be compared for engineering use. The actual values are subject to wide variation depending upon the actual application.

Table 5 Property/cost relationship for typical wall cladding application for various materials

Material	Thickness mm	Weight kg/m^2	Fabricated cost (approximate) $£/m^2$
Precast reinforced concrete	75	180	10
Asbestos cement	6	12	5
grp	6	9	15
grc	9	18	12

Table 4 illustrates that the newer composites are not at present economically viable as structural materials in civil engineering unless some very special performance characteristic is required — for example translucency or ultra light weight. Two factors account for this: firstly, the severe reductions from ultimate to allowable stress and secondly the high cost of the basic material. The allowable stress is low partly due to the inadequate data available; the materials are expensive partly because of limited production.

The wall cladding example illustrates however that where the thickness is controlled by handling requirements, durability (for example to provide adequate cover in the reinforced concrete panel), and low stress wind loading the new composites are much more attractive. For a small increase in cost a dramatic reduction in weight is achieved leading to overall economies.

It therefore looks as though development of resin concrete, fibre resin and fibre cement composites will be along the lines of applications discussed earlier in the paper. The fastest development is likely to be in the field of glass fibre cement (grc) thin sheet materials. This material is stiffer and slightly cheaper than grp and has the great advantage of being incombustable. There would also appear to be performance advantages by the use of continuous fibre reinforced grc rather than chopped fibres currently used.

5 CONCLUSIONS

A number of exciting applications are being developed both for the traditional composite concrete, and for the newer composites fibre reinforced resin and fibre reinforced cement.

In the case of concrete, performance data used by design engineers are taken at two levels depending upon the application. For normal design, Codes of Practice and Specifications, based on information from established test procedure is used. For sophisticated design many decades of past research data are available and techniques have been built up to make use of this.

The new composites suffer from a lack of design data. We urgently require existing data to be gathered together and standardized test procedure to be developed to establish at least the following:

(1) Likely quality variation during practical manufacture, and simple methods of measuring this;

(2) Creep rupture stresses over a range of temperatures −10°C to −70°C, three exposure conditions, dry, wet and outdoor. These must then be combined with (1) to give safe allowable stresses;

(3) Creep data under the conditions for (2) above (could probably be obtained from the same test). This is important in determining long term deflection and buckling problems;

(4) Fatigue data under dry, wet and outdoor conditions. Fatigue due to thermal cycling should also be considered.

Until standard methods of test and simple design manuals are available for the above factors, fibre reinforced composites are limited to useful but non-structural applications in civil engineering.

ACKNOWLEDGEMENTS

The authors would like to thank the directors of Taylor Woodrow Construction for permission to publish this paper, and staff in the Materials Research Laboratory for help in its preparation.

REFERENCES

1 **BRS** Collection of construction statistics, 2nd Edition 1971. (Building Research Station, Watford)

2 *Reinforced Plastics* 18 No 1 (January 1974)

3 **Browne, R.D.** and **Blundell, R.** 'The behaviour of concrete in prestressed pressure vessels', *Nuclear Eng and Des* 20 No 2 (1972)

4 **Browne, R.D.** and **Burrow, R.E.D.** 'An example of the utilization of the complex multiphase material behaviour in engineering design', *Proc of the civil engineering materials*

5 Mead, P.F. 'London Bridge: demolition and construction 1967–73', *Proc Inst of Civil Eng* **54** Part 1 (February 1973) paper 7597

6 Stubbs, S.B. 'Concrete structures in deep waters', paper presented at *Offshore Scotland conferences*, Aberdeen (1973)

7 Browne, R.D. and Domone, P.L. 'Concrete for surface and underwater structures', paper presented at *Materials for underwater technology symposium*, Admiralty Materials Laboratory (May 1973)

8 Oliver, P.C. and Roach, E.C. 'The use of reinforced plastics in civil engineering with particular reference to Dubai international airport terminal', *8th International Reinforced Plastics Conf, Brighton,* the British Plastics Federation (1972)

9 Makowski, Z.S. 'Plastics Structures', reprinted from *Systems, Building and Design*, available from Scott Bader, Woolaston, Wellingborough, Northants

10 McCurrich, L.H. and Kay, W.M. 'Polyester and epoxy resin concrete', paper presented at *Resins and concrete symposiums,* the Plastics Institute, North Eastern section, Newcastle (1973)

11 McCurrich, L.H. and Adams, M.A.J. 'Fibres in cement and concrete', current practice sheet in *Concrete Magazine* (April 1973)

12 Nervi, P.L. 'Ferrocement, its characteristics and potentialities' *L'Ingegnere* (January 1951)

13 Romauldi, J.P. and Mandel, J.A. 'Tensile strength of concrete affected by uniformly distributed and closely spaced short lengths or wire reinforcement', *J American Concrete Institue* (June 1964)

14 'Fibre reinforced cement composites', *Technical Report* 51.067, The Concrete Society, London (1973)

15 Lonkard, D.R. 'Applications of wire reinforced concrete', paper presented at *Symposium on fibrous concrete, USA and UK,* The Concrete Society, Birmingham (1972)

16 Dave, N.J. paper presented at *ACI Symposium* (October 1973)

17 Hodgson, A.A. 'Fibrous Silicates', European Printing Co (1965)

18 'Health: dust in industry', *Technical Data Note* 14, Department of Employment and Productivity, HM Factory Inspectorate

19 Thomas, J.A.G. 'Glass Fibre Reinforced Cement', *Composites* (June 1971)

20 Walkus, I.R. and Kowalski, T.G. 'Ferrocement: a survey', *Concrete* (February 1971)

21 Neville, A.M. 'Properties of concrete', Pitman (1963)

22 *CP 110, Code of practice for the structural use of concrete* British Standards Institute (Novemebr 1972)

23 Krenchal, H. 'Fibre reinforcement', *Akademist Forlag*, Copenhagen (1964)

24 Howe, R.J. and Owen, M.J. 'Cumulative damage in chopped strand mat polyester resin laminates', *8th International Reinforced Plastics Conf Brighton,* British Plastics Federation (1972)

25 Steel, D.J. 'The creep and stress rupture of reinforced plastics', *Plastics Institute Transactions and Journal* **33** No 107 (October 1965)

26 Gibbs and Cox Inc., 'The Marine Design Manual', McGraw Hill (1960)

27 McCurrich, L.H. 'A glass reinforced plastic roof designed for mass production', *Proc 5th International Reinforced Plastics Conference* British Plastics Federation, London (1966)

28 Kabelka, J. 'The behaviour of a polyester laminate under long term loading', *Proc 5th International Reinforced Plastics Conference* British Plastics Federation, London (1966)

29 'Jointing with mastics and gaskets', *BRE Digest* 37, Building Research Station, Watford (1963)

30 'Reinforced plastics cladding panels', *BRE Digest* 161, Building Research Station, Watford (January 1974)

31 Johnston, C.D. and Sidwell, E.H. 'Testing concrete in tension and compression', *Magazine for Concrete Research* **20** No 65 (December 1968)

32 Hannant, D.J. 'Steel fibre reinforced concrete', *Proc of a conference on prospects for fibre reinforced construction materials* Building Research Establishment (1972)

33 Hannant, D.J. and Edginton, J. 'Steel fibre reinforced concrete, the effect on fibre orientation of compaction by vibration', *Materiaux et Constructions* **5** No 25 (1972)

34 Hughes, B.P. 'Fibre reinforcement for concrete', *Symposium on fibrous concrete USA and UK* the Concrete Society, Birmingham (September 1972)

35 Lankard, D.R. 'Steel fibres in mortar and concrete', *Composites* (March 1972)

36 Dixon, J. and Mayfield, B. 'Concrete reinforced with fibrous wire', *Concrete* (March 1971)

37 Shah, S.P. and Rangan, B.V. 'Fibre reinforced concrete properties', *ACI Journal* (February 1971)

38 Harris, B., Varlow, J., and Ellis, C.D. 'The fracture behaviour of fibre reinforced concrete', *Cement and Concrete Research* **2** No 4 (1972)

39 Allen, H.G. 'The purpose and methods of fibre reinforcement', *Proc of a conference on prospects for fibre reinforced construction materials* Building Research Establishment (1972)

40 Aveston, J. and Cooper, G.A. 'Single and multiple fracture', *Proc of the conference on properties of fibre composite materials, National Physical Laboratory, 1971* IPC Science and Technology Press (1972)

41 *Fibreglass Data Sheet* FA 430 (May 1971)

42 Allen, H.G. 'The fabrication and properties of glass reinforced cement', *Composites* (September 1969)

43 Biryukovich, D.L. 'Glass Fibre Reinforced Cement', *Kiev Budivelnik, Civil Engineering Research Association Translation* No 12 (1965)

44 Grimer, F.J. and Ali, M.A. 'The strengths of cements reinforced with glass fibres', Building Research Station, Watford (October 1969)

45 Maries, A. and Tseung, A.C.C. 'Factors influencing the strength of cement/glass fibre composites. Structure, solid mechanics and engineering design', *Proc of the Southampton 1969 civil engineering materials conference* published by Wiley Interscience

46 Waller, J.A. 'Carbon cement composites', *Civil Engineering and Public Works Review* (April 1972)

47 Majumdar, A.J. and Ryder, J.F. 'Glass fibre reinforcement of cement products', *Glass Technology* **9** No 3 (June 1968)

COMMENTS

1 On Table 5, by A.A.K. Whitehouse (Scott Bader Co Ltd)

I refer to Table 5 and mention that grp cladding only 3 mm thick is now being used instead of the 6 mm quoted, with significant reductions in cost.

Reply by Mr McCurrich

This is an interesting comment and would certainly lead to a reduction in cost though it would not of course be pro rata to the thickness.

2 On fire performance, by C. Urbanowicz (Cape Universal)

I would like to make a comment relating to the conference programme, regarding the lack of general interest shown in a

L. H. McCurrich* and M. A. J. Adams†

very important composite property, namely that of fire performance. This gap in the fundamental knowledge of the behaviour of building materials exposed to fire has been the indirect cause of the continued loss of property and indeed, loss of life. Some work has been done but this has mostly been confined to overall evaluation on the relevant British Standards.

It can be demonstrated, however, that the mechanisms governing crack arrest by use of fibres in a loaded composite such as fibre reinforced cement are also applicable to the stress induced thermally during a fire. These stresses, in a material without fibres, cause large-scale cracking and combined with spalling and loss in strength, result in material collapse. The use of selected fibres prevents such a failure.

Reply by Mr McCurrich

These comments about the fire performance of materials are certainly most important and relevant to this conference: brief mention of fire performance was given in our paper (page 88) and to the advantages which the fibre reinforced cement materials can offer in this respect. It is quite true that fibre reinforcement can ehlp prevent spalling of concrete in fire conditions, this has been confirmed in the use of fibres for refractory concretes and also some fire tests carried out on the use of fibre cement sheet materials used as permanent formwork for columns. The fire resistance of the column has been appreciably extended by the fibre cement sheet material providing increased protection to the steel.

QUESTIONS

1 On use of wire rather than rod and bar reinforcement, by Dr Cooper (Atlas Copco – Switzerland)

Does Mr McCurrich anticipate any problems or advantages for wire reinforced cement over conventional rod and bar reinforcement in demolition?

Reply by Mr. McCurrich

Dr Cooper has raised an interesting comment with regard to demolition of concrete structures especially as this problem is now becoming increasingly significant as some of our older concrete buildings especially in urban areas are in need of modification or demolition. Some of the early work on fibre reinforcement was carried out to provide explosion resistant structures and fibres are effective in preventing concrete spalling. In view of this it does seem likely that fibre reinforced concrete will be more difficult to demolish than conventional unreinforced concrete. For main structural applications however it is unlikely that wire fibre reinforcement will replace conventional rod reinforcement but will be used in special cases where improvement in certain properties is required.

2 On fatigue of reinforced concrete in offshore structures, by P. Bristoll (Kominkliske Shell Laboratorium, Amsterdam)

Would Mr McCurrich please expound on his comments on the fatigue of concrete with respect to offshore structures.

Reply by Mr McCurrich

Quantitative data on the fatigue of concrete with regard to off-shore structures is not at present available for publication: it is however an important aspect of the material technology for these applications and Taylor Woodrow have been involved in detailed studies on the effect of fatigue on reinforcing rods under the three conditions; exposure to salt water, exposure to salt water in a concrete environment, and exposure to air. These tests have to be carried out to ensure that the performance of the final structures will be satisfactory with regard to long term durability. With regard to fatigue of concrete itself as opposed to the reinforcing rods within concrete, the material is generally designed to be in compression and the working stresses are appreciably below those which cause fatigue damage.

Fibre reinforced cements — scientific foundations for specifications

J. AVESTON, R.A. MERCER and J.M. SILLWOOD

This paper reviews the theory of multiple cracking and crack suppression in brittle matrix composites and shows how it can be applied to cement reinforced with carbon fibre and steel wire. The extension of the theory to include the strength, crack spacing and modulus of composites for the case where the fibres are in a random planar array is discussed and an analytical expression for the flexural strength of a material which obeys the theory is derived.

MAIN SYMBOLS (Others are defined in text)

E	Young's modulus
N	Number of fibres crossing unit cross section
V	Volume fraction
a	Scaling factor equal to $E_m V_m / E_f V_f$
γ_m	Surface work of fracture of the matrix
ϵ	Strain
η	Efficiency factor
σ	Stress
τ	Average shear strength of fibre-matrix interface
d	Fibre diameter
l	Fibre length
l_c	Shortest fibre which can be fractured in the composite
r	Fibre radius
x'	Length needed for continuous fibres to transfer load $\sigma_{mu} V_m$ per unit area of composite

Subscripts:
- c Composite, or critical fibre length
- f Fibre
- m Matrix
- u Ultimate stress or strain

1 INTRODUCTION

Although the first patent for wire reinforced cement [1] dates from before the general use of reinforced concrete, and indeed the use of fibres to toughen bricks and pottery can be traced to the very beginnings of civilisation, it is only in the last decade that the principles governing the fibre reinforcement of brittle matrices have begun to be understood. As a matrix, portland cement has some extremely attractive properties, eg it is about six times stiffer and one hundred times cheaper than resin but it has one over-riding disadvantage — a very low failure strain.

In a composite the stiffness and hence the stress at which a brittle matrix such as cement will crack can be increased by using high modulus fibres but in general the cracking strain of the matrix remains unaltered. Moreover the increase in stiffness with the small volume fraction of random fibres normally used in cement is not of much practical value. The idea of increasing the matrix cracking strain by using closely spaced fibres was first suggested by Romualdi and Batson [2] who predicted that the cracking strain should be inversely proportional to the square root of the fibre spacing as long as the fibres remain bonded to the cement. Unfortunately other workers have been unable to confirm the large cracking strains predicted by this theory.

More recently Aveston, Cooper and Kelly [3] considered the case of a sliding frictional bond between the fibre and matrix and showed that an enhanced cracking strain should still be observed but the effect would be much less than that predicted by Romualdi. Subsequently Aveston and Kelly [4] using a shear-lag approach to the bonded case showed that the fibre-matrix interface is unlikely to be able to withstand the shear stress thrown onto it when the matrix cracks, with the result that at least some debonding will occur and the large predicted matrix failure strains are not likely to be achieved in practice. Kelly [5] has since shown that the bonded theory and that of Romualdi lead to essentially the same result; both predict a cracking strain inversely proportional to the square root of the fibre spacing and both predict improbably large cracking strains. By the same token they are both inapplicable to fibre reinforced cement (frc) and we see why the Romualdi theory has to some extent fallen into disrepute.

As designers in frc usually aim to keep within the uncracked region and use the additional post cracking strength to take care of accidental overload including impact there is clearly a need to establish whether the cracking strength of a brittle matrix can be increased to a practically significant extent, and if so, how this point and the remainder of the stress strain curve can be predicted in terms of measurable properties of the fibre, matrix and interface. In this paper we review the essentials of the Aveston, Cooper, Kelly (ACK) theory, then show how the theory can be used to predict the matrix failure strain and other important properties of carbon fibre and steel wire reinforced cement, and finally indicate how it can be extended to include composites reinforced with short fibres and to specimens stressed in flexure.

National Physical Laboratory, Teddington, Middlesex

2 CONTINUOUS ALIGNED FIBRES

When a brittle matrix composite containing continuous or long (relative to l_c) fibres is loaded in tension, then according to the ACK theory it will crack when the strain has reached the normal failure strain of the matrix or a value

$$\epsilon_{mu} = \left[\frac{12\tau \gamma_m E_f V_f^2}{E_c E_m^2 r V_m} \right]^{\frac{1}{3}} \quad (1)$$

whichever is the greater. The stress strain curve to this point will be linear with a slope for practical purposes given by the mixtures rule

$$E_c = E_m V_m + E_f V_f = E_m + (E_f - E_m) V_f \quad (2)$$

If the strength or resistance to pull out of the fibres bridging the crack is then sufficient to withstand the total stress on the composite, ie

$$\sigma_{cu} = \left(1 - \frac{l_c}{2l}\right) \sigma_{fu} V_f \geqslant \sigma_{mu} V_m + \epsilon_{mu} E_f V_f \quad (3)$$

the additional load thrown onto the (continuous) fibres will be transferred back into the matrix over a distance

$$x' = \frac{E_m V_m}{V_f} \cdot \frac{\epsilon_{mu} r}{2\tau} \quad (4)$$

and the matrix will become transversed by a series of cracks spaced between x' and $2x'$. Gale [6] has pointed out that the actual spacing is the same as the minimum average spacing between cars of length x' parked at random in a given space which he has shown statistically to be equal to $1.364 \pm 0.002\ x'$.

The additional stress in the fibres will therefore vary linearly between $\sigma_{mu}(V_m/V_f)$ at the crack to $\sigma_{mu}(V_m/V_f)(1 - 1.364/2)$ at a distance $(1.364/2)x'$ from the crack and the total strain at the limit of multiple cracking will be

$$\epsilon_{mc} = \epsilon_{mu}(1 + 0.659a) \quad (5)$$

Further loading will merely result in the fibres stretching and slipping relative to the blocks of matrix which can take no further share of the load so that the tangent modulus will become $E_f V_f$ and the composite will eventually fail at a stress

$$\sigma_{cu} = \left(1 - \frac{l_c}{2l}\right) \sigma_{fu} V_f \quad (6)$$

and strain

$$\epsilon_{cu} = \left(1 - \frac{l_c}{2l}\right) \epsilon_{fu} - 0.341a\ \epsilon_{mu} \quad (7)$$

In principle therefore a single value of τ obtained from the measured crack spacing, C, using equation (4) setting $C = 1.364\ x'$ and a value of γ_m, eg by the method described by Davidge and Tappin [7] should be sufficient to calculate ϵ_{mu} from Equation (1) and hence, using Equations (5)–(7), the whole stress strain curve to failure.

The theory was tested using cement paste reinforced with either carbon fibres (frc) or continuous steel wire (wrc). The

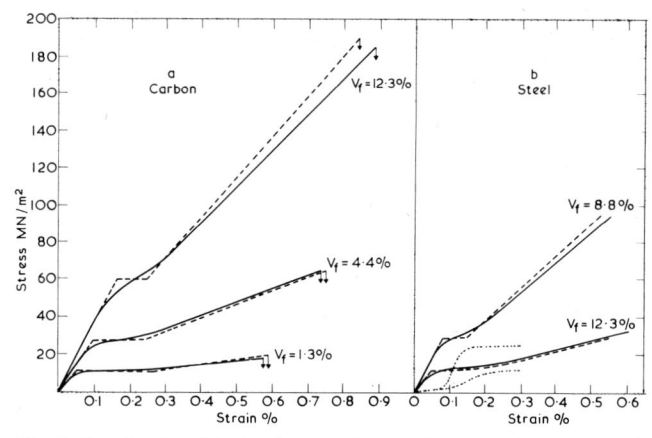

Fig. 1 Tensile stress/strain diagrams for continuous carbon and steel wire reinforced cement. Full lines are the experimental curves, broken lines are the curves predicted by the ACK theory and the dotted lines are the integrated acoustic emission (arbitrary units) of the wrc specimens

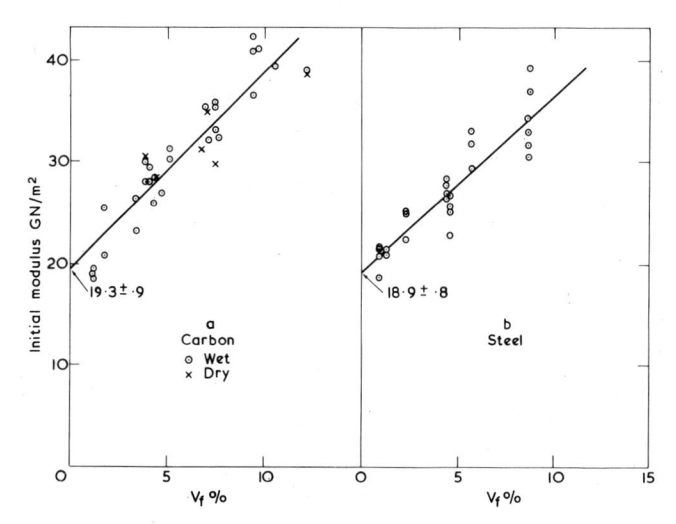

Fig. 2 Initial Young's modulus of continuous carbon and steel wire reinforced cement. Lines are least squares fits

experimental stress strain curves for typical specimens are shown in Fig. 1 (continuous lines) together with those predicted by the ACK theory (broken lines). The extent to which the salient points of all the tests fit the theory is indicated in more detail in Figs. 2, 4, 5, 6 and 7.

The initial slopes of the stress strain curves (Fig. 2) show considerable scatter which arises from the difficulty of making accurate strain measurements on cement in direct tension but the value of $E_f = 192\,\text{GN/m}^2$ and $E_m = 19\,\text{GN/m}^2$ from the least squares slope and intercept of the wrc line using Equation (2) agrees reasonably with the independently measured values of $196\,\text{GN/m}^2$ and $20\,\text{GN/m}^2$. The corresponding value of $E_f = 214\,\text{GN/m}^2$ for the carbon fibre is outside the manufacturer's range of 180–$200\,\text{GN/m}^2$ and probably arises from the loss of water from the thin layers of cement paste that were painted onto the fibre during lay-up of the specimens, with the result that the water-cement ratio of the matrix will decrease and hence its modulus will increase with increasing V_f. The discrepancy in E_f of $\sim 20\,\text{GN/m}^2$ would only require a reduction in the water/cement (w/c) ratio from 0.5 to 0.45 at the highest V_f and as there is no real doubt about the practical applicability of the mixture rule to the initial modulus we need not pursue the point. We are concerned with the contentious question of where these curves depart from linearity.

The failure strain of the unreinforced matrix in flexure computed assuming elastic behaviour was about 0.06% but for direct tension (of dog's bone specimens loaded through flexible wire rods cast into the ends in an effort to provide axial loading) was only about 0.02%. The experimental curves in Fig. 1 leave no doubt that the strain, as well as the stress, at the limit of proportionality increases with V_f and that for the 12.3% cfrp at least the apparent strain at first crack is not only greater than that of the unreinforced matrix tested in tension ($\sim 0.02\%$) but is actually greater than what can be taken as the upper limit of 0.06% from the flexural test of the matrix. However a precise test of Equation (1) is more difficult. The matrix in practice cracks over a range of stress and so it is not possible to measure x' corresponding to the initial value of ϵ_{mu} and hence the value of τ, because the crack spacing is infinite at this point. Moreover as implied above the value of ϵ_{mu} at this point is likely to be lower than the true value because of bending stresses, which will disappear as multiple cracking proceeds. The final crack spacing, which is what is measured, will lie between the extremes of x' corresponding to the initial value of ϵ_{mu} and $2x'$ where in this case x' corresponds to a final value of ϵ_{mu} at the limit of cracking. We have therefore taken it as equal to $1.364 \bar{x}'$, where $\bar{x}'$ is the mean transfer length corresponding to the mean matrix failure strain $\bar{\epsilon}_{mu}$. For the wrc the maximum acoustic emission coincided with the minimum slope of the stress strain curves and, since there can be no wires breaking at this stage, we have taken the stress at this point to be the mean matrix cracking stress $\bar{\epsilon}_{mu}$ and the elastic strain in the matrix at this point $\bar{\epsilon}_{mu} = \bar{\sigma}_{mu}/E_c$ as the mean matrix failure strain.

Figures 3a and 3b show the (decorated) cracks in typical wrc and cfrc specimens and Fig. 4 shows the mean crack spacings plotted against $\epsilon_{mu} V_m/V_f$ to obtain τ from Equation (4). The least squares slopes give $\tau = 6.8 \text{MN/m}^2$ for wrc and $\tau/r = 5.2 \times 10^{11} \text{Nm}$ (the effective value of r is not known) for cfrc. The observed values of $\bar{\epsilon}_{mu}$ shown by the points in Figs. 5a and b though again showing considerable scatter are generally in good agreement with the theoretical curves computed from Equation (1) using the above values of τ and τ/r, values of E_f, E_m and E_c reported in this paper and a value of $\gamma_m = 4 \text{J/m}^2$ determined by Brown [8].

The wrc specimen usually failed by pull-out of the wires from the block of matrix remaining in the grips but those which broke did so at the theoretical uts of $\sigma_{fu} V_f$. The cfrc eventually failed in a brittle manner at about $0.7 \sigma_{fu} V_f$ (Fig. 6).†
Finally the slopes of the stress strain curves at the completion of matrix cracking (Fig. 7) are within 5% of $E_f V_f$ for both systems and the displacement of this part of the curve due to the inflexion in the stress strain diagram as a result of multiple cracking (Fig. 1) is also close to the predicted value.

3   RANDOM LONG FIBRES

The simplest expression for the modulus of a composite containing a random two dimensional (2-D) array of fibres is the mixtures rule [Equation (2)] with the term in V_f multiplied by an efficiency factor $\eta = \frac{1}{3}$ as proposed by Krenchel [9]. More rigorous expressions have been derived — Christiansen

---

† Taking σ_{fu} as the manufacturers value for an impregnated tow, ie 2.3 GN/m^2

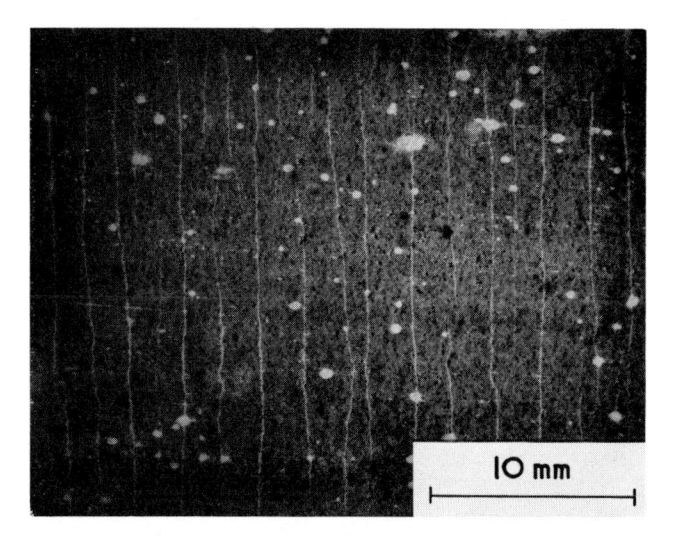

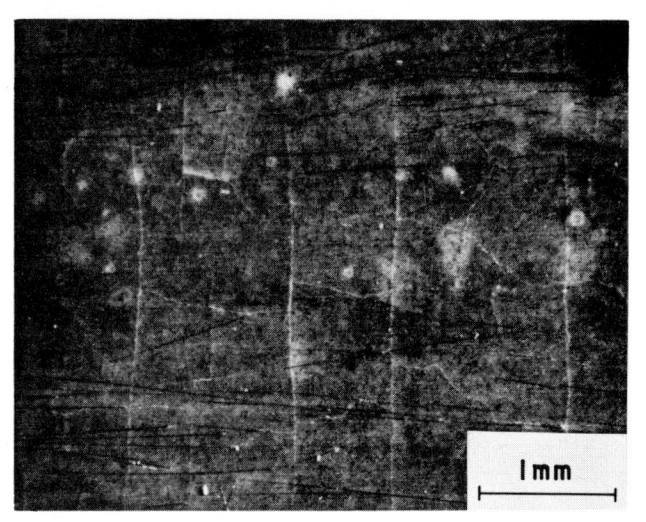

Fig. 3 Multiple cracking in steel wire reinforced cement, (top) and carbon fibre reinforced cement, (bottom)

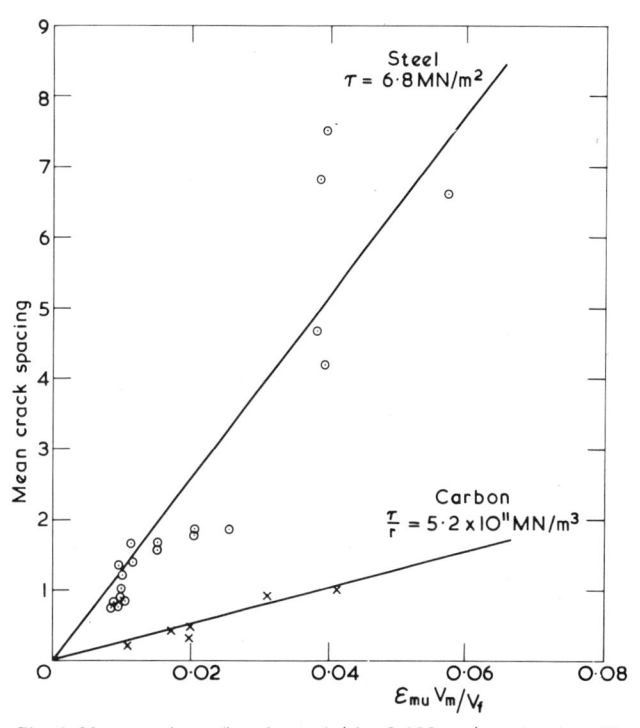

Fig. 4 Mean crack spacings in steel ($d = 0.132$ mm) and carbon fibre reinforced cement. Lines are least squares fits

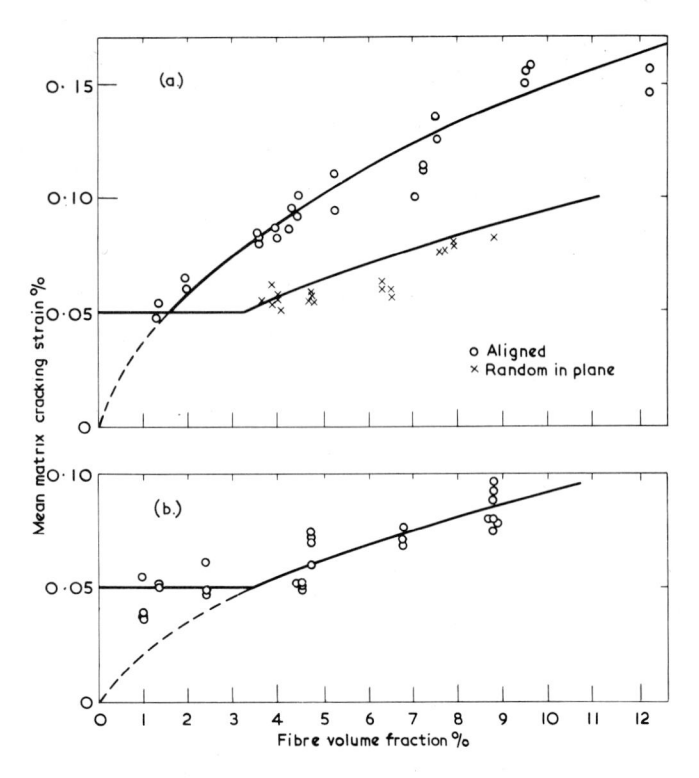

Fig. 5 Mean matrix cracking strain of steel wire ($d = 0.132$ mm and carbon fibre reinforced cement. Lines are the values computed from Equation (1) using the values of τ from Fig. 4 and the measured value of $\gamma_m = 4 J/m^2$

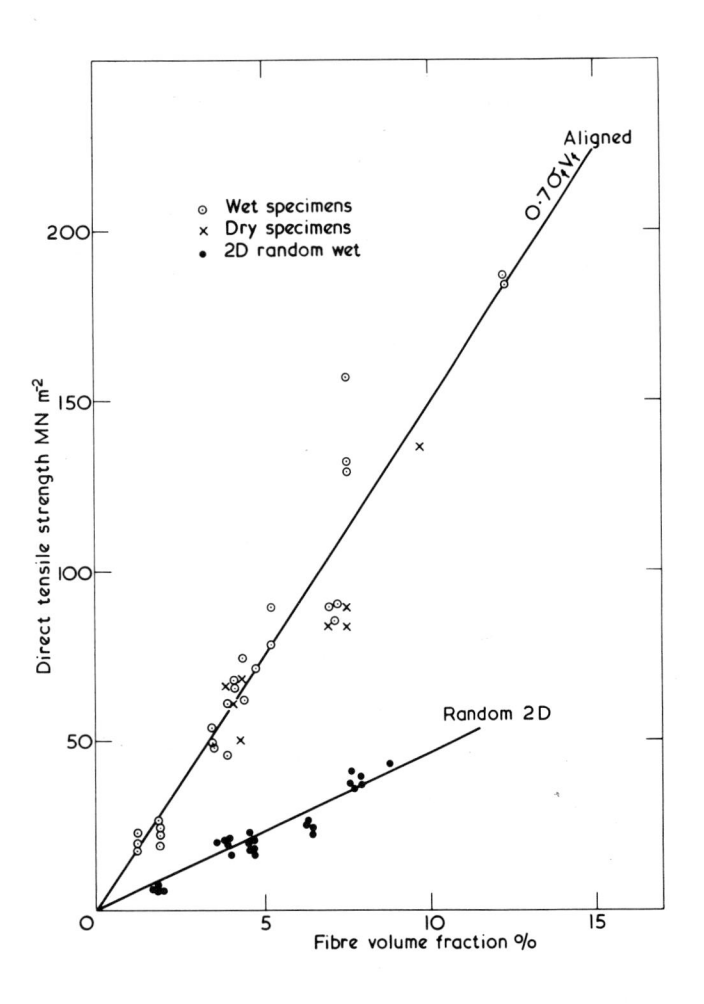

Fig. 6 Ultimate tensile strength of carbon fibre reinforced cement

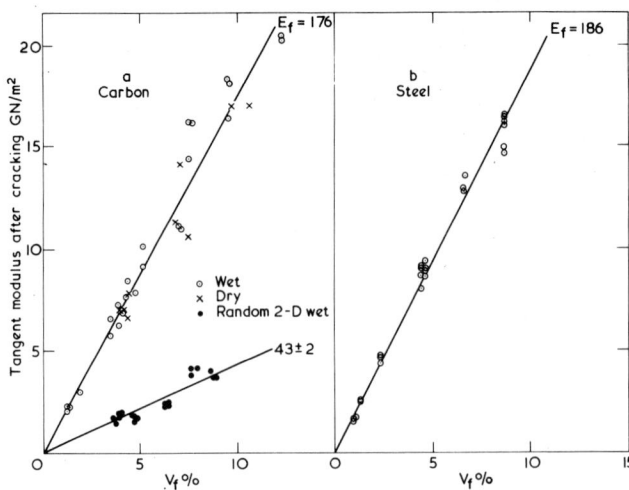

Fig. 7 Tangent moduli of carbon fibre (a) and steel wire (b) reinforced cement specimens after completion of multiple cracking

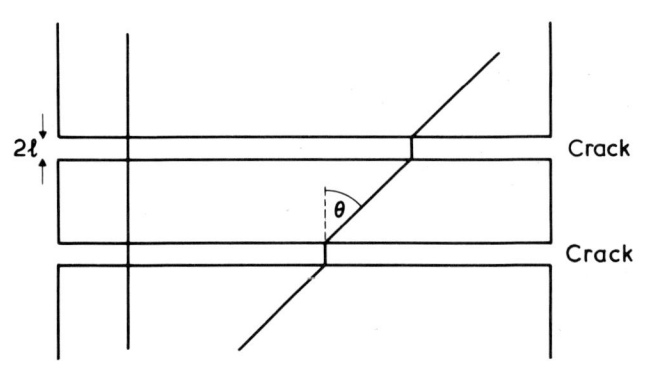

Fig. 8 Model assumed for fibres crossing a crack

and Waals [10] but for fibre reinforced cement where V_f is small the difference between them and the simple mixtures rule modified by η is likely to be less than the experimental error and is of no practical importance.

When the matrix has cracked the modulus will be determined by the fibres extending and slipping through the blocks of matrix as in the aligned case but it will be less because (a) there are fewer fibres crossing a crack and (b) fibres at an angle θ to the normal are strained less than the fibres normal to the crack, ie parallel to the applied stress, for a given crack extension. Thus if the crack opens a small amount to produce a uniform increase in strain $\Delta \epsilon_f$ and load $\Delta \epsilon_f E_f \pi r^2$ throughout the fibres normal to the crack (see Fig. 8) the corresponding values for fibres at angle θ will be $\Delta \epsilon_f \cdot \cos \theta$ and $\Delta \epsilon_f E_f \pi r^2 \cos \theta$ because the extension is equal to the crack opening and is independent of θ. The number of fibres at angle θ is $N \cos \theta \, d\theta$ where $N = V_f / \pi r^2$ and hence the total increase in load per unit area of composite over all the fibres for an increase in strain $\Delta \epsilon_f$ in the fibres normal to the crack, which is equal to the strain in the composite is

$$\Delta \sigma_c = \frac{2}{\pi} \int_0^{\pi/2} N \Delta \epsilon_f E_f \pi r^2 \cos^2 \theta \, d\theta =$$

$$= \frac{\Delta \epsilon_f E_f V_f}{2} \quad (8)$$

hence

$$E_c = \frac{\Delta \sigma_c}{\Delta \epsilon_c} = \frac{\Delta \sigma_c}{\Delta \epsilon_f} = \frac{E_f V_f}{2} \quad (9)$$

Similarly when the load in the fibres normal to the crack has reached the breaking load $\pi r^2 \sigma_{fu}$ the total load per unit area of composite is

$$\sigma_{cu} = \frac{2}{\pi} \int_0^{\pi/2} N\pi r^2 \sigma_{fu} \cos^2\theta \, d\theta = \frac{\sigma_{fu} V_f}{2} \quad (10)$$

and this will be the strength of the composite if it breaks when the most highly stressed fibres break.

The crack spacing for a matrix with a single valued breaking strain according to this model has been shown [4] to be equal to $\pi/2$ times that of a corresponding aligned composite. This arises because the sum of the resolved tensile force due to fibre-matrix shear and the 'pulley' force exerted by the fibre as it is bent through an angle θ at the crack surface at the cracking stress of the composite is independent of θ. However when the load is increased the 'pulley' force will also increase but this will be offset by a decreasing frictional force as the fibres contract laterally with the result that the ultimate crack spacing is difficult to quantify.

The strength will also be influenced by two opposing factors: the tensile stress $\sigma_{t\theta}$ which fibres at angle θ can support will be reduced by a factor $\phi = (\sigma_{t\theta} + \sigma_b)/\sigma_{t\theta}$ where σ_b is the bending stress where the fibre enters the crack face, but this bending stress will in turn be reduced by an indeterminate amount by crumbling of the matrix at this point. Matrix crumbling will also tend to reduce the modulus of the cracked composite.

Cox [11] has shown that the stress concentration in a fibre being pulled from the matrix at an angle θ to the fibre axis in the matrix is given by

$$\phi = \frac{\sigma_b + \sigma_{t\theta}}{\sigma_{t\theta}} = 1 + 2\theta \left[\frac{E_f}{\sigma_{t\theta}}\right]^{1/2} \coth\left[\frac{4l}{d_f}\left(\frac{\sigma_{t\theta}}{E_f}\right)^{1/2}\right] \quad (11)$$

where l is the length of fibre protruding from the crack face. In the limit if the fibres are infinitely thin the coth term reduces to unity and we have for the total stress, putting $\sigma_{t\theta} = \sigma_t \cos\theta$,

$$\sigma_{max} = \sigma_b + \sigma_t \cos\theta = 2\theta E_f^{1/2}(\sigma_t \cos\theta)^{1/2} + \sigma_t \cos\theta \quad (12)$$

A plot of Equation (12) for various values of E_f taking σ_t as unity (Fig. 9) shows that fibres at 60° to the load axis are the most highly stressed and the stress concentrations can be as high as 15 for $E_f/\sigma_t = 100$ (eg carbon fibres with $\epsilon_{fu} = 1\%$) and 8 for $E_f/\sigma_t = 25$ (eg glass fibres with $\epsilon_{fu} = 4\%$). Hence if the composite fails when the most highly stressed fibres fail the upper bound to the uts, corresponding to zero bending stress is $\frac{1}{2}\sigma_{fu} V_f$. The lower bound, where the tensile stress in the fibres normal to the crack at the initiation of failure is reduced from σ_{fu} to $\sigma_{fu}/(\sigma_b + \sigma_t \cos\theta)$ is only about $\frac{1}{15}$ of this value for a fibre with $\epsilon_{fu} = 1\%$. In practice we find that the strength of pseudo random cfrc (Fig. 6) is closer to the upper bound, indicating that there is substantial crumbling of the matrix, and this is also consistent with the low values of the modulus.

Derivation of the matrix cracking strain for the random 2-D case is complicated by the non-linear increase in matrix strain with distance from the crack, and the other compli-

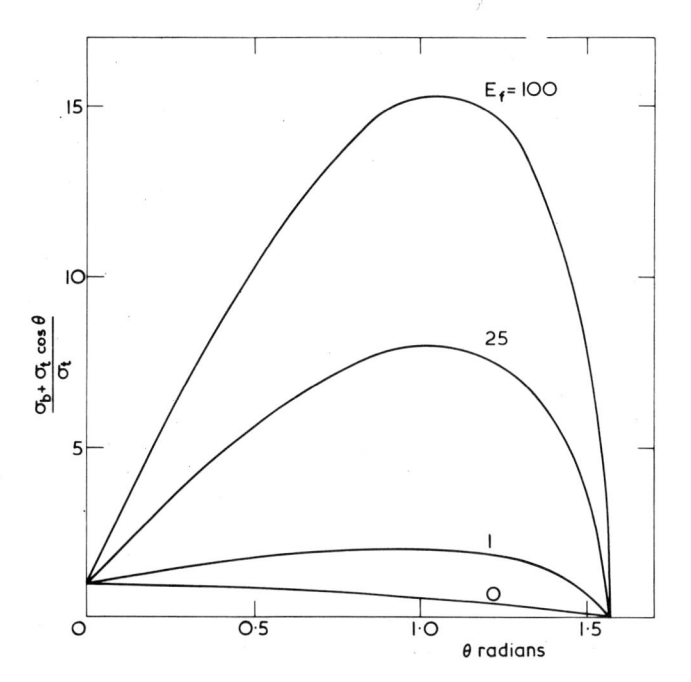

Fig. 9 Total stress (tensile plus bending) in a fibre crossing a crack at angle θ to the normal relative to the stress σ_t in an aligned fibre for various E_f computed from Equation (12)

cations mentioned above raise doubts whether a rigorous treatment is possible. However if we apply the stiffness efficiency factor of $\frac{1}{2}$ to V_f in Equation (1) we see that the cracking strain would be $(\frac{1}{2})^{2/3} \sim 0.63$ of the aligned material, which is reasonably consistent with the few data available so far (Fig. 5a.).

4 SHORT FIBRES

If as before it is assumed that after the matrix has cracked there is a linear transfer of stress from the fibres bridging the crack back into the matrix, then the ACK debonded theory can be modified to predict the mechanical properties of short fibre reinforced cement. As an example the crack spacing is derived in Appendix 1 and the results are given in Fig. 10. They show that the spacing is close to that of the continuously reinforced material until the fibre length approaches $4x'$, ie until the mean transfer length $l/4$ equals the crack spacing.

When $l \leq l_c$, which is usually the case for wrc, and assuming the resistance to pull out of the fibres bridging the crack is sufficient to stand the load on the composite the strength should simply be equal to the frictional force needed to pull out the mean embedded length $l/4$ of fibre multiplied by the number of fibres $N = V_f/\pi r^2$ crossing unit area of crack, ie

$$\sigma_c = 2\pi r \tau \left(\frac{l}{4}\right)\left(\frac{V_f}{\pi r^2}\right) = \frac{V_f \tau l}{d} \quad (13)$$

For the random 2-D case it has been shown that $N = (2/\pi)(V_f/\pi r^2)$ and for the three dimensionally random $N = (\frac{1}{2})(V_f/\pi r^2)$ so that if the pull out force does not vary with angle the strengths should be $2/\pi$ and $\frac{1}{2}$ respectively of the aligned material.

In practice we have found that although the pull-out force F for a single fibre does not vary much with the angle

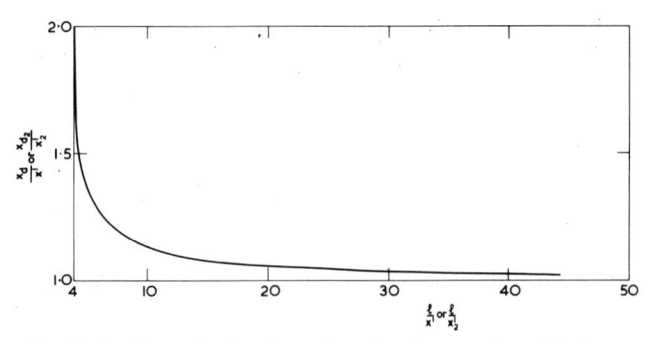

Fig. 10 Crack spacing for short aligned or short random 2-D fibres relative to the corresponding continuous case predicted by Equations (A2) and (A8)

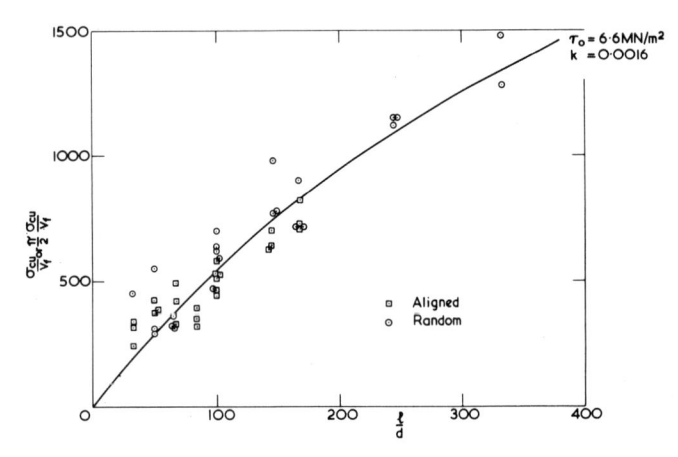

Fig. 11 Tensile strength of discontinuous wire reinforced cement with $V_f = 1.6–1.9\%$. Experimental results for the random 2-D specimens have been multiplied by an efficiency factor $\pi/2$. Line is the least squares fit to Equation (16)

through which the fibre is bent where it enters the material, it increases less than linearly with embedded depth and in the same way the strength of short fibre aligned and random wrc, which under pull out conditions is simply the integrated pull out force also increases less than linearly with fibre length. A simple explanation for this is that the shear strength of the interface decreases as a result of the Poisson contraction of the fibre as the load on the composite is increased. If we assume that the decrease in bond strength is proportional to the stress on the fibre we have

$$\tau = \tau_0 - \frac{KF}{\pi r^2} \qquad (14)$$

and

$$F = 2\pi r \tau l = \frac{2\pi r \tau_0 l}{1 + \frac{2Kl}{r}} \qquad (15)$$

The composite strength will then be equal to the product of the the mean pull out force per fibre between $l = 0$ and $l = l/2$ and the number of fibres $N = V_f/\pi r^2$ per unit area, is

$$\sigma_u = \frac{V_f}{\pi r^2} \frac{2}{l} \int_0^{l/2} \frac{2\pi r \tau_0 l}{1 + \frac{2Kl}{r}} = \frac{V_f \tau_0}{K}\left[1 - \frac{r}{Kl}\log\left(1 + \frac{Kl}{r}\right)\right] \qquad (16)$$

for the aligned case and $2/\pi$ or half of this for the random 2-D and 3-D case.

The observed values of σ_u/V_f and $(\pi/2)\sigma_u/V_f$ were fitted to Equation (16) using a general non linear least squares programme to obtain a value of τ_0, the limiting value of the bond strength at low embedded fibre lengths which should be close to the value of τ obtained independently from the crack spacing. For the aligned wrc the crack spacing gave a mean value of $\tau/\sigma_{mu} = 1.5 \pm 0.2$ using Equation (4). When combined with the value of $\sigma_{mu} = 5.5 \pm 0.8 \text{MN/m}^2$ for those specimens with low aspect ratio wires that failed by single fracture this gives $\tau = 8.3 \pm 2.3 \text{MN/m}^2$ which is reasonably consistent with the value of $\tau_0 = 6.6 \text{MN/m}^2$ obtained from the least squares fit of the strengths to Equation (16). The value of the efficiency factor of $2/\pi$ for a random 2-D composite that fails by pull-out appears to be justified by the fact that the random and aligned points in Fig. 11 scatter about the same line.

With the 3-D random mortar specimens it was found that the higher the wire aspect ratio the lower the volume faction that could be incorporated before the wires 'balled up' in the mixer with the result that the product $V_{fmax}l/d$ was found to have a roughly constant value of three. This meant that many of the specimens (those with $V_f l/d$ less than about one) were no stronger in direct tension than the unreinforced matrix and even the strongest ones, which required such prolonged vibration to compact that they would be of little practical use were only about 50% stronger than the mortar. The flexural strength on the other hand could be more than doubled. All specimens appeared to fail by single fracture although it was difficult to be certain of this because the grips were so soft that sudden failure always occurred.

The experimental tensile strengths (Fig. 13) were fitted to Equation (16) by the method of least squares to yield a value of $\tau_0 = 5.4 \text{MN/m}^2$ which for a water cement (w/c) ratio of 0.5 is consistent with the value of 6.8MN/m^2 for the continuous wire reinforced cement paste (w/c = 0.4) and 8.3MN/m^2 for the short fibre reinforced paste where the w/c ratio was only 0.35.

5 FLEXURAL STRENGTH

When a rectangular beam of elastic material is stressed in flexure the longitudinal stress and strain at a given cross section increase linearly from the surface in compression, through the neutral axis at the centre where the tensile stress is zero, to a maximum at the opposite surface. The stress in the outer surface at failure, computed from the bending moment, assuming elastic behaviour, is known as the modulus of rupture (MOR) and is ideally equal to the tensile strength of the material. In practice even for an elastic material the MOR is usually greater than the uts because a smaller volume of the specimen is stressed and extraneous bending stresses that can arise from mis-alignment of the grips in a tensile test are eliminated. However if, as with frc, the material is elastic to a greater strain in compression than in tension the strain will still vary linearly across a section but the stress in a large part of the tensile zone will be roughly constant at the yield stress and the neutral axis will move towards the face in compression, resulting in an increased moment and a MOR that may be up to some 2–3 times the uts.

The exact relation between the MOR and uts for a brittle matrix composite is derived in Appendix 2. The ratio MOR/uts is found to be a function of a and $\epsilon_{fu}/\epsilon_{mu}$ and to vary (Fig. 12) from a lower bound of unity where $\epsilon_{fu} = \epsilon_{mu}$ and the material is essentially elastic, to a maximum of three at high values of a and $\epsilon_{fu}/\epsilon_{mu}$ where the pseudo ductility from multiple cracking is a maximum. The values in Fig. 12 are for an assumed crack spacing of $2x'$ but the results for the lower bound to the crack spacing x' differ by only about 2–3% and so it would appear that the calculation is not too sensitive to the exact form of the tensile stress strain curve.

For the 3–D wrc which was the only material tested both in flexure and tension the relative value of the MOR and uts of the stronger specimens given by the relative slopes in Figs. 13 and 14 was about 2.7. Although the fibre volume fraction varied from zero up to a maximum of 9.5% for the lowest l/d the majority were in the range 1–2%, corresponding to $a \sim 70$ if we take an efficiency factor of $\frac{1}{6}$. As failure was by fibre pull-out at a stress very little greater than σ_{mu} then the relevant value of $\epsilon_{fu}/\epsilon_{mu}$ (where ϵ_{fu} is the effective fibre strain at pull out in this case) must be close to the lower limit for a given a. The computed ratio we require will therefore be close to the upper bound given by the broken line in Fig. 12 which for $a = 70$ gives MOR/uts ~ 2.6 in good agreement with experiment.

If Fig. 12 can be confirmed over a wider range of a and $\epsilon_{fu}/\epsilon_{mu}$ it should enable the results of flexural tests, which are easy to do, to be used for design purposes or alternatively the maximum bending moment to be computed from the uts provided the compressive strength is known to be sufficient to ensure failure is initiated in the tensile zone. Conversely if the observed MOR/uts ratio is less than predicted it would indicate that failure is initiated in the surface in compression — something that is not always obvious in a flexural test.

6 EXPERIMENTAL

The cfrc was made by hand lay up of Grafil A (Courtaulds Ltd) 10 000 filament tow supplied in the form of a thin veil 100mm wide and cement paste applied with a paint brush.

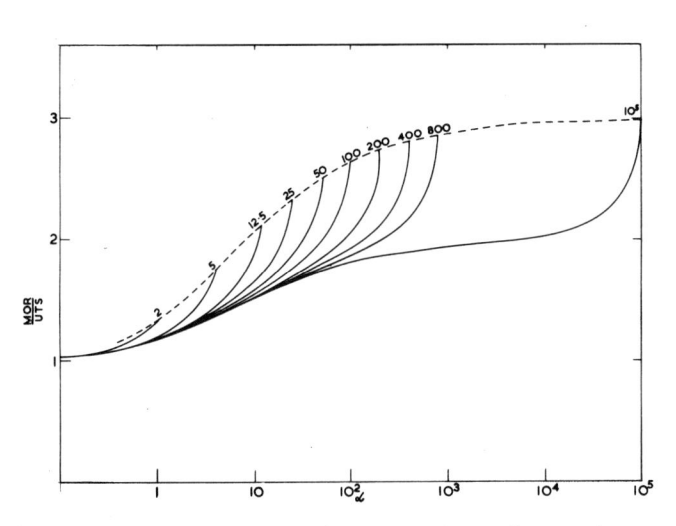

Fig. 12 Ratio of the modulus of rupture to the tensile strength vs $a = E_m V_m / E_f V_f$ computed from Equation (A13) for various values of $\epsilon_{fu}/\epsilon_{mu}$

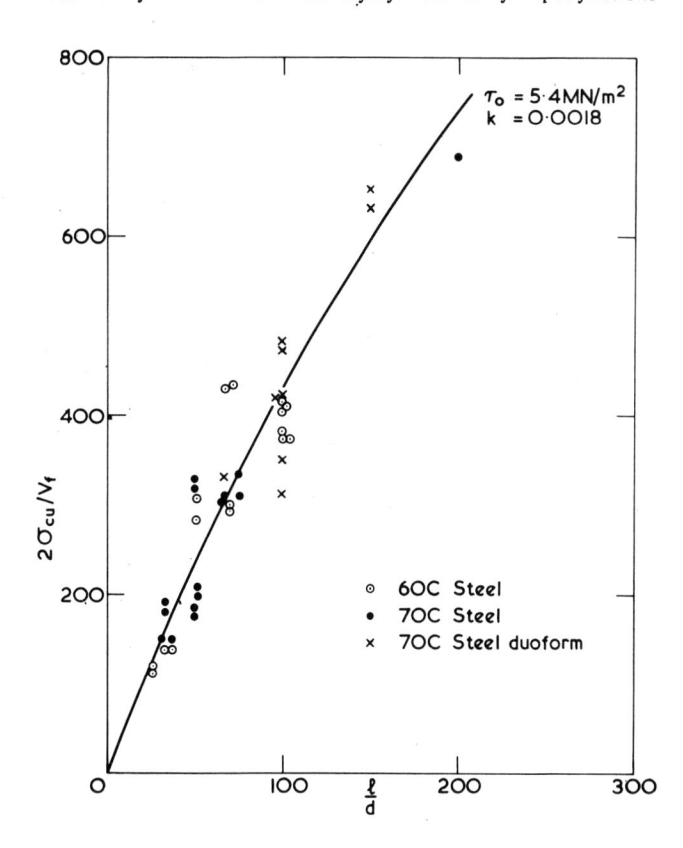

Fig. 13 Tensile strength of 3-D random fibre reinforced mortar

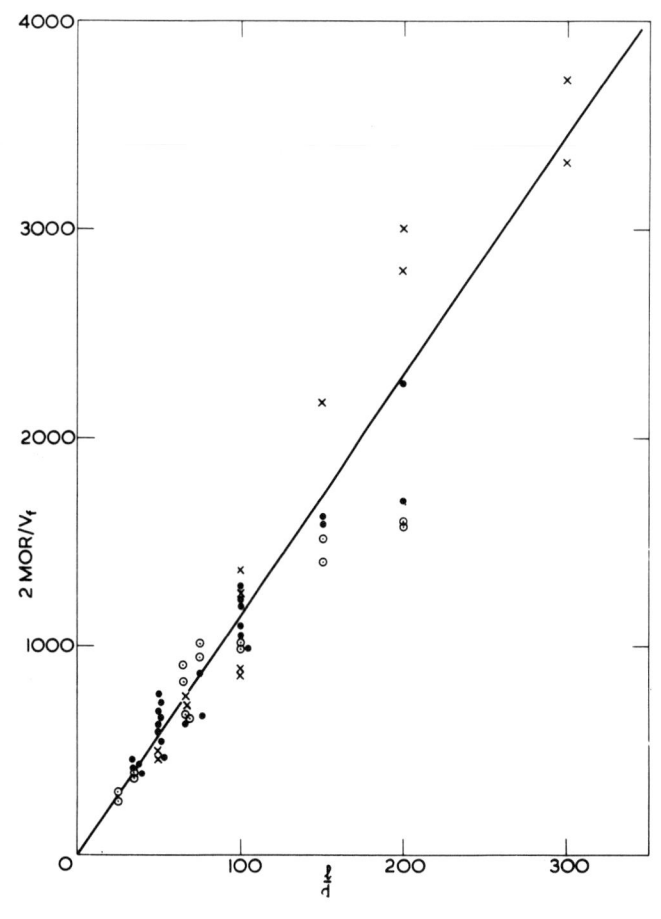

Fig. 14 Flexural strength of 3-D random fibre reinforced mortar

Pseudo random composites were made by rotating the mould through 10° for each layer of carbon: The continuous wrc was made by winding 132μm diameter hard drawn steel wire onto a rectangular mandrel, controlling the pitch of the coil winder and spacing between the layers to give a square array with the required volume fraction. This was then impregnated with cement paste, cured for 24 h and cut from the mandrel using a diamond saw. The matrix of the cfrc was 'Swiftcrete' ultra rapid hardening Portland cement with a w/c ratio of 0.5, and for the wrc ordinary rapid hardening Portland cement w/c = 0.4.

Discontinuous wrc was made by sprinkling alternate layers of random chopped wire and a watery cement paste onto a rectangular filter bed and removing the excess water by suction to give a final w/c ratio of about 0.35. The aligned specimens were made by dropping the wires onto a sheet of corrugated paper backed by magnetic strip, then inverting it over the filter bed and finally removing the magnet to allow the aligned wires to fall onto the filter bed. The wires were then impregnated with the watery slurry, sucked dry and further wires added and the process repeated until the required thickness was built up. Four lengths (12.5, 25, 37.5, and 50 mm) and three diameters (0.152, 0.254 and 0.38 mm) were used to give a range of l/d from 33 to 330. Specimens 300 × 30 × 10 mm were sawn from the cured sheets. A small correction was made to the uts, using the theory of Allen [12], for the loss of effectiveness of wires cut at the edges of the random 2-D specimens.

The 3-D random wire reinforced mortar was made in a conventional pan type concrete mixer using a w/c ratio of 0.5, cement/aggregate ratio of 0.4 and maximum aggregate size of 5 mm. The specimens were cast in moulds 500 × 100 × 100 mm.

All specimens were cured under water (28 days for the mortar, 14 days for cement paste) and then tested wet to avoid problems of shrinkage of the matrix relative to the fibre. Stress strain curves were measured by feeding the average output from a pair of linear variable differential trans transformer type transducers to the x-axis, and the load output from the Instron to the y-axis of an x-y recorder.

7 APPENDIX 1: CRACK SPACING FOR SHORT FIBRES

(a) Aligned fibres

Let x_d be the distance from the first crack for the fibres to transfer the load $\sigma_{mu} V_m$ per unit area of composite back into the matrix. The number of fibres with both ends at a distance greater than x_d from the crack and thus able to transfer their full share of the load is $N(1 - 2x_d/l)$ where $N = V_f/\pi r^2$. The remaining $2x_d N/l$ fibres have one end less than x_d from the crack and transfer load over a mean distance $x_d/2$. We put the total load transferred equal to $\sigma_{mu} V_m$ to give

$$2\pi r \tau x_d N(1 - x_d/l) = \sigma_{mu} V_m$$

or

$$x_d = \frac{1}{(1 - x_d/l)} \cdot \frac{V_m}{V_f} \cdot \frac{\sigma_{mu} r}{2\tau} = \frac{x'}{(1 - x_d/l)} \quad (A1)$$

where x' is given by Equation (4). Hence from (A1)

$$x_d = \frac{1 \pm (l^2 - 4lx')^{1/2}}{2} \quad \ldots (A2)$$

and as $x < 1$ it is the smaller root that is required.

(b) Random 2-D fibres

Let x_{d2} be the distance measured normal to the crack for the load in the matrix to build up to $\sigma_{mu} V_m$. By analogy with the random continuous case [4] the fibres are assumed to all slip over the same distance x_{d2} measured along the fibre so that a discontinuous fibre at angle θ to the crack will contribute its maximum load if both ends are at a distance x_{d2} measured along the fibre from the crack. The number of such fibres at angle θ^* is $N(1 - 2x_{d2}/l) \sin\theta \, . \, d\theta$ and so the total frictional force resolved in the direction of applied stress is

$$f_1 = \frac{2}{\pi} \int_0^{\pi/2} 2\pi r \, \tau \, x_{d2} \sin\theta \, . \, N(1 - 2x_{d2}/l) \sin\theta \, . \, d\theta$$

$$= \frac{\tau x_{d2} V_f}{r} - \frac{2\tau x_{d2}^2 V_f}{rl} \quad (A3)$$

For the remaining fibres with one end less than x_{d2} from the crack the total frictional force is

$$f_2 = \frac{2}{\pi} \int_0^{\pi/2} 2\pi r \frac{\tau x_{d2}}{2} \sin\theta \, . \, \frac{2x_{d2}}{l} N \sin\theta \, d\theta$$

$$= \frac{\tau x_{d2}^2 V_f}{rl} \quad (A4)$$

hence the total frictional force $f_1 + f_2$ is

$$F_1 = \frac{\tau x_{d2} V_f}{r} - \frac{\tau x_{d2}^2 V_f}{rl} \quad (A5)$$

Each of the n_1 fibres with both ends at a distance greater than x_{d2} from the crack will exert a pulley force (see Ref. [4]) $p_1(1 - \sin\theta)$ normal to the crack where p_1 is the load in the fibre, and the n_2 fibres with one end less than x_{d2} from the crack will exert a mean pulley force $p_2(1 - \sin\theta)$ so that the total pulley force is

$$F_2 = \frac{2}{\pi} \int_0^{\pi/2} n_1 (\sin\theta) \, p_1 \sin\theta \, d\theta +$$

$$+ \frac{2}{\pi} \int_0^{\pi/2} n_2 (1 - \sin\theta) \, p_2 \sin\theta \, d\theta$$

$$= \frac{2}{\pi} (1 - \pi/4)(n_1 p_1 + n_2 p_2) \quad (A6)$$

For large α, $n_1 p_1 + n_2 p_2$ is equal to $(\pi/2) \sigma_{mu} V_m$, the load thrown onto the fibres in a random 2-D composite when the matrix cracks so from Equation (A6)

$$F_2 = \sigma_{mu} V_m (1 - \pi/4) \quad (A7)$$

Putting $F_1 + F_2$ equal to $\sigma_{mu} V_m$ and remembering that the crack spacing for a continuous random 2-D composite

---

*Note: θ is defined here as the angle between the fibre and crack, to be consistent with Ref. [4].

x_2' is equal to $(\pi/2)x'$ we get

$$x_{d_2} = \frac{l \pm (l^2 - 2\pi lx')^{1/2}}{2} = \frac{l \pm (l^2 - 4lx_2')^{1/2}}{2} \quad (A8)$$

and see by comparison with Equation (A2) that the discontinuous fibre crack spacing varies with l in the same way for aligned and random composites.

8 APPENDIX 2: FLEXURAL STRENGTH

Consider the beam in Fig. 15a. The stress strain curve of the material is assumed to be as predicted by the ACK theory for a crack spacing of $2x'$ (Fig. 15b) in tension and linear in compression. Plane sections are assumed to remain plane so that

$$\epsilon = \frac{y}{r}, \quad \epsilon_1 = \frac{h_1}{r} \quad \text{and} \quad \epsilon_2 = \frac{-h_2}{r} \quad (A9)$$

The sum of the normal forces acting at any cross section is zero, ie

$$b \int_{h_2}^{h_1} \sigma \, dy = 0 \quad (A10)$$

where b is the breadth of the beam, and the moment of the same forces with respect to the neutral axis is equal to the bending moment M (see Timoshenko [13]),

$$M = b \int_{-h_2}^{h_1} \sigma y \, dy = 0 \quad (A11)$$

from Equation (A9) $dy = rd\epsilon$ and substituting into Equation (A10) we get

$$\int_{-h_2}^{h_1} \sigma \, dy = r \int_{\epsilon_2}^{\epsilon_1} \sigma \, d\epsilon = 0$$

Hence the position of the neutral axis is such that the areas under the tensile and compressive stress strain curves are equal. Putting ϵ_1 equal to $\epsilon_{fu} - a\epsilon_{mu}/2$ we get

$$\frac{E_c \epsilon_2^2}{2} = \frac{E_c}{2} \epsilon_{mu}^2 + E_c \epsilon_{mu} \frac{a\epsilon_{mu}}{2} +$$

$$+ \left(\epsilon_{fu} - \frac{a\epsilon_{mu}}{2} - \frac{a\epsilon_{mu}}{2} - \epsilon_{mu} \right) \frac{(E_c \epsilon_{mu} + \epsilon_{fu} E_f V_f)}{2}$$

$$\text{or} \quad \epsilon_2^2 = \frac{\epsilon_{fu}^2}{1+a} \quad (A12)$$

We put the sum of the absolute values of the maximum strain equal to

$$\Delta = \epsilon_{fu} \frac{a\epsilon_{mu}}{2} - \epsilon_2 = \frac{h_1}{r} + \frac{h_2}{r} = \frac{h}{r}$$

and substitute from (A9) $dy = rd\epsilon$ into Equation (A11) to get

$$M = hr^2 \int_{\epsilon_2}^{\left(\epsilon_{fu} - \frac{a\epsilon_{mu}}{2}\right)} \sigma \epsilon \, d\epsilon$$

$$= \frac{bh^2}{\Delta^2} \int_{\epsilon_2}^{0} E_c \epsilon^2 \, d\epsilon + \int_{0}^{\epsilon_{mu}} E_c \epsilon^2 \, d\epsilon +$$

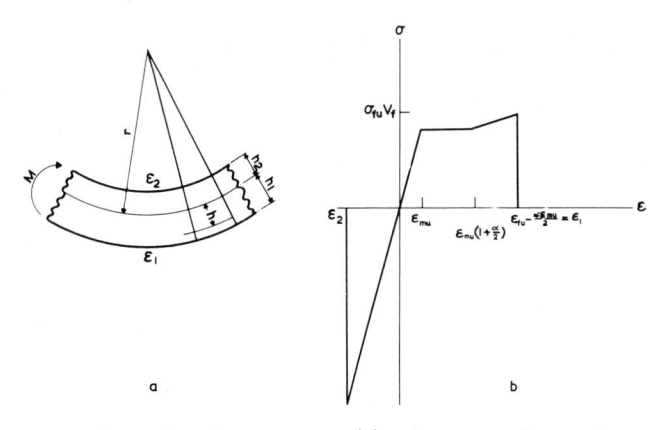

Fig. 15 Flexural specimen geometry (a) and corresponding tensile and compressive stress strain curve assumed in deriving Equation (A14)

$$+ \int_{\epsilon_{mu}}^{\epsilon_{mu}(1+a/2)} E_c \epsilon_{mu} \epsilon \, d\epsilon +$$

$$+ \int_{\epsilon_{mu}(1+a/2)}^{\left(\epsilon_{fu} - \frac{a\epsilon_{mu}}{2}\right)} E_f V_f \epsilon^2 \, d\epsilon +$$

$$+ \int_{\epsilon_{mu}(1+a/2)}^{\left(\epsilon_{fu} - \frac{a\epsilon_{mu}}{2}\right)} \frac{E_m V_m \epsilon_{mu} \epsilon}{2} \, d\epsilon$$

$$= \frac{bh^2}{\Delta^2} \left\{ \frac{E_c \epsilon_{mu}^3}{3} - \frac{E_c \epsilon_2^3}{3} + \frac{E_c \epsilon_{mu}}{2} (\epsilon_{mu}^2 (1+a/2)^2 - \epsilon_{mu}^2) + \right.$$

$$+ \frac{E_f V_f}{3} \left[\left(\epsilon_{fu} - \frac{a\epsilon_{mu}}{2} \right)^3 - \left(\epsilon_{mu}(1+a/2) \right)^3 \right] +$$

$$\left. + \frac{\epsilon_{mu} E_m V_m}{4} \left[\left(\epsilon_{fu} - \frac{a\epsilon_{mu}}{2} \right)^2 - \left(\epsilon_{mu}(1+a/2) \right)^2 \right] \right\}$$

$$(A13)$$

The modulus of rupture (MOR) is defined as $6M/bh^2$ and the uts is equal to $\epsilon_{fu} E_f V_f$ and hence

$$\frac{\text{MOR}}{\text{uts}} = \frac{6M}{\epsilon_{fu} E_f V_f bh^2} = \frac{6}{\epsilon_{fu} \Delta^2} \times$$

$$\times \left\{ \left[(1+a) \epsilon_{mu}^3 \left(\frac{a^2}{8} + \frac{a}{2} + \frac{1}{3} - \frac{\epsilon_2^3}{3\epsilon_{mu}^3} \right) \right] + \right.$$

$$+ \frac{1}{3} \left[\left(\epsilon_{fu} - \frac{a\epsilon_{mu}}{2} \right)^3 - \epsilon_{mu}^3 (1+a/2)^3 \right] +$$

$$\left. + \frac{a}{4} \epsilon_{mu} \left[\left(\epsilon_{fu} - \frac{a\epsilon_{mu}}{2} \right)^2 - \epsilon_{mu}^2 (1+a/2)^2 \right] \right\} \quad (A14)$$

The condition for multiple cracking

$$\sigma_{fu} V_f \geq \sigma_{mu} V_m + \epsilon_{mu} E_f V_f$$

or $$\quad (A15)$$

$$a \geq \frac{\epsilon_{fu}}{\epsilon_{mu}} - 1$$

where σ_{fu} is the stress to either break or pull out the fibres, fixes the upper limit to the ratio MOR/uts shown by the broken line in Fig. 12 where Equation (A14) is plotted as a function of a for various values of $\epsilon_{fu}/\epsilon_{mu}$.

The analogue of Equation (A14) for the lower bound to the crack spacing x' is

$$\frac{MOR}{uts} = \frac{6}{\epsilon_{fu}\Delta^2}\left\{(1+a)\epsilon_{mu}^3\left(\frac{9a^2}{32}+\frac{3a}{4}+\frac{1}{3}-\frac{\epsilon_2^3}{3\epsilon_{mu}^3}\right)+\right.$$
$$+\frac{1}{3}\left[\left(\epsilon_{fu}-\frac{a\epsilon_{mu}}{4}\right)^3-\epsilon_{mu}^3\left(1+\frac{3a}{4}\right)^3\right]$$
$$\left.+\frac{a\epsilon_{mu}}{8}\left[\left(\epsilon_{fu}-\frac{a\epsilon_{mu}}{4}\right)^2-\epsilon_{mu}^2\left(1+\frac{3a}{4}\right)^2\right]\right\} \quad (A16)$$

The values of MOR/uts computed for crack spacings of x' and $2x'$ using (A15) and (A16) differ at the most by 4%.

ACKNOWLEDGEMENTS

The authors are grateful to Dr A. Kelly and Dr D. McLean for helpful discussions and to Dr B. Gale for the general least-squares programme used to fit Equation (16).

REFERENCES

1. **Hyatt, T.** *UK Patent Specification* 2701 (30 June 1876)
2. **Romualdi, P.J.** and **Batson, G.B.** *Proceedings of the American Society of Civil Engineers* 89 No EM3 147 (1963)
3. **Aveston, J., Cooper, G.A.** and **Kelly, A.** 'The properties of fibre composites', *Conference proceedings, National Physical Laboratory* published by IPC Science and Technology Press Ltd (1971) paper 2, p 15
4. **Aveston, J.** and **Kelly, A.** *J Mat Sci* 8 352 (1973)
5. **Kelly, A.** paper in this conference
6. **Gale, B.** Private communication
7. **Davidge, R.W.** and **Tappin, G.** *J Mat Sci* 3 165 (1968)
8. **Brown, E.C.** Private communication
9. **Krenchel, H.** Fibre reinforcement, Akademick Forlag Copenhagen (1964)
10. **Christensen, R.M.** and **Waals, F.M.** *J Comp Mat* 6 518 (1972)
11. **Cox, H.L.** Private communication. See also [3] p 71
12. **Allen, H.G.** *J Phys D; Appl Phys* 5 331 (1972)
13. **Timoshenko, S.** 'Strength of Materials', 3rd edition (D. Van Nostrand Co Ltd) p 366

QUESTIONS

1 Concerning Fig. 1, by R.W. Davidge (AERE Harwell)

I would like to raise two points concerning Fig. 1, the load deflection curves for aligned composites of carbon fibre reinforced cement, and in particular the third region where matrix cracking is complete and all the load is taken on the fibres.

(a) If the deformation here is still elastic why does the extrapolation of the theoretical curve give a positive intercept on the stress axis rather than pass through the origin?

(b) Is the fact that experimental slopes are lower than the theoretical ones a reflection on the fabrication techniques used? Presumably hand lay up from veil material leads to both non-perfect alignment and distribution of fibres. The method we use at Harwell involves filament winding under tension and this gives very good alignment and distribution. [**Briggs, A., Bowen, D.H.** and **Kouck, J.** *Int Conf 'Carbon fibres, their place in modern technology'* The Plastics Institute (London 1974)].

Reply by Mr Aveston

In the third region of the stress-strain curve the average load borne by the matrix (for crack spacing $2x'$) remains constant at $\sigma_{mu}V_m/2$ but all the *additional* load is taken by the fibres. The slope of the stress strain curve is therefore $E_f V_f$ and its equation is

$$\sigma = E_f V_f \epsilon + \sigma_{mu} V_m/2$$

The final slopes give $E_f = 176\,GN/m^2$ for Grafil A (Fig. 9) which is at the lower end of the range $175-205\,GN/m^2$ quoted by the manufacturers. If the alignment had been perfect I imagine a slightly higher modulus would have been observed in this region.

2 On asbestos reinforced cement products, by C. Urbanowicz (Cape Universal)

The conference consisted mainly of papers dealing with the properties and applications of the newer composites such as grp, cfrp and grc. I would like to make the point that very little research work of a comparable standard has been done or been published on the improvements of properties of the old established composites still widely used such as asbestos cement, quoted in Mr McCurrich's paper as having total consumption second only to concrete.

It would be interesting to see the results of some basic research by universities and other research institutions into the mechanisms of fibre reinforcement of such products which already possess a vast market.

Reply by Mr Aveston

Some basic research on the reinforcement of cement by asbestos has been published by H.G. Allen [*Composites* (June 1971)] and more recently at NPL we have done some contract research on this material. Although these results are on a commercial in-confidence basis, it is fair to say that they are broadly consistent with the general theory of brittle matrix composites outlined in our paper.

COMMENTS

1 On increase of tensile failure strain of Portland cement reinforced with glass fibre bundles, by Miss V. Laws (Building Research Establishment)

Mr Aveston and co-authors have shown that the tensile failure strain of cement paste can be increased by the incorporation of fibres, namely steel and carbon. We have carried out a similar study to establish whether an increase can be obtained when glass fibre bundles are added to cement. The matrix was ordinary Portland cement, mixed at 0.5 water/cement ratio. The fibres were continuous and aligned in the direction of the stress, and the composites were cured at 90% RH for 7 or 28 days.

The Young's modulus of the matrix, deduced from the measured composite modulus, increased by a factor of approximately 2½ over the range of volume fibre fractions used.

The figure shows the measured matrix failure strains as a function of volume fibre fraction. The solid line is the Aveston, Cooper and Kelly prediction, based on the deduced values of the matrix modulus. Although the scatter is wide, the ACK curve is a reasonably good fit.

I have not drawn the 'base line', ie the line showing the failure strain of the unreinforced matrix. The matrix modulus increased with volume fibre fraction; possibly due to loss of water by evaporation during composite fabrication. Information on the effect of a decreased water/cement ratio on the failure strain in tension of cement paste is not available, but data in compression (Spooner, 1972) shows that this could be a significant factor contributing to the observed increase in matrix failure strain with volume fibre fraction.

REFERENCE

1 **Spooner, D.C.** *Mag Concrete Res* **24** No 79 (1972) pp 85–92

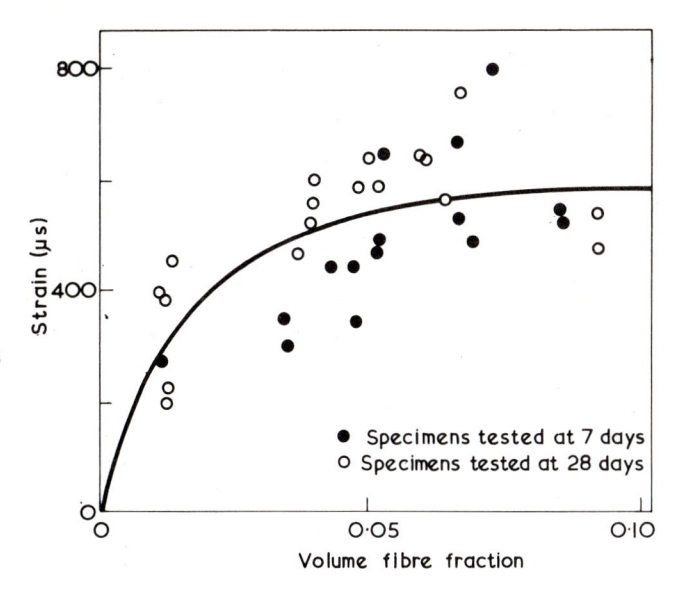

Matrix failure strain of glass fibre reinforced cement paste. The full circles refer to composites tested at 7 days, the open circles to those tested at 28 days. The curve is the prediction using Equation (1) of the paper by Aveston et al.

Some engineering considerations of fibre concrete

R.N. SWAMY and C.V.S. KAMESWARA RAO

The development of any new building material is an interdisciplinary process. In addition to a basic materials science approach, engineering considerations are essential in developing a material system that can be used in the construction industry. Some engineering aspects with respect to properties, testing and specification for design are discussed.

1 INTRODUCTION

The very nature of the construction industry and its close relationship with the community and environment necessitates an inevitable time-lag between the development of a new building material in the laboratory and its practical application. Although considerable progress is being made in the understanding of the basic mechanics of strength and deformation of fibre reinforced cement composites, there is still a long way to go to establish acceptable methods of testing and specification of the properties required in the design process. The basic requirements of the composite in the fresh and hardened states are briefly discussed in this paper.

2 RHEOLOGY OF THE FRESH COMPOSITES

One important factor affecting the overall cost of the material and its performance in the hardened state is the mix formulation and handling characteristics of the material whether used in situ or as a precast unit. A clear distinction should be made in this respect between fibres in a cement matrix and fibres in a concrete mix. Coarse aggregates are cheap, and, like fibres, add to the stability and stiffness of the matrix, and their presence is therefore beneficial particularly in load-bearing components. A more significant argument for their incorporation is that aggregates also create heterogeneity and remove the apparent brittleness of the matrix and thereby mobilise the intrinsic properties of the matrix in resisting load. Their presence, however, creates complex interparticle friction, and very much influences the mobility and compactibility of the fresh composite (Fig. 1). The conventional consistency tests are not reliable for fibre composites but the compactibility tests give a relatively more reliable measure of the rheological properties of the composite. The geometry and volume fraction of both the fibres and aggregates have to be carefully controlled if the properties of the composite are to be optimised. Conventional mix design techniques may also be readily extended to fibre concrete by treating the fibre content as an equivalent aggregate content in relation to shape and surface area.

3 TESTING AND SPECIFICATION

The testing methods and the design specifications should be related to both short- and long-term performance. The latter is just as important to non-load-bearing components, and the ability to accommodate movements due to moisture and temperature changes within and between structural components should be clearly established. Non-load-bearing elements made of traditional materials, such as cladding panels, have failed because of their inability to resist stresses redistributed and transferred on to them due to movements

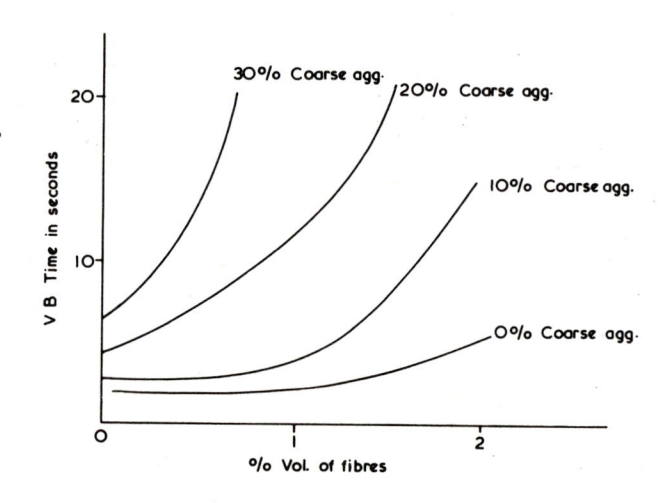

Fig. 1. Fibre-aggregate interaction on the compactibility of steel fibre reinforced concrete

Department of Civil and Structural Engineering, University of Sheffield, Yorkshire, England.
This is an original contribution which was not presented at the conference.

occurring in the rest of the structure. Fibre composites are ideally suited to accommodate such unforseen movements which cannot always be readily designed for, and data need to be established to enable these movements to be quantified. In other words, strength is not the only criterion for design, and the ability to deform without deterioration in serviceability is probably far more significant.

Two factors need to be considered in the development of test methods and design specifications. The significant ductility of fibre concrete, and in particular, its superior strain capability, need to be considered in the testing of the material, and in establishing relations between the various tensile strengths. Few concrete components are subjected to a purely tensile stress field, and the inherent difficulties of obtaining such a stress in a test makes the standardization of such a test questionable, although desirable, particularly when one considers the arguments still raging about the age-old compression test.

Secondly, design equations should be simple and reflect the physical processes involved in the behaviour of the material. Cement-based matrices are neither elastic nor plastic but they display a kind of inelasticity that cannot be represented by a simple mathematical model. Unlike the predictions based on conventional assumptions, the unique property of the concrete matrix which is increasingly being recognised in building codes is its ability to transfer stresses across microcracks. This unique property of the cement matrix is enhanced by the presence of fibres, and needs clearly to be recognised in design specifications.

4 DESIGN FOR FLEXURAL STRENGTH

The superior crack arrest properties of the fibre composite require that the mode of failure should also be considered in relating the design specifications to the type of test. The presence of fibres modifies the stress distribution in the flexural tensile test, for example (Fig. 2). Noting that at failure the matrix is cracked and that the fibres pull-out, the

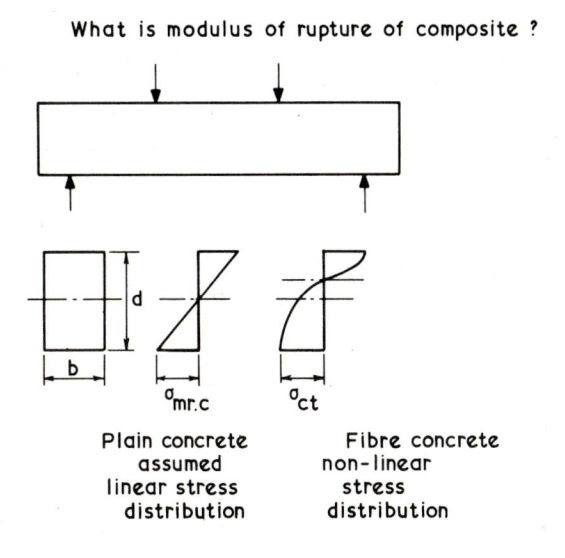

Fig. 2. Modulus of rupture stress distribution at failure

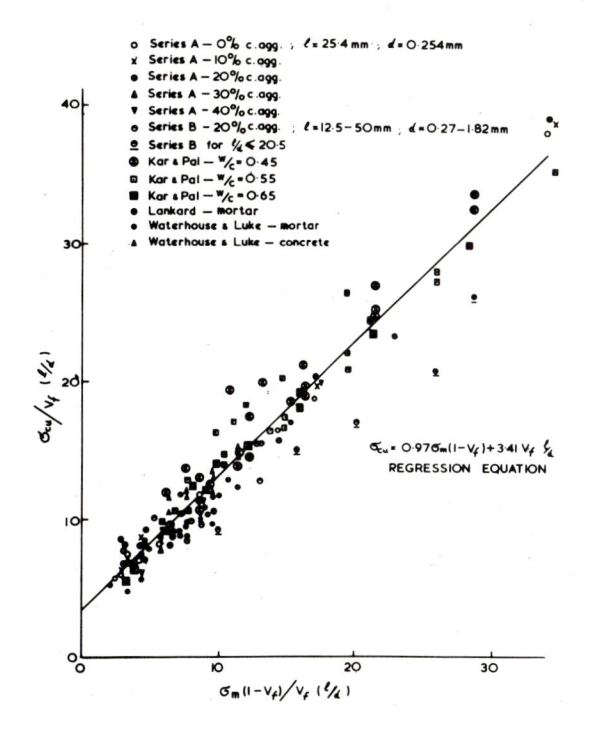

Fig. 3. Crack control-composite mechanics equation for ultimate flexural strength of steel fibre concrete

flexural strength σ_f of the fibre concrete with random steel fibres can be shown to be:

$$\sigma_f = A\sigma_m V_m + 0.82 \, \tau \, V_{f\,(l/d)}$$

where A is a constant, σ_m is the modulus of rupture of the matrix, τ the interfacial bond strength, V_f the volume fraction of the fibre, V_m the volume fraction of the matrix, and l and d define the length and diameter of the fibre respectively. The advantage of a solution of this type is its simplicity in establishing a lower-bound equation from which a design equation with suitable partial safety factors can be easily derived. The equation not only establishes the contribution of the cracked matrix to the composite strength, but also expresses in simple terms the physical processes involved in the fracture of the composites.

Fig. 3 shows the excellent correlation between the above equation and data from three continents comprising a wide range of variables of matrix properties, aggregate size and volume, fibre volume and fibre content, type of cement and curing regime.

5 CONCLUSIONS

In a contribution restricted by space only general ideas can be aired but the main emphasis in the paper is the need to link the laboratory to the construction practice and the materials science to practical engineering design considerations. Concrete is probably the most complex of all construction materials but the versatility of the fibre concrete makes it a better component where conventional concrete can only be used inefficiently. It is, however, emphasized that it is no universal panacea for all problems. It is also important to recognize that in engineering the economics of construction should be related to performance and not merely to cost of materials alone.

Tensile stress/strain characteristics of glass fibre reinforced cement

B.A. PROCTOR, D.R. OAKLEY and W. WIECHERS

Simple strip specimens of glass fibre-reinforced cement, containing about 4% (by volume) glass fibre, were tested in direct tension. Strain measurements were by extensometer and strain gauge. The stress/strain curve showed four distinct regions, the first with an initial tangent modulus of about 25 GN/m² in which the material behaved as an uncracked composite, secondly a transition zone, thirdly a region of extensive crack development and finally a crack opening region with a tangent modulus approximately predicted by fibre values.

1 INTRODUCTION

The work which is very briefly described in this paper forms part of an extensive and continuing investigation into the mechanical properties of glass fibre-reinforced cementitious materials being carried out at the Pilkington Research Laboratories. Following the comments and analysis of Allen [1] and Nair [2] indicating the limitations of flexural testing with this type of material, a considerable effort has been devoted to detailed observations of the stress/strain curve in direct tension.

The results reported below were obtained with material produced by the spray/suction de-watering method devised at Building Research Establishment [3] and containing 5½% to 6% glass fibre by weight of the green de-watered board (approximately 4% by volume). The glass was in the form of chopped strands of 38 mm length, disposed approximately randomly in the plane of the board, each strand containing about 204 filaments of 13 μm diameter.

The matrix was a rapid hardening Portland cement (Velocrete) and after dewatering to a water cement ratio of 0.32 and curing in a fog room at 95% RH and 21 °C for seven days, samples were stored in the laboratory until testing.

2 TEST METHODS, RESULTS AND DISCUSSION

Simple strip specimens 150 × 25 × 8 mm were found to be satisfactory; a 12 mm long strain gauge was attached to the centre portion of the sample and gave accurate strain measurements for initial modulus values. The complete stress/strain curve was obtained from a clip-on extensometer with a 25 mm gauge length. Two additional strain gauges, attached one to each face of the specimen, were connected in opposition and operated at high gain to give a sensitive (differential) indication of the initiation of micro-crack propagation at very low stresses. The specimens were pulled to failure in an Instron Testing Machine at strain rates in the range 10^{-3}/minute to 10^{-2}/minute.

The stress/strain behaviour of a variety of slightly different types of material is shown in Fig. 1. There is an initial 'elastic' region common to all samples, followed by an extensive pseudo-ductile region in which the samples appear to be divided into two groups reflecting a degree of fibre anisotropy which was introduced by the particular spray equipment used in their manufacture and subsequently confirmed qualitatively by microscopic examination. Samples cut transversely from the original board appeared to have fewer fibres oriented along their axes, and deformed and failed at lower stresses than the longitudinal samples. Within these two groups the behaviour of samples tested after 1, 3, 6 and 12 months storage appeared very similar, although there was some drop in strain to failure after 1 month.

The most complete picture of stress/strain behaviour is shown by the 1 month old, longitudinal samples (eg Fig. 2) although most features can also be recognised in the curves of older transverse and longitudinal samples. Fig. 2 is a reproduction of the actual extensometer trace of a typical sample and four main regions of the curve have been identified.

Region I has an initial tangent modulus of about 25 GN/m² which is within the range of that for neat cement paste. Simple 'Law of mixtures' considerations would indicate that these low fibre contents should only add about 5% to the stiffness of the cement matrix; on the basis of a modified

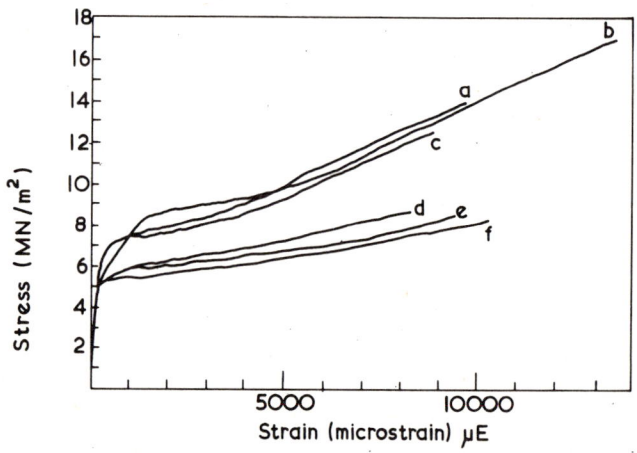

Fig. 1 Tensile stress/strain curves for laboratory-stored grc samples: longitudinal samples—(a) 6 months old; (b) 1 month old; (c) 1 year old; transverse samples—(a) 6 months old; (b) 1 year old; (c) 3 months old

Pilkington Brothers Research and Development Laboratories, Lathom, Ormskirk, Lancashire, England

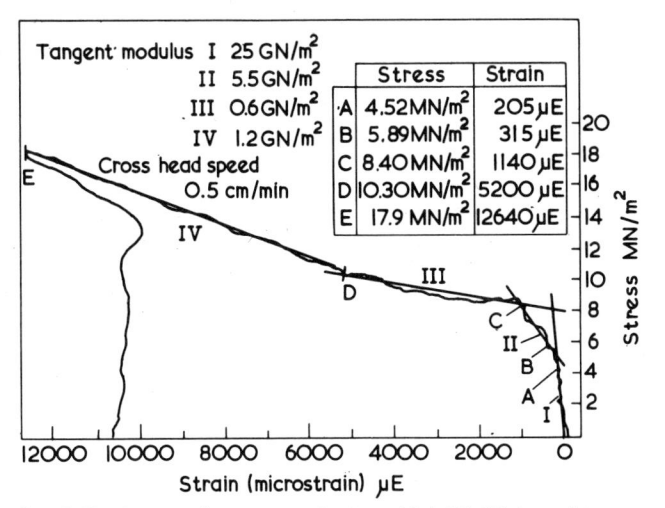

Fig. 2 Tensile stress/strain curve for board LA 77: 28 days old

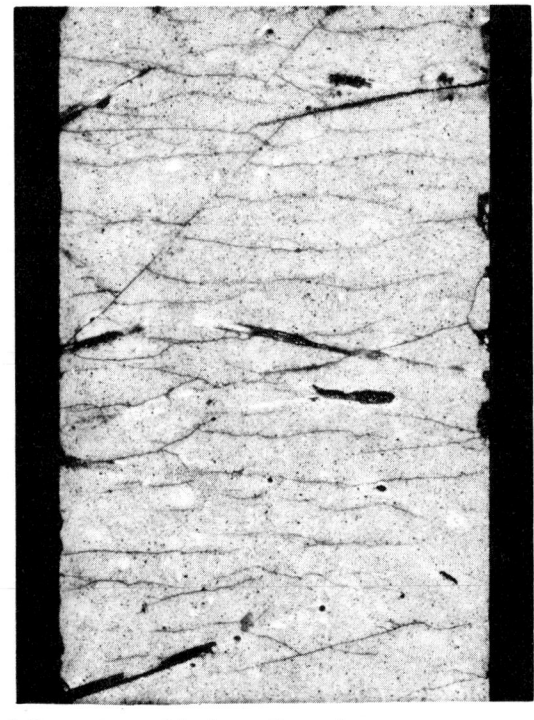

	Stress	Strain
A	4.52 MN/m^2	205 µE
B	5.89 MN/m^2	315 µE
C	8.40 MN/m^2	1140 µE
D	10.30 MN/m^2	5200 µE
E	17.9 MN/m^2	12640 µE

Tangent modulus I 25 GN/m^2; II 5.5 GN/m^2; III 0.6 GN/m^2; IV 1.2 GN/m^2. Cross head speed 0.5 cm/min.

Fig. 3 Transverse cracking in tensile specimen

law of mixtures which takes account of the effects of voidage in transversely oriented strands Nair [2] has suggested a reduction of 5 to 10% in modulus. Material variability renders it difficult to distinguish between these predictions. The more detailed strain gauge work also indicated that there was some micro-crack propagation and slight changes in modulus within this Region I.

The distinct second Region, II, is so far unexplained. It appears most clearly and consistently in this particular type of material (1 month old longitudinal) but may also be detected in other samples. The transition portion ABC culminates in a sharp 'bend-over point', C, followed by extensive pseudo-ductile behaviour due to multiple cracking of the cement matrix as discussed by Allen [1] and Aveston et al [4]. By point D at the end of Region III the crack development has been completed and the whole stressed length of the specimen is covered with fine transverse cracks as illustrated in Fig. 3. Individual spacings range from about 0.2 to 5 mm with a most frequently occurring value of about 1.1 mm. Comparison of these crack spacings with directly measured strand bond strengths suggests that the efficiency of fibres in controlling cracking is little affected by changing from an aligned to a pseudo random planar array (ie $\eta \approx 1$ for crack spacing).

It is also interesting to note that Spurrier and Luxmoore [5] suggest a gradual transition from Regions III to IV in two dimensional randomly reinforced materials; in these samples however the changes associated with point D at the end of the crack development region appeared to be quite distinct.

The final Region, IV, corresponds to crack opening with elastic fibre stretching, after failure the cracks close again almost completely outside the localised failure zone. The final failure involves considerable fibre pull-out but the fibre strengths utilised correspond closely with values measured in separate 'strand-in-cement' tests and it is possible that failure is actually initiated by fibre breakage. Efficiency factors for the use of fibre strength and stiffness in Region IV are close to the values predicted by Laws et al [6], (ie $\eta \approx 0.2 - 0.35$).

The stress/strain curve over regions III and IV corresponds, qualitatively, in a quite detailed way with the multicracking behaviour of continuous, aligned fibre, brittle-matrix composites described by Aveston et al in Reference [4]. Due to the complexities introduced by the incorporation of fibres in multi-filament strands, variable penetration of cement into the strand and random (or near random) fibre orientation in a plane, it has not been possible to establish a fully quantitative interpretation of behaviour at the present time.

3 CONCLUSION

The full tensile stress/strain curve of fibre reinforced cementitious materials provides valuable experimental data for the establishment of an understanding of material behaviour. Simple strip specimens have been found to be adequate and strain gauging was a useful supplement to strain measurements with an extensometer.

Qualitatively the 'random' spray/suction-dewatered grc material behaved in a multi-crack manner having an initial modulus equal to that of cement paste and extensive multi-cracking and crack opening regions with large strain to failure and high energy absorption.

ACKNOWLEDGEMENTS

The authors wish to thank the Directors of Pilkington Brothers Limited and Dr D.S. Oliver, Director of Group Research and Development, for permission to publish this article.

REFERENCES

1 **Allen, H.G.** *J Comp Mat* 5 (April 1971) pp 194–207

2 **Nair, N.G.** Private communication

3 **Steele, B.R.** 'Glass fibre reinforced cement', *Proc Int Building Conf, 'Prospects for fibre reinforced construction materials', 1971* Department of the Environment (1972)

4 **Aveston, J., Cooper, G.A.** and **Kelly, A.** 'Single and multiple fracture', *NPL Conf, 'The properties of fibre composites'* IPC Science and Technology Press Ltd (1971).

5 **Spurrier, J.** and **Luxmoore, A.** *Fibre Science and Technology* 6 (1973) pp 281–298

6 **Laws, V., Lawrence, P.** and **Nurse, R.W.** *J Phys D: Appl Phys* 6 (1973) pp 523–537

Modification of grc properties

A.J. MAJUMDAR

The properties of glass fibre reinforced cement (grc) are dependent on age and environmental conditions. As part of a research programme aimed at establishing the age-strength relationship for grc, the effect of altering certain fabrication variables such as fibre parameters and matrix formulations on grc properties is being investigated. The effect of adding a second fibre is also being studied. A very brief summary of the work in progress at the Building Research Establishment in these areas is given in this paper.

The mechanical properties of fibre cement and concrete are beginning to be understood in terms of theoretical models [1,2,3,4] and in the work of Aveston, Mercer and Sillwood [4] presented at this conference good agreement between theory and experiment has been obtained for composites containing steel wire and carbon fibre. These are short-term results however. The long-term properties of certain types of fibre cement composites can differ significantly from the short-term ones due to the effects of the inherent reactivity of the cement paste and a rigorous application of the theory to specify properties is not possible yet for all cases. An accurate knowledge of the properties of the fibre after they have lived with the matrix for some time is not available and little is known about the changes which take place at the interface between the fibre and the matrix over a period of time. When the reinforcement is in the form of a fibre-bundle, as is the case with asbestos, glass and carbon fibres, the determination of all interfacial properties poses very difficult experimental problems.

The effect of age and environment on the properties of high-alumina cement composites containing two different types of glass fibre is shown in Fig. 1. Both composites were prepared by the BRE spray-suction method [5]. Each had approximately 4 volume % of 34 mm long fibres which were distributed uniformly in the matrix in a two-dimensional random array. In the case of E-glass, point I and II refer to stress-strain properties at failure obtained with one-year and four-year old specimens, whereas for the AR-glass, the corresponding ages were slightly longer. It may be mentioned here that the AR-glass used in this study was the first experimental batch of zirconia containing glass fibre produced commercially for research at BRE. Substantial developments have taken place since then and Pilkingtons have launched a special fibre Cem-FIL* for cement reinforcement.

Building Research Establishment, Department of the Environment

* Registered trade mark of Pilkington Brothers Limited

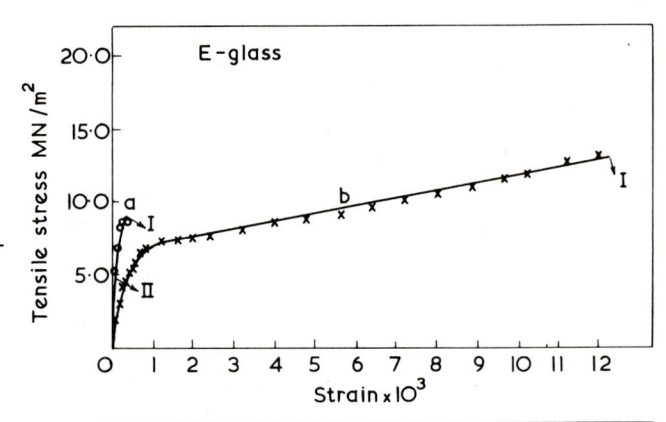

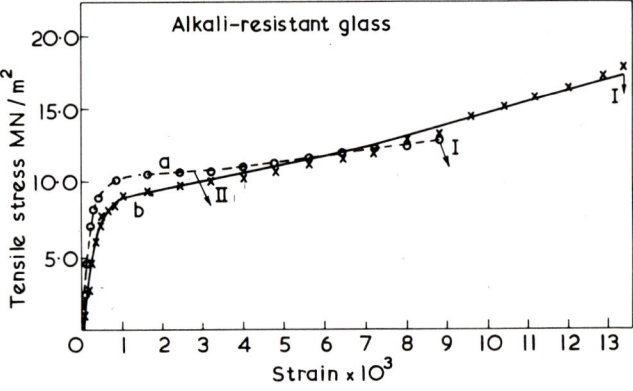

Fig. 1 The effect of environment on the stress-strain behaviour of glass fibre high-alumina cement: (A) E-glass (a) water stored, (b) air stored; (B) Alkali-resistant glass (a) water stored, (b) air stored

Three important points emerged from the study with high-alumina cements.

1. The modulus of the composite and the stress at the limit of proportionality (LOP) are higher in the case of water-stored samples for both glasses. This has to do with the degree of hydration of the cement.

Table 1 Stress-strain properties of glass fibre cement: fibre length 34 mm

Sample	Fibre volume (%)	Number of fibres per strand	Fibre diameter (μm)	At LOP		At failure	
				Stress (MN/m^2)	Strain (μ-strain)	Stress (MN/m^2)	Strain (μ-strain)
grc (28 days old)	4.0	200	13.5	7.0	300–350	14.0	9 000
grc (28 days old)	8.0	200	13.5	11.0	300–350	18.0	7 500
grc (28 days old)	4.0	50	13.5	10.0	300–350	18.0	12 000
grc (28 days old)	4.0	200	10.0	8.5	300–350	16.0	8 000
grc + 10% pvp (one year old)	4.0	200	13.5	10.5	550	13.0	7 000

2 Under dry conditions (40% rh), both glasses are stable in the matrix of high-alumina cement. Results obtained with 4 year old samples show that the failure stress-strain values given in Fig. 1 remain virtually unchanged.

3 In wet conditions, composites containing E-glass lose all its pseudo-ductility after only 12 months, whereas with the first generation AR-glass, a significant proportion of this ductility is retained even after 4 years. This observation can be easily correlated with the microstructure of the fracture surfaces of old, water-stored specimens. In the case of E-glass, fibre fracture on a very extensive scale is observed with virtually no pull-out. For AR-glass, evidence of multiple cracking of the matrix and fibre debonding and pull-out can be clearly seen.

The effects of altering some of the fibre parameters on the properties of glass fibre reinforced Portland cement can be seen in the results given in Table 1. These random 2-D composites were also prepared by the spray-suction method. The properties of the control are listed in the first row. A fibre volume fraction of 0.04 has been chosen for much of the work at BRE with the expectation that such a composite would be economically viable. The effect of doubling the fibre concentration is shown in the second row of Table 1. The next two rows show the effect of increasing the aspect ratio of the reinforcements by reducing the cross-sectional area of the fibre bundle. It may be pointed out that the fibre bundle, which is the effective reinforcement here, is more like a tape than a cylinder. A decrease in the cross-sectional area of the bundle can be achieved either by reducing the number of filaments per bundle or by reducing the diameter of individual filaments. Either way, at constant volume fractions the number of reinforcements is increased. Table 1 shows that by these alterations of the fibre parameters, stresses in the composite at both LOP and failure can be susbstantially increased. Regarding the strain at LOP, no statistically significant trend was noticeable.

If it is desirable to increase the strain at LOP, this can be done effectively by the addition to the matrix of a suitable polymer, preferably an emulsion. Polyvinyl propionate (PVP) and some styrene-acrylics have given good results over the period of study. With PVP addition, the retention of large failure strain after one year is also very encouraging (Table 1). For securing these advantages, however, one has to pay a penalty in terms of higher costs. The same is perhaps true of having strands with smaller numbers of filaments.

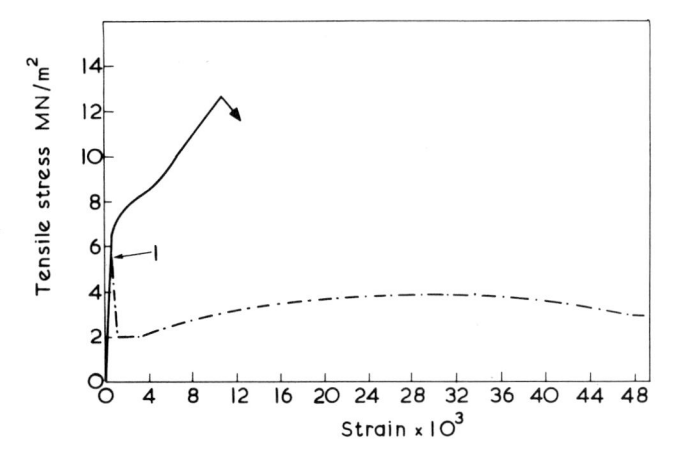

Fig. 2 Tensile stress-strain curves for grc (————) and grc containing 1 volume % polypropylene monofilaments (— · — · — · — · —)

The densities of the various glass fibre cement composites whose properties are listed in Table 1 were not constant. This reflects the differences in the porosity of these materials which must have a considerable influence on the efficiency of reinforcement by the short fibres, thereby controlling the strength of the composites. Such information is not available at present for glass fibre cement but Ali [6] has studied the dependence of critical transfer length of glass fibre strands on porosity, using gypsum plaster composites. Results given in Table 1 indicate that by optimising fibre parameters and matrix formulations, the properties of currently produced glass fibre cement can be further improved.

In some applications of glass fibre cement sheets, for example as roof tiles, a high degree of resistance to impact over a long period of time is required after the component has been placed. For such uses, reinforcement by a mixture of glass and polypropylene fibres is an attractive proposition. The stress-strain curve for such a composite having 1 volume % of polypropylene and which underwent an accelerated form of curing is shown in Fig. 2. In this environment the composite failed initially at the LOP (point I) but the polypropylene fibres were able to support the cracked composite at reduced loads. Even at very high strains, the specimens were not physically separated into two halves. The stress-strain graph of fibre cement composite reinforced by glass alone and cured in air at 20°C is also included in Fig. 2 for comparison.

ACKNOWLEDGEMENT

The work described here is a very brief summary of the on-going research at BRE and in collaboration with Pilkington Brothers Limited. The author acknowledges the help given by several of his colleagues in the provision of experimental results. This paper is published by permission of the Director of BRE.

REFERENCES

1 **Allen, H.G.** *J Phys D: Appl Phys* **5** (1972) pp 331–343
2 **Laws, V., Lawrence, P.** and **Nurse, R.W.** *J Phys D: Appl Phys* **6** (1973) pp 523–537
3 **Kelly, A.** 'Some scientific points concerning the mechanics of fibrous composites', paper in this conference
4 **Aveston, J., Mercer, R.A.** and **Sillwood, J.M.** 'Fibre reinforced cements – scientific foundations and specifications', paper in this conference
5 **Steele, B.R.** 'Glass fibre reinforced cement' *Proc Intern Bldg Exhibition Conf on 'Prospects for fibre reinforced construction materials'* Building Research Establishment (1972) pp 29–39
6 **Ali, M.A.** 'Study of brittle fibre – brittle matrix composite system with particular reference to glass fibre reinforced gypsum', *M Phil Thesis, University of Surrey* (1972)

Ferrocement — a summary of research activity

R.G. MORGAN

Practical application of ferrocement as a hull material is far advanced on the understanding of its behaviour under various loading conditions. A brief summary of recent research activity in studying its properties in tension, compression, flexure and impact is supplemented by a comprehensive list of research studies.

1 INTRODUCTION

Ferrocement can be defined as a strong densely-mesh-reinforced cement mortar skin which portrays high tensile and flexural strength characteristics. In its earliest and simplest form it was used by Joseph Louis Lambot who built a small rowing boat in France in 1848 by plastering a mesh-covered iron framework, and called the material 'ferricement'. Many boatbuilders followed Lambot's techniques up to the early 1910's and then the material reappeared in a more sophisticated form with Pier Nervi's vessels in Italy in 1943. Another lapse followed until about 1960 when several boats were built in New Zealand, Australia and in the United Kingdom. As expertise in construction techniques developed, the material grew in popularity and ferrocement vessels are now being built all over the world.

Nervi carried out the first experiments [1] to assess the properties of the material which he had termed 'ferrocemento' but it is only in the last few years that ferrocement has been seriously researched and generally this has been directed to marine applications. The basic materials for ferrocement construction (cement, sand, rod and mesh) are cheap and no single manufacturer has been able to justify extensive research and development or promote the material. On the other hand the basic ingredients are readily available in most parts of the world, construction methods need little equipment but lots of unskilled labour and the developing countries are now making wide use of ferrocement for a variety of small building units [2].

The major application is for ships built using a variety of constructional methods [3–6] for vessels up to 32 m long, and an attempt has been made at building a 55.5 m X 14.6 m deck cargo barge using post-tensioning techniques [5]. Fishing vessels over 50 m are now being projected and the classification societies [8], universities, national [9] and governmental agencies [10–12] have become involved in recent years as designer and builder moved outside the 'rule of thumb' range. Whatever technique might be employed in laying up the rod and mesh reinforcement it is essential that the steelwork maintains its shape during mortar application and that thorough penetration of meshes be achieved. The mortar must be kept moist and warm for strength to develop in a curing period which could extend into weeks, and the problem of achieving a consistent quality in the material is made more acute because the building process is highly labour intensive and open to wide variations in quality [13]. Ferrocement offers excellent qualities in water tightness, durability, fire resistance, low maintenance, ease of repair and is becoming more attractive in cost in comparison with steel and timber. Its major drawback is lack of knowledge of its mechanical properties and behaviour under various loading conditions.

2 RESEARCH

The major problem is one of interpretation of results (size of specimen, load application) and correlation with the actual behaviour of the material in a ship's hull in operation. Also ferrocement lacks a universally accepted definition [14,15] and can be made up of a variety of meshes (chicken wire, woven cloth, welded mesh) and rods (mild or high tensile steel) in different quantities and sizes, with mortars ranging in strength from 30 to 80 N/mm^2.

2.1 Tension

Several research papers [12,16–18] report that the ultimate tensile strength is dependent on the strength of the reinforcement in the direction of load and the onset of cracking relies on the dispersion of the steel within the matrix. This important parameter, called specific surface (k), is defined as the surface area of the reinforcement per unit volume of composite with Bezukladov [12] giving a minimum requirement of 0.2 mm^{-1}.

Correct choice of mesh and rod can produce ferrocement skins taking up to 30 N/mm^2 in uniaxial tension while limiting the crack sizes to an acceptable 0.050 mm width.

2.2 Compression

Rao and Gowdar [19] show that the compressive strength depends primarily on the mortar and is independent of the specific surface or the volume of reinforcement when loading in the direction of the mesh layers. This is because at failure the mortar splits longitudinally and the meshes buckle. Increasing k beyond about 0.35 mm^{-1} reduces compressive strength as complete penetration becomes a problem and meshes form layers of weakness.

2.3 Flexure

Simple bending tests on plate specimens show that increasing the specific surface of the steel increases the moment at which cracks first appear, and again, as in uniaxial tension, the crack pattern and width can be calculated and are dependent on the mesh sizes near the surface. Many research workers [3,10,11,20–24] have shown that an adaptation of classical reinforced concrete theory predicts well the uncracked, cracked and ultimate load behaviour of ferrocement in flexure.

Tancreto and Haynes [21] suggest that a 6 mm sized mesh with wire diameter of 0.63 mm is the most suitable reinforcement from considerations of strength, workability and cost.

Lecturer in Civil Engineering, University of Bristol, Bristol, Avon, England

2.4 Modulus of elasticity

Test results show that the law of mixtures holds in calculating E values in tension, compression and flexure [18] but only a little work has been done on creep [12] and evaluation of Poisson's ratio [12,18].

2.5 Impact

The holing of several ferrocement vessels has raised grave doubts about the impact resistance of ferrocement although there is nothing to suggest that other materials would have fared better under similar conditions. Several tests have been used to assess what is an 'acceptable' impact failure from measuring flow of water after damage to examination of cracking patterns and indentation depths on the struck surface. Again several researchers [10,11,25–27] using drop weights or ballistic pendulums generally agree that high tensile strength and a large specific surface is desirable to minimise cracking and leakage in ferrocement after impact damage.

3 CONCLUSIONS

Summaries of research work [5,10,11,28] and development in design and constructional techniques [29,30] show that knowledge of the properties and behaviour of ferrocement still lags a long way behind the use of the material. More intensive studies are needed to give to the naval architect and structural engineer reliable design data.

Tancreto and Haynes [21] state 'The limited knowledge of ferrocement properties is probably the most prevalent factor preventing the widespread application of the material especially in designs traditionally reserved for noncementitious materials', whereas the Concrete Society/RINA joint working party on ferrocement vessels reported in 1973 [31] that it had 'found no reason to suppose that the present ferrocement boat industry is likely to suffer because of the lack of results of experimental research... The prime need at present is for authoritative guidance on the design, construction and supervision necessary to ensure satisfactory performance'.

Hopefully the designer, builder and researcher will collaborate to fulfil the great potential of ferrocement as a structural material.

REFERENCES

1 **Nervi, P.L.** 'Ferrocement: its characteristics and potentialities', *L'Ingenere* ANIAI, Italy No. 1 (1951) (C and CA Translation No Cj 60 7/56)

2 **National Academy of Sciences.** 'Ferrocement: applications in developing countries', *Board of Science and Technology for International Development (BOSTID)*, Office of Foreign Secretary, National Academy of Sciences, Washington DC (February 1973)

3 **Scott, W.G.** 'Ferrocement for Canadian fishing vessels' Vol 1 *Project Report* No 42, Industrial Development Branch Fisheries Service, Department of the Environment, Ottawa (August 1971)

4 FAO Papers presented at 'Seminar on the design and construction of ferrocement fishing vessels', (Session III Construction methods and costs) Wellington, New Zealand 1972. *FAO Fisheries Report* No 131, Vol II. Food and Agricultural Organisation of the United Nations, Rome (June 1973)

5 **Morgan, R.G.** 'Concrete floating and submerged structures', *Report* 54,010, The Concrete Society, London (1973)

6 UNO, 'Boats from ferrocement', *United Nations Industrial Development Organisation* No 10/85, New York (1972)

7 **Nontanakorn, D.** 'Construction of a 32 m ferrocement barge and selected application of ferrocement in Thailand', *FAO Seminar* New Zealand (1972)

8 Lloyds Register of Shipping, 'Tentative requirements for the construction of yachts and small boats in ferrocement', London (1967) – being currently revised

9 **Mahafrey, P.I.** 'Ferrocement', *Technical Note* 21, Cement and Concrete Association of Australia (April 1972)

10 **Greenius, A.W.** and **Smith, J.D.** 'Ferrocement for Canadian fishing vessels', Vol 2, *Report* No 48 (June 1972) – see Ref 3.

11 **Greenius, A.W.** 'Ferrocement for Canadian fishing vessels', Vol 3, *Report* No 55 (August 1972) and Vol 4, *Report* No 64 (1973) – see Ref. 3

12 **Bezukladov, V.F. et al.** 'Ship hulls made of ferrocement', Leningrad (1968) *Navships Translation* 1148

13 **Kowalski, T.G.** and **Walkus, B.R.** 'Concrete technology in the quality control of ferrocement vessels', *FAO Seminar*, New Zealand (October 1972)

14 **Howard, D.J.** 'Basis for a definition of ferrocement', *FAO Seminar* New Zealand (October 1972)

15 **Bigg, G.W.** 'A definition of ferrocement', *FAO Seminar* New Zealand (October 1972)

16 **Walkus, R.** 'State of cracking and elongation of ferrocement under axial tensile load', *Buletinul Institutului Politcnic*, Din Iasi, Poland, Part I Vol XIV (XVIII) (1968), Part II Vol XVI (XX) (1970)

17 **Naaman, A.E,** and **Shah, S.P.** 'Tensile tests of ferrocement', *J American Concrete Institute* 68 No 9 (September 1971) pp 693–698

18 **Lee, S-L. etal** 'Mechanical properties of ferrocement', *FAO Seminar* New Zealand (October 1972)

19 **Rao, A.K.** and **Gowder, C.S.K.** 'A study of behaviour of ferrocement in direct compression', *Cement and Concrete* (October–December 1969) pp 231–237 (not seen by author – reported in Ref. 29)

20 **Chang, W.F. et al.** 'Flexural behaviour of ferrocement panels', *American Society of Civil Engineers Ocean Engineering Conference, Civil Engineering in the Oceans,* Vol II (December 1969) pp 1023–1044

21 **Tancreto, J.E.** and **Haynes, H.H.** 'Flexural strength of ferrocement panels', *Tech Report* R.772, Naval Civil Eng Lab, Port Hueneme, California 93043 (August 1972)

22 **Buchner, T.E.** 'Investigation into the intensity and distribution of cracking in ferrocement panels subject to flexure', *FAO Seminar* New Zealand (October 1972)

23 **Johansen, J.K.** 'Ferrocement for ship hull construction – an alternative solution of the strength calculation quandry?', *Schiff und Hafen* 25 Part 4 (1973) pp 327–331

24 **Logan, D.** and **Shah, S.P.** 'Moment capacity and cracking behaviour of ferrocement in flexure', *J American Concrete Institute* 70 No 12 (December 1973) pp 799–804

25 **Shah, S.P.** 'Ferrocement as a new engineering material', *Report* No 70–11, College of Engineering, University of Illinois at Chicago Circle (November 1970)

26 **Ellen, P.E.** 'Practical ferrocement design – reinforced and post-tensioned', *FAO Seminar* New Zealand (October 1972)

27 **Pedersen, H-P.** 'Panels of ferrocement, glassfibre reinforced polyester and plywood subjected to dynamic point loads', *FAO Seminar* New Zealand (October 1972)

28 **Shah, S.P.** and **Srinivasan, M.G.** 'Strength and cracking of ferrocement', *FAO Seminar* New Zealand (October 1972)

29 **Bigg, G.W.** 'An introduction to design for ferrocement vessels', *Project Report* No 52 (January 1972) – see Ref. 3

30 **Moar, H.M.** 'Ferrocement – the design of its properties to meet the requirements of marine usage', *FAO Seminar* New Zealand (October 1972)

31 **Concrete Society/RINA.** 'Preliminary report of the Concrete Society Working Party', *The Naval Architect* No 4 (October 1973) pp 115–116

The characterization of reinforced thermoplastics for industry and engineering uses

C.M.R. DUNN and S. TURNER

The deformation and strength characteristics of short fibre reinforced thermoplastics are illustrated by results on reinforced polypropylene, nylon and polytetramethylene terephthalate. The mechanical properties of mouldings are anisotropic and vary from point to point, but it is suggested that, even so, a relatively simple system of evaluation should suffice for most purposes. It is proposed that upper and lower bounds of strength and resistance to creep in tension are measured on specimens cut from simple mouldings and that these data are supplemented by tests in flexure and torsion at small strains.

1 INTRODUCTION

The assessment of short-fibre thermoplastics composites by physical tests is a far more complex matter than the assessment of isotropic unreinforced thermoplastics, mainly because there are several extra variables and partly because there is no generally applicable theory. The fibres become oriented during the processing and confer anisotropy on the properties. This in itself is a severe complication but the actual situation is more difficult because the alignment of the fibres is imperfect, variable from point to point and often very different from what simple reasoning would suggest.

The distribution of the fibre alignment, and its variability, cannot be accommodated simply as test variables because the specimen is a structure and its behaviour is determined by the separate properties of the matrix and the filler and by the geometrical conformation of the two. Thus a test on such a specimen does not yield a 'property' that can be transposed unambiguously into a prediction of service performance. This characteristic, coupled with the anisotropy, entails a far larger number of tests than would suffice for the proper evaluation of an unreinforced sample, and it is imperative that a working compromise between comprehensiveness and economy of effort be reached. There are two distinct, though related targets; the rapid assessment of a property for diagnostic or materials selection purposes and the comprehensive evaluation necessary to provide data for the prediction of service performance. In general, the first target is achieved by a range of short-duration tests, of which the most common are those familiarly known as strength and modulus tests, and the second target is achieved by studying those same properties as a function of elapsed time. The properties are sensitive to the temperature, to the chemical environment, to the fabrication conditions and to other variables. A knowledge of the effects of these various factors is a requisite for any serious design exercise but in this paper we make merely passing reference to them, to avoid encumbering the text and because they are partly mitigated by the fibres at the cost of much greater anisotropy.

Imperial Chemical Industries Limited, Plastics Division, Bessemer Road, Welwyn Garden City, Hertfordshire, England

There are many practical applications for which fibre-reinforced thermoplastics are preferred to metals and even some for which they are the only candidates. However, as with unfilled thermoplastics, their suitability for engineering applications is limited by their relativley low strengths and moduli which, despite the fibres, are very much below those of most metals. These particular properties therefore merit the greatest attention, and this paper concentrates on the associated anisotropy, which bestows benefits on the one hand and disadvantages on the other. The published literature on the properties of short-fibre composites has largely failed to discuss the problem, less from any deliberate policy of silence than from an inadequate basis of test results. This paper attempts to rectify the omission but without imposing the array of compliances that are formally required for a full discussion. It is based on recent studies to establish the minimum test programme necessary to provide a *sufficient* evaluation of a short-fibre composite.

There is essentially very little novel in the system of tests that we propose. It has been found that the strain to yield (or fail) is a useful quantity and therefore the use of an extensometer, which can also provide modulus as a function of strain from the same test, is advocated. The use of the ratio of tensile modulus to shear modulus, as described by Bonnin, Dunn and Turner [1], is also recommended; but otherwise the solution lies in the careful choice of simple mouldings and a limited number of specimen geometries.

2 TEST SPECIMENS

The overriding problem in the testing of composites is the choice of appropriate test specimens. It has been standard practice to use ASTM Type 1 tensile bars, or specimens cut from them, for tensile strength and tensile modulus tests and, with a few exceptions, this has had to suffice for most evaluation exercises. The data obtained from such tests present fibre-reinforced materials at their best but the results of Bonnin et al [1], show how misleading a single datum can be. For instance, the fibres in a typical commercial grade of reinforced nylon 66 enhance the 100 s tensile modulus measured in flexure by a factor of more than 6 but the 100 s

Table 1 100 s tensile creep modulus (GN/m^2) at 20°C

Specimen		Strain	
		0.002	0.010
Control, no glass.	Tray, radial	1.73	1.31
	Tray, tangential	1.50	1.23
0.075 volume fraction of glass fibre, average length 50 μm			
	Tray, radial	2.20	1.54
	Tray, tangential	2.29	1.66
	ASTM Bar	2.35	1.77
0.080 volume fraction of glass fibre, average length 200 μm			
	Tray, radial	2.35	1.86
	Tray, tangential	3.09	2.36
0.080 volume fraction of glass fibre, average length 500 μm			
	Tray, radial	2.58	2.08
	Tray, tangential	3.54	2.92
	ASTM Bar	4.45	3.70

*These values are approximate, since accurate measurement of length and its distribution is difficult and tedious.

shear modulus (torsion about the axis of the original bar) is increased by a factor of only 2.5. These factors will depend to some degree on the moulding conditions, as they affect the fibre orientation, but the general picture remains unaltered, the tensile modulus is increased by a larger factor than the shear modulus. This is a straightforward manifestation of the anisotropy caused by fibre alignment and is alone sufficient to raise doubts about the validity of an evaluation programme based solely on the ASTM tensile bar. There are further difficulties; the volume fraction of glass fibres is not uniform through a cross-section and the pattern and the degree of fibre alignment in the bar is not necessarily typical of what would be found in a more practical moulding.

The results of some recent experiments illustrate the latter points rather dramatically. Modulus measurements were carried out on ICI creep specimens [2] cut from ASTM bars and the bases of injection-moulded seed boxes. The seed boxes were gated at the centre on the base and specimens were taken with their axes either along radii or at right angles to them. The polymer was a propylene homopolymer and contained about 0.080 volume fraction of glass fibres which were well coupled to the matrix. There were three compounds, with average fibre lengths 50 μm, 200 μm and 500 μm*. Modulus data are given in Table 1.

The radial and tangential values for the unreinforced polymer are unexceptionable and similar in magnitude to those reported by Ogorkiewicz and Weidmann [3]. The presence of fibres, however, reverses the order of merit, radial specimens having the lower modulus. The anisotropy grows worse as the fibre length increases and so does the anomaly. The probable explanation is that the fibres are not predominantly aligned with their axes lying along radii from the sprue, as might be expected from the most elementary argument; many lie in the tangential direction and a few in the transverse (through thickness) direction. The pattern of orientation varies through the thickness, also, and therefore the moduli measured in tension differ from those measured in flexure and both are affected by the removal of surface layers of material. A specimen cut from an ASTM bar moulded from the sample containing the longest fibres had a modulus measured in flexure of 4.23 GN/m^2 when the tension and compression faces of the test specimen were the surfaces of the original moulded bar, a modulus of 4.62 GN/m^2 after a layer 0.16 mm thick had been machined from each surface and 4.17 GN/m^2 after a further 0.99 mm had been machined from one face to expose a core of rather different appearance. Clearly, for the sample with average fibre length 500 μm, there is poor agreement between the modulus of a specimen cut from an ASTM bar and that of specimens cut from the seed box and hence, presumably, poor correlation of such a modulus with the overall 'stiffness' of an article.

The enhancement factors [4]** calculated from the averages of the radial and tangential data are 1.40, 1.69 and 1.90 in order of increasing fibre length. The 1.40 for the compound with the shortest fibres is significantly better than the value of 1.20 that is obtained from the well known equation for spherical inclusions

$$\frac{G_{\text{comp}}}{G_{\text{matrix}}} = 1 + 2.5\phi + 14.1\phi^2 \qquad (1)$$

where: G_{comp} = shear modulus of composite; G_{matrix} = shear modulus of matrix; ϕ = volume fraction of fibres. The specimens chosen for this investigation are reasonably typical of the seed box as a whole and the average value is probably a good first order approximation, via the geometry, to the stiffness of the article. However, it would be inconvenient for several reasons to use such mouldings as the starting point of a general evaluation, and there is therefore a need for a standard plaque that combines the practical validity of the seed box and the geometrical simplicity of the ASTM tensile bar. Our tentative solution has been to use three mouldings, the ASTM bar, an edge-gated disc and a flash-gated square plaque. The primary intention was that the ASTM bar would be the source of 'along flow' specimens, the disc would be the source of very weak 'across flow' specimens, by analogy with the practice for impact testing [5], and the plaque would provide specimens with a range of fibre orientations. To some degree this expectation has been fulfilled and a self-consistent set of data on strength and modulus emerges, but there are distinct skin and core effects which complicate the picture and emphasise the inescapable point that a test specimen of a composite will not yield basic physical property data because it is itself a structure. Thus, even when the results are derived from several types of specimen, one must question the relevance of the data to specific situations. Essentially, the performance of one structure has to be predicted from the performance of another rather than from basic properties data.

3 DEFORMATION AND STRENGTH UNDER STRESSES OF SHORT DURATION

3.1 Tests in tension

Reference has been made already to so-called 'along-flow' and 'across-flow' specimens. In a sense, this usage implies a regularity of fibre orientation that does not exist in short

---

**Enhancement factor is defined as the ratio of the modulus of the composite to the corresponding modulus of the matrix at the same strain.

fibre composites, as is demonstrated by the data quoted in the previous section. A serious study of the relationship between properties and fibre orientation requires a more quantitative nomenclature, but short-fibre composites fabricated by standard thermoplastics processing machinery do not provide a unique fibre axis as a reference. In this paper we arbitrarily assume that the 'along-flow' specimen from an ASTM tensile bar defines a 0° specimen, corresponding to what would be obtained from a sheet containing a uniaxial array of continuous fibres. With slightly less justification, a 0° direction is defined as shown in Fig. 1 for the rectangular plaque. It transpires that a 0° specimen from a plaque has a lower tensile modulus and a lower strength than the ASTM bar or a specimen cut from it. This conforms to the standard picture of fibre orientation in relation to mould geometry and is in concord with Dimmock's results on 0° specimens cut from different parts of a plaque moulded from glass fibre-reinforced nylon 66 [6]. There was a continuous decrease of modulus from the edge to the centre, 10% decrease for one grade and 30% for another. In our work the 0° specimens are always typical of the material near the edge. The 0° specimen for the edge-gated disc is also defined in Fig. 1.

Although these definitions are far less precise than those possible with a uniaxial array of continuous fibres, the results, which are typified by Figs. 2a and 2b, are surprisingly regular and reasonable, with the one unexpected feature that the disc is not very anisotropic. Modulus and strength both fall to about half as the specimens change from 0° to 90° for the plaque and the ASTM bar. The datum for the 90° specimen from an ASTM bar is rather speculative, since it had to be obtained from a specimen of questionable geometry, but it conforms to the general pattern of the data and to the yield stress of the matrix.

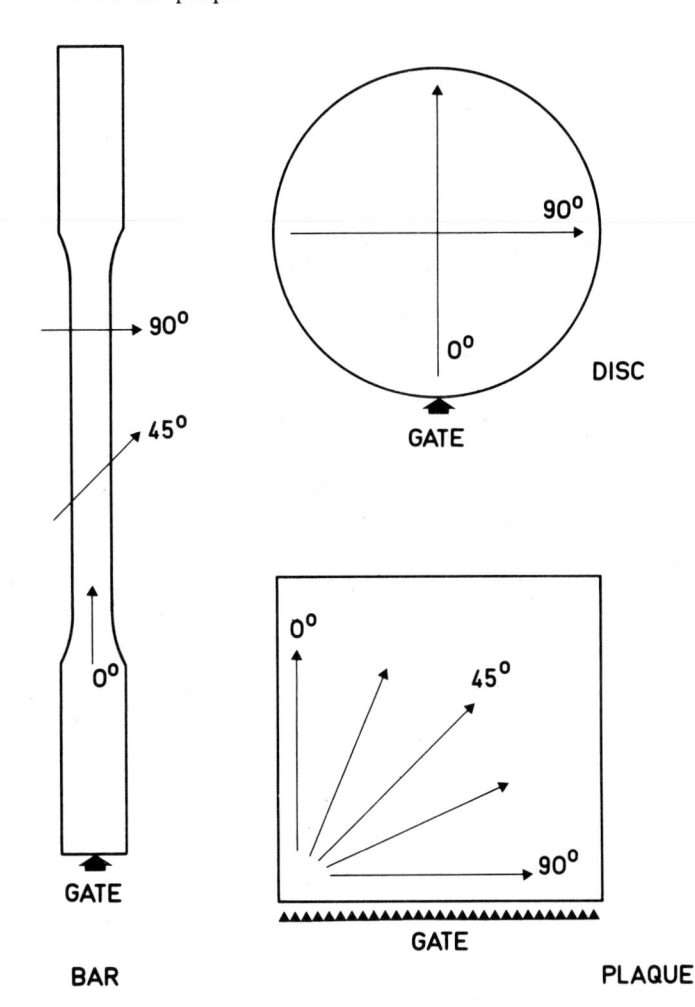

Very similar results have been obtained for other fibre-reinforced thermoplastics. The discs always show a relatively low anisotropy and, in the case of polypropylene, the results are in reversed order to what would be intuitively expected, as were the modulus data in Table 1 for specimens cut from seed boxes. There is a relatively straightforward explanation for this reversal which is given briefly in the Discussion.

It might seem that the plaque would suffice as the sole source of specimens, but the bar is so firmly established as a

× ASTM TYPE 1 BAR
⊙ PLAQUE 150 mm x 150 mm
△ DISC 150 mm DIA.

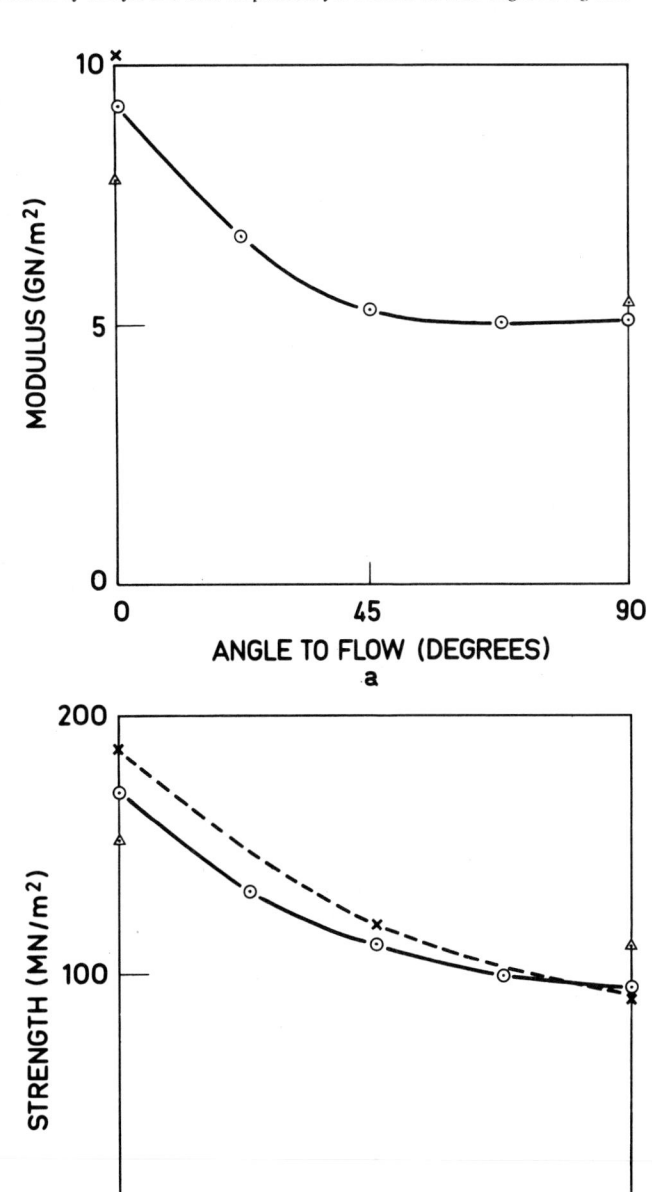

Fig. 2 Anisotropy of modulus and strength: glass fibre-reinforced nylon 66 at 20°C

Fig. 1 Definition of specimens

Table 2 Tensile strengths at 20°C (MN/m^2)

Specimen	Glass fibre-reinforced nylon 66	Glass fibre-reinforced polypropylene
0° bar	187	80
0° plaque	169	–
0° disc	150	58
90° disc	112	79
90° plaque	95	–
90° disc, notched (tip radius 2 mm)	107	–
90° disc, notched (tip radius 250 µm)	84	55
0° disc, notched (tip radius 250 µm)	–	43

test specimen that its exclusion from any evaluation system would be both undesirable and unwise. The disc is also well established as a source of test specimens, the 90° ones of which, wrongly as it now transpires, were used to give a measure of the lowest possible strength in service. It can be deduced from the results that the fibre orientation is fairly random in the plane of the disc, which may correspond better on average to the orientation in practical mouldings than does that in the plaque. Thus for most evaluation purposes tests on 0° bar and appropriate disc specimens may suffice though there are certain hazards of service for which the latter specimens would give optimistic data, eg when the component is so designed that a sharp angle coincides with unfavourable fibre orientation. Such situations are probably accommodated satisfactorily by data from tensile tests on notched specimens. Results for a commercial grade of glass fibre-reinforced nylon, some of which are abstracted from Fig. 2b, and for a commercial grade of glass fibre-reinforced polypropylene are given in Table 2.

One might expect 90° specimens with a notch to be very weak, but the relatively modest reductions that are observed in the strength suggest either that there are a significant number of fibres aligned in a direction conducive to a crack-stopping role or that only a few favourably aligned fibres are needed to delay cracks propagating from a notch.

Apart from the inverted order of merit for the polypropylene disc results, the strength and modulus data are very straightforward and the only imperative requirement for their successful utilization is that the datum chosen should be appropriate to the application. However, past practice in this respect has been far from satisfactory, most of the data presented for such use having been derived from 0° bar specimens. Another analogous deficiency has been that very few results have been published on deformation modes other than tension.

3.2 Tests in flexure and torsion

Ogorkiewicz and Weidmann recently published results for unreinforced polypropylene showing how the shear modulus † and the tensile modulus varied with the angle between the specimen axis and an arbitrary direction of flow [3]. In

† 'Shear modulus' used throughout this paper refers to a modulus measured by torsion about an axis in the plane of the bar, plaque or disc.

agreement with theory, the curve for the shear modulus passed through a maximum at approximately the same angle as that at which the curve for the tensile modulus passed through a minimum. Short fibre composites behave similarly, but the ratios of their tensile modulus to their shear modulus are usually far greater than those for unreinforced polymers, ie the tensile moduli are enhanced by the fibres far more than the shear moduli. Bonnin et al [1] measured the 100 s tensile modulus in flexure $[E_f(100)]$ and the 100 second shear modulus in torsion $[G_t(100)]$ and proposed that the ratio $E_f(100)/G_t(100)$ should be used as a crude measure of anisotropy. The ratio lies between 2.5 and 3.0 for 0° bar specimens of most unreinforced thermoplastics and may be as high as 3.5 for 45° plaque specimens. For such values of the ratio, nothing can be deduced about the degree of anisotropy without an independent measure of Poisson's Ratio but, in contrast, reinforced thermoplastics have typical values of $5.5 \rightarrow 7.0$ for 0° bar specimens, clearly indicative of high degrees of anisotropy. As might be expected, the ratio for some specimens depends on the fibre volume fraction, see Table 3 for examples.

A specimen cut with its axis in the tangential direction from a sprue-gated tray, 280 mm in diameter, moulded from a commercial grade of the same nylon containing 0.18 volume fraction of fibres, corresponds approximately to the 90° disc specimen and has a ratio of 2.57. A specimen cut with its axis in the radial direction, and certainly not corresponding closely to the 0° bar specimen, had a ratio of 3.70. The ratios for specimens cut from a pin-gated tray were very nearly the same.

The results given in Table 3 show that the $E_f(100)/G_t(100)$ ratio for shear and tensile deformation of the 90° specimens is independent of the fibre volume fraction, but give no indication of the degree of enhancement of $E_f(100)$ and $G_t(100)$. This deficiency is rectified by Fig. 3 in which enhancement factor is plotted against volume fraction fibres. The tension data for 90° disc specimens and all the shear data constitute a lower bound and the tension data for 0° bar specimens form what is presumably an upper bound. Glass fibre-reinforced polypropylene shows similar bound characteristics.

4 DEFORMATION AND STRENGTH UNDER STRESSES OF LONG DURATION

Modulus and strength data of the type discussed in the previous sections are not necessarily meaningful in relation to those applications demanding a load-bearing capability over extended periods. In such cases modulus has to be replaced by either a family of creep curves or a family of

Table 3 $E_f(100)/G_t(100)$ for nylon 66 glass fibre composites (20°C)

Volume fraction of glass fibre	0° bar specimens	90° disc specimens
0	2.72	2.69
0.145	4.65	2.69
0.27	5.90	2.59
0.40	6.48	2.48

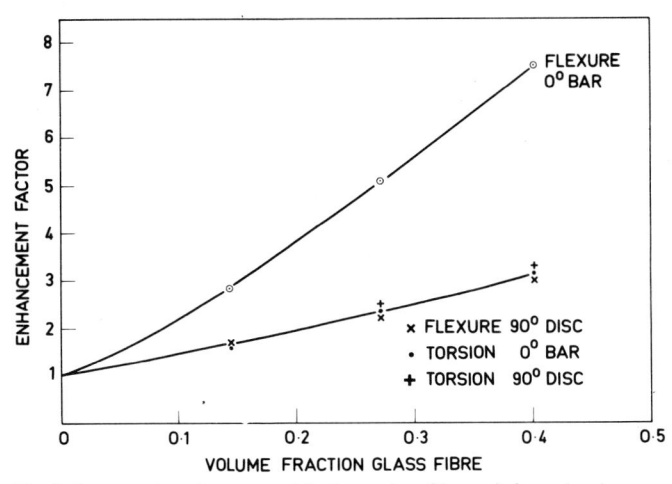

Fig. 3 Low strain enhancement factors: glass fibre-reinforced nylon 66 at 20°C

stress relaxation curves and strength has to be replaced by creep rupture (static fatigue) and dynamic fatigue data. Thus a very large experimental effort is necessary for a complete evaluation if the sample is anisotropic, unless by chance the time dependencies of some of the stiffness coefficients and the compliances are similar to one another. Darlington and Saunders [7] have shown that the creep behaviour of specimens cut with their axes at different angles from a highly anisotropic sheet of unreinforced polyethylene varies widely. Less extreme variations than theirs might be expected for specimens cut similarly from a plaque moulded from a short fibre composite, firstly because the imperfect orientation of the fibres gives a relatively low degree of anisotropy and secondly because the stiffening mechanism is entirely different in the two cases. The latter is an important point; a high degree of molecular orientation radically changes the inherent viscoelastic response time of the material whereas the incorporation of a filler merely dilutes the overall effect, changing the amplitude but not the response time.

In fulfilment of this expectation, it transpires that the creep behaviour of our 0° bar specimens and 90° disc specimens are rather similar, provided proper allowance is made for the dilution effect. If the creep curves are plotted as log strain vs log time the dilution effect appears as a vertical shift. Very close similarity is found between the creep of the 0° bar and the 90° disc creep specimens of glass fibre-reinforced polytetramethylene terephthalate (PTMT) at 140°C, for instance. The curves are plotted in Fig. 4, where the constancy of the shape denotes that the time-dependencies are nearly identical. Reinforced nylon exhibits the same regularity at high temperatures.

There is not such a close correspondence between the creep behaviour of 0° bar and 90° disc specimens at room temperature though the divergence is not serious, see Fig. 5. The latter has virtually the same creep rate as a corresponding specimen of unreinforced PTMT. This is consistent with a simple model of a series assembly of elastic (glass) and viscoelastic (matrix) components for the 90° disc tests and with a part series/part parallel model for the 0° bar tests.

The one obvious difference between the two test conditions cited above is that 140°C is above the glass-rubber transition temperatures (T_g) and 20°C is below it. Above T_g, creep behaviour in partially crystalline plastics is regular and simple, with creep strains up to 0.05 or more representable as the product of a function of time and a function of stress.

Below T_g, the variables in the creep equation are not readily separable, possibly as a consequence of a lower yield strain in the glassy polymer than in the rubbery polymer. Thus the slightly different time dependence for the 0° bar and 90° disc specimens at 20°C might be caused by incipient plastic deformation near the fibres, which would be greater in the series model 90° disc tests.

The different dilution effects for 'along flow' and 'across flow' tests give very different and sometimes widely diverging strain vs log time curves (ie standard creep curves) when the two specimens are subjected to the same stress. Our view is that this is the information needed for design claculations, such pairs of curves constituting an upper, or stiffer, and a lower, or weaker, bound to the load-bearing behaviour of a component subjected to that stress, see Fig. 6. The stiffer bound is probably a genuine upper limit to the load bearing capability of mouldings in service, but there are some uncertainties about the weaker bound because the transverse creep properties of a highly aligned system will approximate to those of the matrix, which is clearly too pessimistic for those situations in which the fibres have a fair degree of random-in-the-plane alignment. On the other hand, the weaker bound must not be too optimistic and we propose to validate it by comparison with the properties of specimens cut from practical mouldings though this step is less simple in practice than it is in principle.

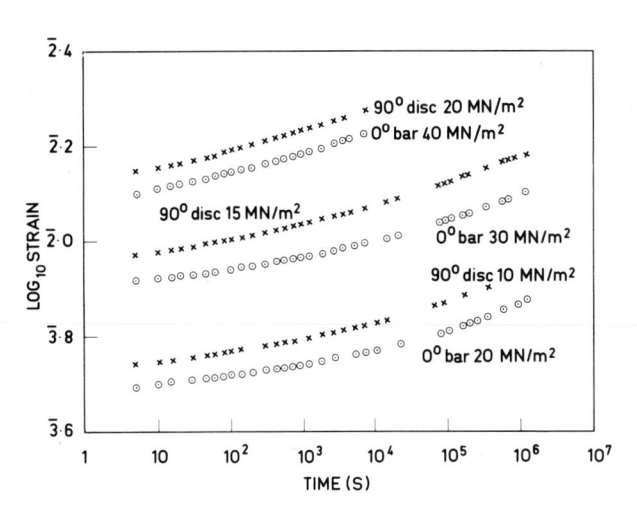

Fig. 4 Tensile creep, glass fibre-reinforced PTMT at 140°C: viscoelastic response is dominated by matrix; dilution effect varies with fibre orientation.

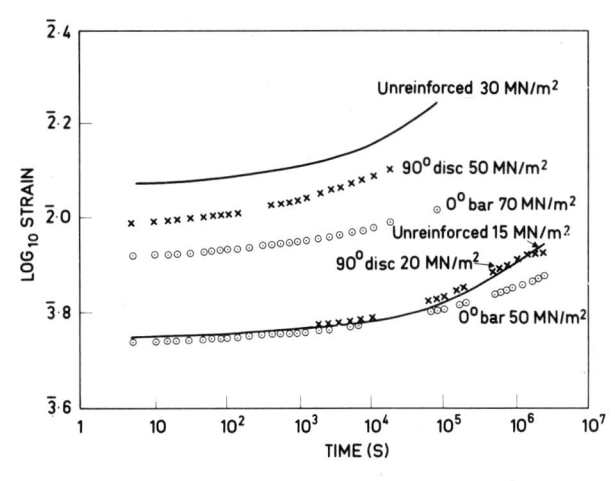

Fig. 5 Tensile creep, glass fibre-reinforced PTMT at 20°C: unreinforced PTMT creeps at a slightly higher rate than the 90° disc specimen

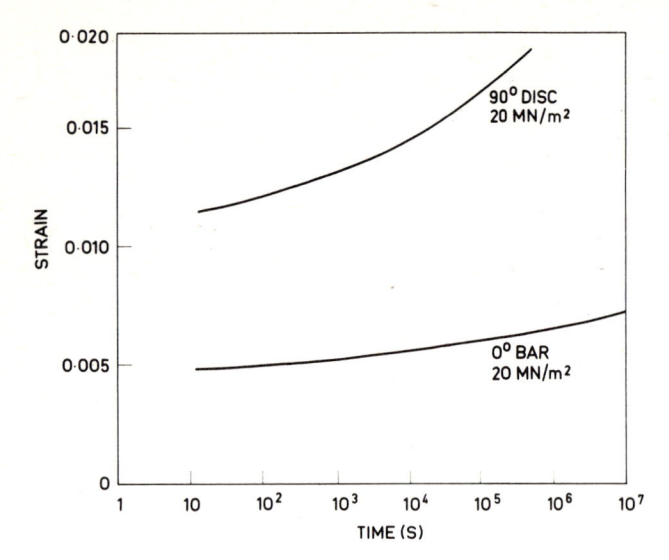

Fig. 6 Bounds to deformation: glass fibre-reinforced PTMT at 100°C

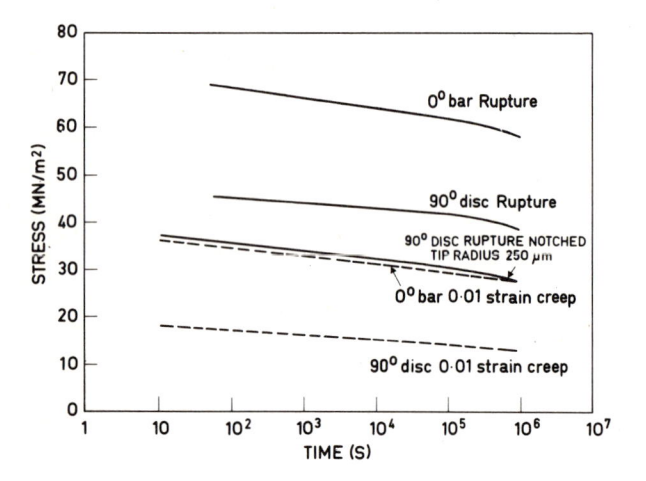

Fig. 7 Creep rupture: a glass fibre-reinforced PTMT at 100°C

These creep curves have to be supported by creep rupture data. The tests are not difficult to perform and the apparatus is far less elaborate than that needed for creep. However, a larger number of experiments are needed for an evaluation of creep rupture than suffices for deformation studies, firstly because failure data show inherent scatter and secondly because an investigation of notch sensitivity, using specimens with different notch geometries is almost obligatory. A notch tip radius of 250 μm seems to be the most useful and a minimum programme seems to require unnotched 0° bar and 90° disc specimens (to accord with the practice for creep), and notched 90° disc specimens. Some results for a glass fibre-reinforced PTMT are plotted in Fig. 7. The specimen geometries were those recommended by Gotham [8].

The relationship between these rupture curves and the creep behaviour is readily apparent if the latter are in the form of isometric curves (stress time sections at constant strain across a family of creep curves). Thus, from Fig. 7 it is obvious that for PTMT the strain at failure is approximately the same for failures at long times as for failures at short times, except for the dip in the curve at the longest times. This dip, an indication of lower ductility, is characteristic of a matrix rather than a composite and its detection is the main objective of creep rupture tests. In the absence of any major change in slope the upper and lower bounds for the long term strength are very simply related to the bounds for the short-term strength and one can then argue that the main considerations in the selection of an overall design strain are the anisotropy and the effect of stress concentrations, with elapsed time often playing a minor role.

It is now well known that lifetimes of thermoplastics, particularly glassy amorphous ones, under fluctuating stress are often significantly shorter than lifetimes under steady stress. Cessna has reported on the dynamic fatigue resistance of glass-fibre reinforced polypropylene [9] and that type of information, or data on creep rupture under intermittent stress as described by Gotham and Turner [10], is a desireable supplement, roughly equivalent in importance to data on the creep rupture of notched specimens under steady stress. Our own experimental study of dynamic fatigue is in its infancy. We have some results from experiments involving flexure, but they are difficult to interpret and of no general significance because of the skin-core distribution of fibre orientation. They are to be supplanted by studies using tension/compression and intermittent tension.

5 DISCUSSION

The use of bounds advocated in the preceding sections represents a significant economy of testing effort when compared with a proper formal evaluation. It can be argued that the bounds are too widely separated, that the weaker one corresponds so nearly to the properties of the matrix that it implies that the fibres confer no advantage whilst the stronger one refers to an extreme case unlikely to arise very often in mouldings. The counter argument is that accurate design practices for anisotropic systems involve such complexities that they are rarely justifiable for purely commercial reasons and therefore a designer needs only approximate data.

A reasonable working compromise for deformation data could be for him to use the mean value of the bounds. This is probably an unsafe practice where strength is the prime consideration, since in general a moulding will break when its weakest point is stressed to its critical level. However, one need not assume a pessimistic position here, since the mould can often be gated in such a way that the fibre orientation is favourable in relation to the direction of the service stresses and the bounds can be calibrated against the results from mouldings or specimens cut from them. Such calibration will serve to highlight the one significant flaw in the system of data discussed here, ie the effect of thickness. As has been demonstrated by the microscopy of sections, by inference from the experimental results and by deliberate experiments, the fibre orientation in the skin layer is very different from that in the core. Very roughly, the fibres tend to lie parallel to lines radiating from the gate in the skin layers and tangential to such lines in the core. There is also some transverse orientation in the core. The proportions of skin and core orientation depend on the mould temperature, the cylinder temperature, the viscosity of the molten polymer, etc, but clearly under constant conditions the proportion of skin orientation will decrease as the specimen thickness increases. Thus data obtained on specimens of a specific thickness will not necessarily be valid for other thicknesses. In particular thick sections will tend to fail in tension at a lower stress than thin sections if the situation corresponds to 0° bar specimens and at a higher stress if the situation corresponds to 90° plaque specimens. A design calculation for flexure is relatively simple, requiring the two bound values in appropriate proportions, but the converse process in which observation of a flexural deformation is translated into a modulus is far less satisfactory.

The procedures and results described in this paper are applicable to, and typical of, all types of short-fibre composite. The examples that have been quoted all refer to glass-fibre reinforced systems in which the coupling between the matrix and the fibre is good, but very similar results are obtained for other types of reinforcement, eg asbestos, though the creep behaviour at long times and high strains tends to be inferior for poorly-coupled systems. As a compensation to this inferiority, poorly coupled systems are often tougher than their well-coupled counterparts, possibly because fibre pull-out contributes as an energy absorbing mechanism.

Other types of reinforcement, platelets such as mica, glass spheres, etc, give less anisotropy and, in general, lower reinforcement factors and enhancement factors at high strains than fibres with a high aspect ratio.

The results quoted here as examples were derived for standard, commercial compounds with volume fractions of between 0.08 and 0.15. Higher volume fractions improve the modulus at low strains more or less in accordance with the parallel model, but the strength is not increased correspondingly because interactions between the more crowded fibres cause the enhancement factor to decrease rapidly with increasing strain and induce failure at lower strain. In glass fibre-reinforced nylon compounds the strain at failure in a conventional tensile test is about 0.026 for 0° bar specimens at a volume fraction of 0.18 (the standard commercial compound) and falls steadily to 0.016 at a volume fraction of 0.40, ie there is a significant reduction of the ductility.

The application of results such as these to design practice is at present tentative rather than established. We suggest that our system of evaluation provides sufficient data for most design purposes and that the development of its use now depends more on the designer than on the materials scientist.

6 ACKNOWLEDGEMENT

The authors wish to acknowledge the care and skill with which C. Toates carried out the many essential experiments upon which this paper is based.

REFERENCES

1. **Bonnin, M.J., Dunn, C.M.R.** and **Turner, S.** 'A comparison of torsional and flexural deformations in plastics', *Plastics and Polymers* **37** (1969) pp 517–522

2. **Mills, W.H.** and **Turner, S.** 'Tensile creep testing of plastics', *Symposium on developments in materials testing machine design,* Manchester, England (September 1965) paper 23

3. **Ogorkiewicz, R.M.** and **Weidmann, G.** 'Anisotropy in polypropylene injection mouldings', *Plastics and polymers* **40** (1972) pp 337–342

4. **Turner, S.** 'Creep studies in plastics', *Applied polymer symposia 17. Mechanical properties and design in polymers* (1971) pp 213–140

5. **Vincent, P.I.** 'Impact tests and service performance of thermoplastics', *Plastics Institute Monograph*

6. **Dimmock, J.** 'The deformational characteristics of anisotropic glass-filled thermoplastics', M.Sc Thesis, Imperial College of Science and Technology, University of London (1968)

7. **Darlington, M.W.** and **Saunders, D.W.** 'The tensile creep behaviour of cold-drawn low density polyethylene', *J Phys D* **3** No. 4 (1970) pp 535–549

8. **Gotham, K.V.** 'A formalised experimental approach to the fatigue of thermoplastics', *Plastics and Polymers* **37** (1969) pp 309–319

9. **Cessna, L.C. Jr.** 'Cyclic creep test data for gr polypropylene', *SPE J* **28** (1972) pp 28–33

10. **Gotham, K.V.** and **Turner, S.** 'Procedures for the evaluation of the long term strength of plastics and some results for polyvinyl chloride', *Poly Eng and Sci* **13** (1973) pp 113–119

QUESTIONS

1 On compressive creep properties, By N.A. Gardiner, (Railko Ltd)

Many 'engineering' thermoplastics are used in compression. Most creep work has been done in tension and shear – what work has been done on compressive creep? Do the tensile and shear results give any indication of what may be expected in compression?

Reply by Dr Turner

Creep in compression has not received much attention. The data that are available show that the resistance to deformation is greater in compression than in tension and therefore conservative design calculations for compression can be based on tensile data. The difference between the moduli in tension and compression is small at low strains but may be of the order of 20% at 0.020 strain. The compressive yield strength is usually higher than the tensile yield strength. We have seen no evidence of anomalies in failure modes under compression.

2 On lower torque-out values obtained with glass filled nylon 66 relative to plain nylon, by K.B. Armstrong (British Airways – Overseas Division)

A few tests at British Airways showed that a 10 VNF thread tapped in glass filled nylon 66 had a lower torque-out value than that of a similar thread in plain nylon. Do you have an explanation for this?

Reply by Dr Turner

This phenomenon is probably attributable to the lower modulus of the unreinforced nylon. Under the cutting action of the tap the unreinforced nylon will deform more than the reinforced nylon and therefore less material will be cut away. Subsequently the deformed material will recover, gripping the tap as it does so and ultimately leaving a smaller hole. A possible contributory factor is that the two materials have different surface textures.

Time-dependence and anisotropy of the stiffness of fibre-plastics composites

R.M. OGORKIEWICZ

Study of the deformational characteristics of composites consisting of glass fibres and thermoplastics or thermosetting resins indicates that their tensile stiffness increases or remains constant with time relative to that of their plastics matrices. It also shows that the number of observations required to determine the anisotropy in stiffness can be minimised by using the theory of orthotropic materials. Considerable anisotropy has been found in shear, as well as in tension, and in some directions the shear stiffness of the composites relative to that of their matrix materials is comparable to the relative stiffness in tension.

1 INTRODUCTION

Reinforced plastics are known as being considerably stiffer, as well as stronger, than unfilled plastics. This knowledge is, however, largely based on their performance under tensile loads of short duration and to a far lesser extent on their behaviour under other types of loading. There is, therefore, a need to examine further whether, or to what extent, their performance under other types of loads might vary from that under uniaxial tension applied for relatively short periods of time.

Even under tensile loads of short duration the relative stiffness of reinforced and unreinforced plastics is known to vary, with strain and with temperature [1]. Moreover, data quoted for fibre reinforced plastics are commonly confined to their performance in the direction of any alignment of the fibres, which may well be very different from that in other directions.

Any such variations are of considerable importance to the users of reinforced plastics. In particular, they need to know more about any variation in their performance with the duration of loading and about their anisotropy, under tensile and also other types of loading. It should be of interest, therefore, to consider some pertinent results extracted from a broad-based programme of work on the deformational characteristics of composites made of glass fibres or spheres and thermoplastics or thermosetting resins.

2 TIME-DEPENDENCE OF STIFFNESS

Because plastics are visco-elastic, reinforced plastics creep to a significant extent, except for composites with continuous fibres where tensile loads can be transmitted solely by the fibres. Otherwise, their moduli decrease with the duration of loading and it is important to know whether the time rate of change of the modulus of reinforced plastics is higher or lower than that of unreinforced plastics, as this shows whether the effectiveness of the reinforcement increases or decreases with time.

Some early work with glass fibre reinforced thermoplastics suggested that the effectiveness of the reinforcement decreases with the duration of loading [1]. However, more

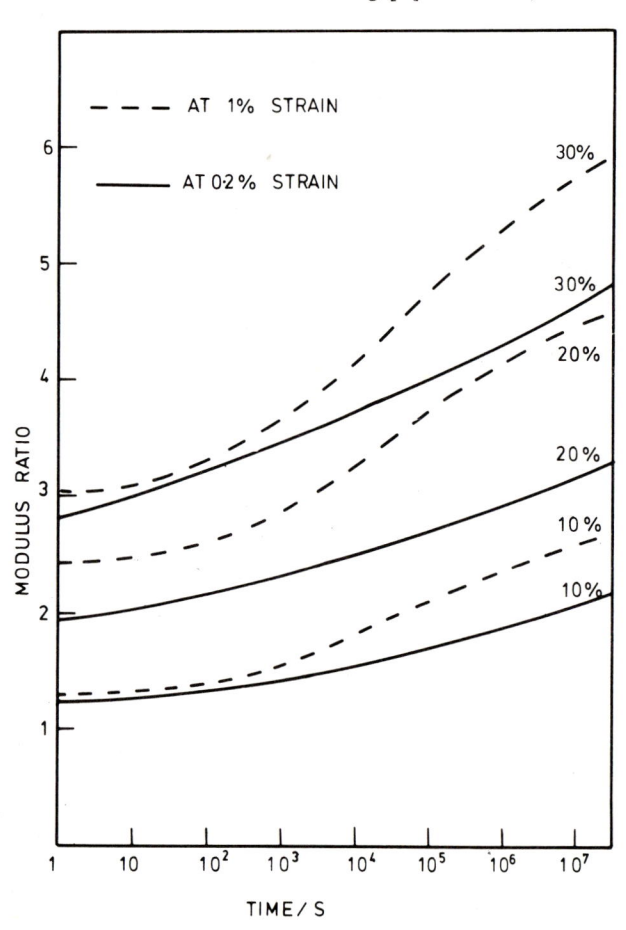

Fig. 1 Variation with time of the ratios of the 100s tensile creep moduli of a polypropylene with 10, 20 and 30 per cent, by weight, of glass fibres to that of the unfilled polypropylene at 0.2 and 1.0 per cent strain

Imperial College of Science and Technology, London SW7 2BX, England recent results do not bear this out. For instance, ratios of the tensile moduli of glass sphere-filled to unfilled polypropylene subjected to constant loads for periods of up to 3×10^6 seconds or approximately 4½ weeks, are independent of time at small (0.2 per cent) strains and at first increase with time at 1 per cent strain [2]. Ratios of the tensile modulus of a unidirectional glass fibre-epoxy laminate at 45° and 90° to the alignment of the fibres to that of the epoxy resin were also found to be substantially constant [3]. Moreover, in the case of glass fibre reinforced polypropylene the ratios of the moduli increase at both 0.2 and 1 per cent strains, as shown in Fig. 1.

All this indicates that the effectiveness of the reinforcement remains constant or even increases with time and that the stiffness of reinforced plastics relative to that of unreinforced plastics is at least as high under long as it is under short loading times.

3 ANISOTROPY

The variation in the stiffness of fibre reinforced plastics with direction poses the problem of making a sufficient number of tests to determine it. Results obtained with several reinforced plastics have shown, however, that the variation in both tensile and shear moduli in a plane can be adequately described by equations based on the theory of orthotropic materials, which call for only three experimental values of modulus, namely at 0°, 45° and 90° to the alignment of the fibres [2,4]. Thus, the variation in the tensile modulus with the angle θ, measured from the direction of the fibre alignment, can be adequately described by the following equation:

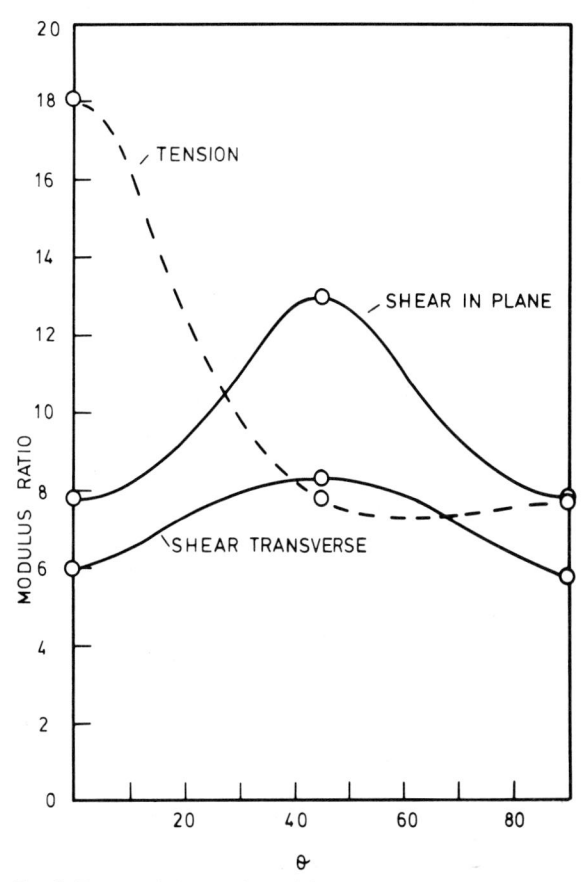

Fig. 3 Ratios of the tensile and shear moduli of a glass fibre-epoxy laminate at different angles to the direction of the alignment of the fibres to the modulus of the unfilled epoxy

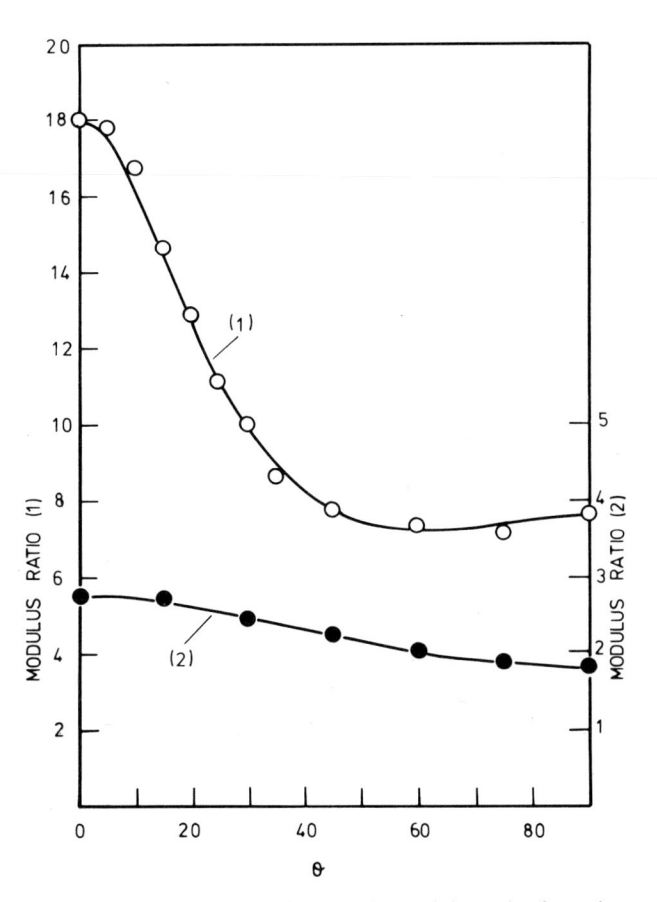

Fig. 2 Experimental values of the tensile modulus ratios for a glass fibre-epoxy laminate (1) and a glass fibre reinforced polypropylene (2) at different angles to the direction of the alignment of the fibres compared with the curves obtained from Equation 1

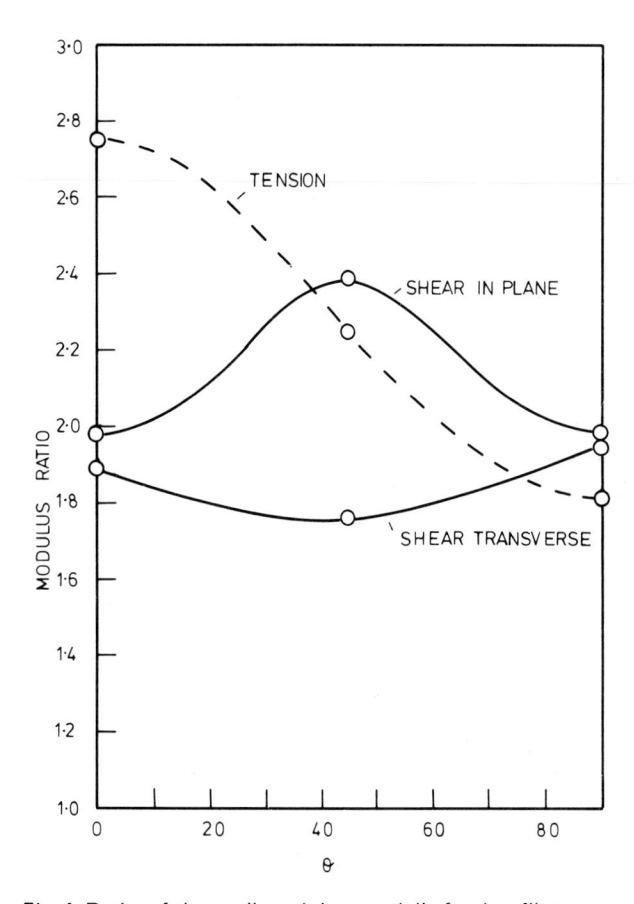

Fig. 4 Ratios of the tensile and shear moduli of a glass fibre reinforced polypropylene at different angles to the direction of the alignment of the fibres to the modulus of unreinforced polypropylene.

$$\frac{1}{E_\theta} = \frac{\cos^4\theta}{E_0} + \frac{\sin^4\theta}{E_{90}} + \left(\frac{4}{E_{45}} - \frac{1}{E_0} - \frac{1}{E_{90}}\right) \sin^2\theta \cos^2\theta \tag{1}$$

where E_θ is the modulus at any angle θ and E_0, E_{45} and E_{90} are respectively the moduli at $0°$, $45°$ and $90°$ to the direction of the alignment of the fibres.

The fit of experimental results to Equation 1 is illustrated in Fig. 2 in terms of the ratios of moduli and is evidently close.

4   SHEAR STIFFNESS

The stiffness of fibre reinforced plastics relative to that of unreinforced plastics has been thought to be much lower under shearing loads than under tension. The ratio of shear moduli of the reinforced and unreinforced plastics is certainly lower than the ratio of the tensile moduli in some directions but in others it is as high or even higher [5,6]. This is particularly so in the plane of the alignment of the fibres and typical results are illustrated in Figs. 3 and 4, which show the ratios of tensile moduli and of shear moduli in the plane of the fibre alignment and normal to it for a glass fibre epoxy laminate and a glass fibre reinforced polypropylene, respectively.

REFERENCES

1   **Ogorkiewicz, R.M.** and **Turner, S.** 'Mechanical characteristics of reinforced thermoplastics', *Plastics and Polymers* **39** (1971) pp 209–213

2   **Weidmann, G.W.** and **Ogorkiewicz, R.M.** 'Time-dependence of the tensile stiffness and anisotropy of a reinforced thermoplastic', *J Mat Sci* **9** (1974) pp 193–200

3   **Weidmann, G.W.** and **Ogorkiewicz, R.M.** 'Tensile creep of a unidirectional glass fibre-epoxy laminate', *Composites* **5** No 3 (1974) pp 117–121

4   **Ogorkiewicz, R.M.** 'Orthotropic characteristics of glass-fibre-epoxy laminates under plane stress', *J Mech Eng Sci* **15** No 2 (1973) pp 102–108

5   **Ogorkiewicz, R.M.** 'Anisotropy in asbestos-filled pvc', *Composites* **3** No 5 (1972) pp 230–233

6   **Ogorkiewicz, R.M.** 'Stiffness and strength of polyester dough moulding compounds', *Composites* **4** No 4 (1973) pp 162–166

The characterisation of glass-reinforced polycarbonate by means of creep-rupture and tensile experiments

J. HUGO

Creep rupture strength and tensile strength at break at different speeds of grips separation in the range of 23° to 130°C have been measured on short glass fibre reinforced polycarbonate. The data have been approximated by use of parametric equations of the 'time-temperature' type. A system of tensile tests is proposed for characterisation of materials as well as of the anisotropy resulting from moulding operation.

1 INTRODUCTION

The anisotropy of mechanical properties of the mouldings made from reinforced thermoplastics may influence the applicability of engineering properties data determined on standard test specimens. There are two areas where the knowledge of realistic properties data should be improved: evaluation of new composites during their development and determination of mechanical properties for a rational choice of materials and design of the parts as well. For both areas the strength limit seems to be a very important property.

The strength limit of polymeric composites is a typical temperature and time dependent property the determination of which requires very much experimental work and time. A relative high variation of rupture times due to differences in properties of single specimens leads to testing of numbers of the specimens. This is the only way, however, to achieve statistically valuable results. Concerning the anisotropy, to evaluate the material completely seems to be a task the economy of which may be disputable.

In order to simplify the system of evaluation we carried out an experimental programme in which several glass fibre reinforced thermoplastics are tested under different conditions of loading in a wide temperature range. The following preliminary results were achieved on commercial type of the glass-reinforced polycarbonate, (Macrolon GV30–Beyer).

2 RESULTS AND DISCUSSION

In Fig. 1 two sets of experimental points are presented. On the right side the points show the times to rupture at different stresses and temperatures of the injection moulded tensile specimens after the action of constant load (loading time approximately 5 s). On the left side the points show the tensile strength at break determined by the tensile testing machine at different speeds of the grips separation. On the time scale there are the times from the start of loading to the break of the injection moulded specimens.

Independent to the fact that the number of creep rupture points is still statistically insufficient, we generalised both sets of data by the parametric equation of Manson and Haferd excluding 130°C (Fig. 2). We found that the quality (concerning variation) of tensile data was much better than that of creep strength data. Every tensile experiment was repeated 5 to 30 times.

Trying to describe the behaviour at normal temperatures from the experiments at elevated temperatures (up to 130°C) by means of different parametric equations we found that

1) the agreement of the calculated data with experimental data is far better for tensile strength experiments than for creep-rupture experiments;

2) generally the Manson-Haferd equation performs better approximation than the equations of Larson-Miller and Sherby-Dorn.

Further we found that, although we have not yet finished the computer program for nonlinear relationship 'log time – log stress', we were able to estimate the creep rupture behaviour from tensile data at elevated temperatures using the equation of Sherby and Dorn. Fig. 3 presents the calculated creep strength values derived from tensile data for 130°, 70° and 40°C. We believe (keeping in mind the differences in the damage cummulation due to different loading regimes as well as the complex influences of the relaxation spectra of the polymer matrix) that the tensile testing at different speeds and temperatures is an economic mean for the estimation of some engineering properties of thermoplastic composites.

Based on these results an experimental programme has been carried out evaluating the influence of eventual anisotropy due to moulding. The specimens taken from mouldings were tested for tensile strength at different speeds and temperatures. Fig. 4 is presented as an example describing the difference between injection moulded specimens and specimens randomly taken from a compression moulded plate (400 × 400mm). Isothermal curves were obtained approximating the experimental data by means of

---

National Research Institute for Materials, Opletalova 25, Praha 1, Czechoslovakia

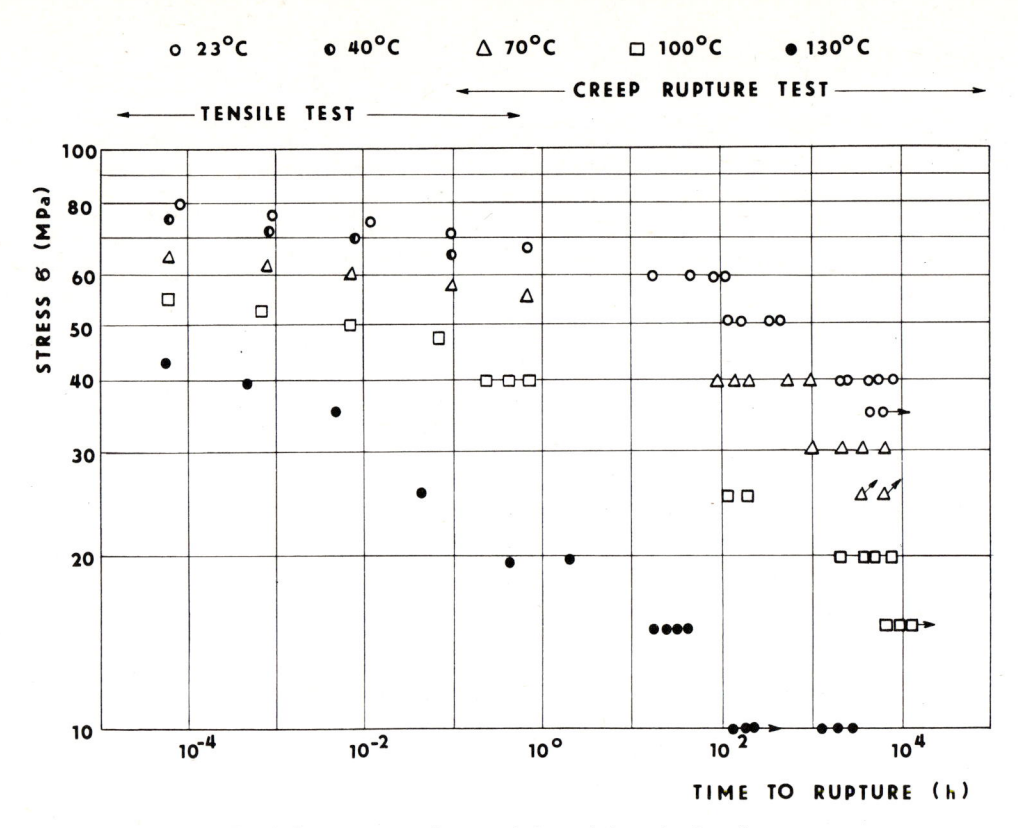

Fig. 1 Stress-rupture diagram of glass-reinforced polycarbonate

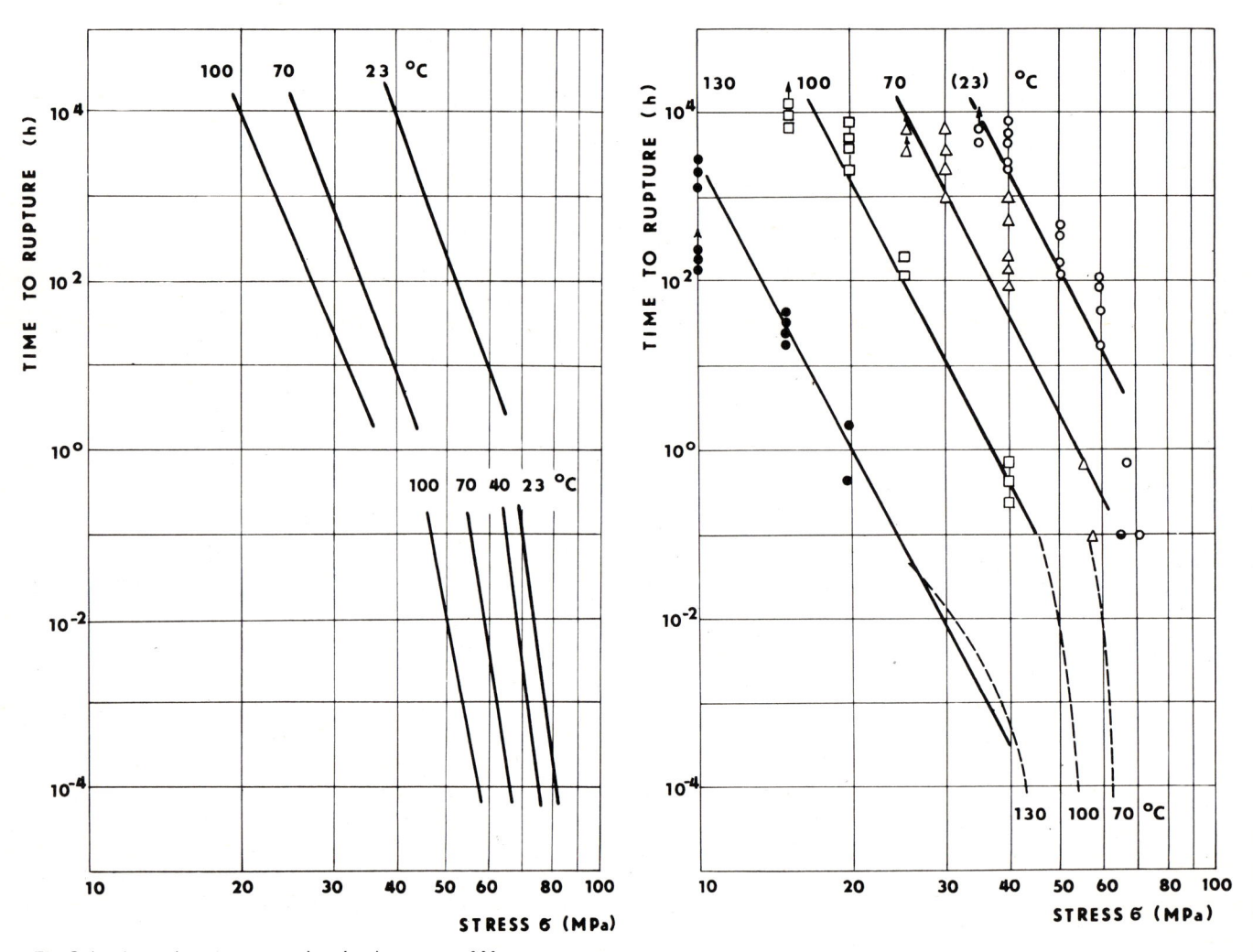

Fig. 2 Isothermal curves; approximation by means of Manson-Haferd equation

Fig. 3 Isothermal curves calculated from tensile test data

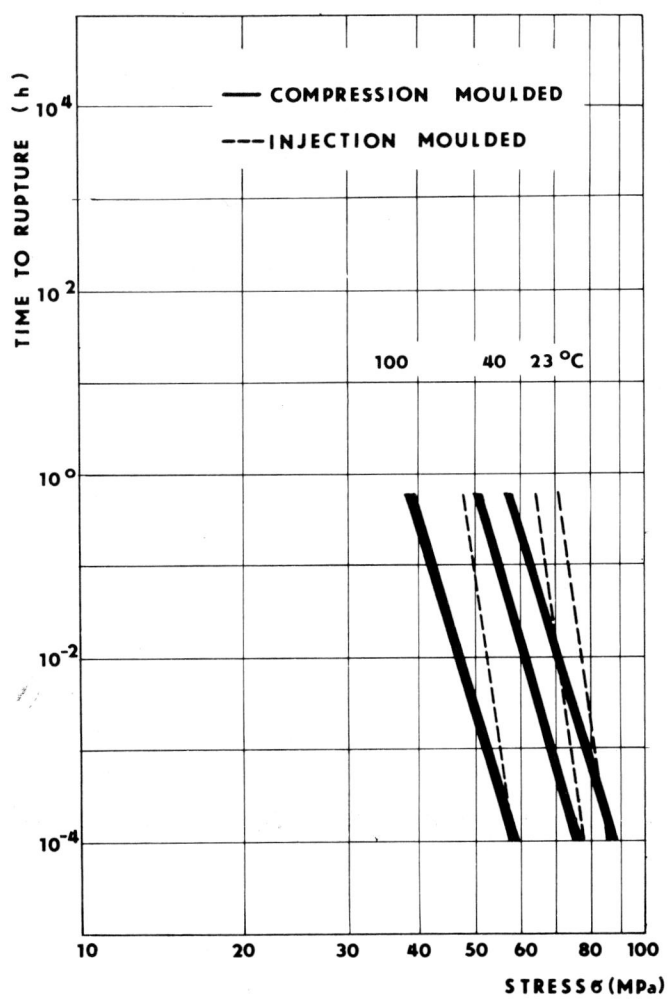

Fig. 4 Isothermal curves for tensile data of specimens prepared by different methods

the Sherby-Dorn equation. The slope of isotherms, being sensitive to reinforcing effect of short glass fibres and/or to time dependent fracture mechanisms of the material, is a kind of more general information than the mean value of tensile strength and their variation determined at standard conditions. Moreover, the change of the creep rupture strength may be estimated within an economical time period.

ACKNOWLEDGEMENT
The author thanks Mr J. Šváb for the active experimental co-operation

APPENDIX
Parametric equations for regression analysis:

Manson and Haferd

$$\log t_B = A_0 + A_1 T + A_2 \log \sigma \cdot T + A_3 \log \sigma \quad (1)$$

$$T = °C$$

Dorn

$$\log t_B = A_0 + A_1 \frac{1}{T} + A_2 \log \sigma \quad (2)$$

$$T = K$$

where T = temperature; σ = stress; t_B = time to rupture.

QUESTIONS

1 On reference for equations and use of Sherby-Dorn equation, by K. Thomas (NPL Teddington)

(1) What are the references for the equations of Manson-Haferd, Larson-Miller, Sherby-Dorn?

(2) Having concluded in the middle of page 123 that '... generally the Manson-Haferd equation performs better approximations than the equations of Larson-Millar and Sherby-Dorn', why is the Sherby-Dorn equation used for estimating creep rupture behaviour from tensile data (Figs. 3 and 4)? It doesn't seem very satisfactory to use one equation for fitting some of the data, and a different equation for fitting the remainder.

Reply by J. Hugo

1. The references are as follows:

1 **Manson, S.S.** and **Haferd, A.M.** 'A linear time-temperature relation for extrapolation of creep and stress-rupture data', *NASA Tech Note* 2890 (1953)

2 **Orr, R.L., Sherby, O.D.** and **Dorn, J.E.** 'Correlations of rupture data for metals at elevated temperature', *Trans ASM* **46** (1954) pp 113–128

3 **Larson, F.R.** and **Miller, J.** 'A time-temperature relationship for rupture and creep stresses', *Trans ASME* **74** (1952) pp 765–775

For further information see also:

Dorn, J.E. 'Mechanical behaviour of materials at elevated temperatures', (McGraw-Hill 1961)

Mendelson, A., Roberts, E. Jr. and **Manson, S.S.** 'Optimization of time-temperature parameters for creep and stress-rupture with application to data from German cooperative long-time creep programme', *NASA Tech Note* D-2975 (1965)

'Methods used for the analysis of tensile and creep-rupture data', Document ISO/TC 17/WG 10/ETP-SG

Harvey, R.P. and **May, M.J.** 'The application of time-temperature parameters for the prediction of long term elevated temperature properties using computerised techniques', paper presented at the *ASM Materials Engineering Congress*, Detroit, Michigan, USA (October 1968)

Goldhoff, R.M. *Materials in Design Engineering* (April 1959) pp 93–97

2. Computing tensile and creep-rupture data as two independent sets of points, the fit of calculated curves with the experimental data is better using the Mansen-Haferd (MH) equation than for the Larson-Miller and Sherby-Dorn equations. On the contrary, the prediction of creep-rupture behaviour from tensile strength data is evidently not very successful because 'log stress–log time' in the MH relation is linear. It is linear in the Dorn equation as well: however, the slope of experimental 'curves' at higher temperatures (near to the glass–rubber transition of polymeric matrix) influences the slope of calculated curves so far that these curves coincide more or less with creep-rupture points.

It is evident that this operation is still very empiric. The way to reach a more exact approximation may lie among others in finding a proper nonlinear relationship 'log stress–log time' which in the form of a parametric equation could be used for all tensile strength (or yield strength) and creep-rupture data in the wide range of temperatures.

Characterisation of fibre composites using ultrasonics

G. DEAN

An ultrasonic technique is reviewed for determining symmetry and measuring the elastic properties of composite materials. Methods for interpreting data to indicate fibre concentration and defect content in continuous fibre systems and the degree of orientation in short fibre systems are discussed and illustrated with experimental results.

1 INTRODUCTION

The presence of directions of preferred fibre orientation in a composite will give rise to anisotropy in elastic properties. The fibre configuration and packing arrangement determine the type of anisotropy, that is material symmetry, and hence the number and identity of the modulus components that must be measured in order to fully characterise the material for its elastic properties. Ultrasonics offers a method for determining material symmetry and all of these components from measurements of wave velocity propagating along known directions in the composite. Since a small diameter ultrasonic beam can be used, measurements may be made on specimens of small size which require little preparation and high accuracy may be achieved for many of the components on material of good quality.

This paper considers the dependence of measurements upon fibre fraction, fibre orientation and the presence of voids, and it is shown how fibre fraction and the degree of orientation may be assessed in uncharacterised material. Where these factors vary within a specimen or component this will be reflected by a variation in velocity from one position to another. Results on inhomogeneous material may then be interpreted to indicate the nature and extent of the inhomogeneity.

2 BASIC PRINCIPLES

The propagation of plane elastic waves in an homogeneous material whose dimensions are large so that the boundary has no influence on the mode of propagation is described by the equations:—

$$(c_{ijkl} n_j n_l - \rho V^2 \delta_{ik}) p_k = 0 \qquad (1)$$

See Reference [1] for derivation. Solutions to these equations give the displacement vector components, p_k, for a wave with phase velocity V and wave normal, n, travelling in a material with stiffness components, c_{ijkl}, and density, ρ. The condition for non-zero values for p_k is:

$$|c_{ijkl} n_j n_l - \rho V^2 \delta_{ik}| = 0 \qquad (2)$$

This equation is cubic in V^2 and the solutions for all wave normal directions may be plotted to give a velocity surface consisting of three sheets, indicating that, along any direction, three types of wave can be propagated whose velocities are given by the lengths of the radius vector in that direction to each of the sheets. For materials of arbitrary symmetry, these sheets are labelled quasi-longitudinal, q-L, or quasi-transverse, q-T, implying that, in general, waves are neither purely longitudinal or transverse but have displacement vectors directed predominantly along or normal to the wave normal.

3 COMPOSITES WITH HEXAGONAL SYMMETRY

A section of the velocity surface for a uniaxial, continuous, carbon fibre composite is shown in Fig. 1 in a plane containing the fibre axis, 3-axis. The material is transversely isotropic so the section is a plane of symmetry and the full surface may be obtained by rotating the figure about the fibre axis. The only non-zero components of the stiffness matrix for this symmetry are, $c_{11} = c_{22}$, c_{33}, $c_{44} = c_{55}$, $c_{66} = \frac{1}{2}(c_{11} - c_{12})$ and $c_{13} = c_{23}$ which give solutions to Equation 2 for the

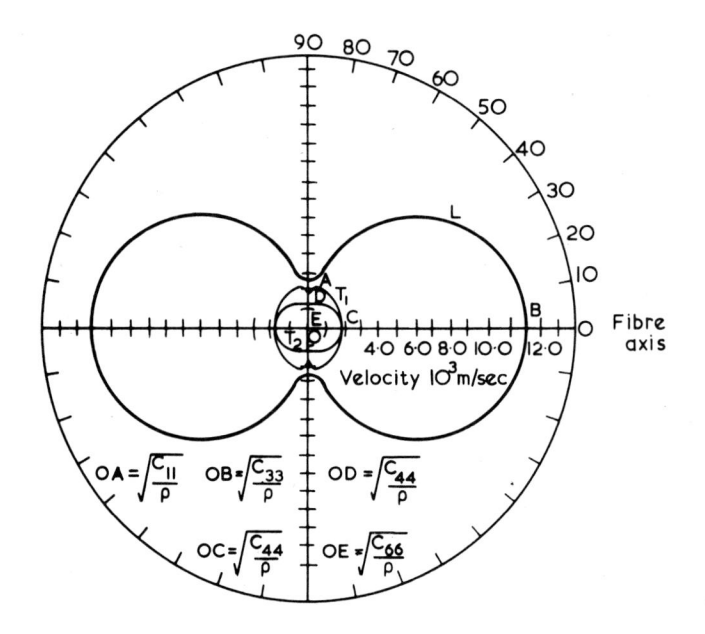

Fig. 1 Section of velocity surface for carbon fibre composite

Division of Materials Applications, National Physical Laboratory, Teddington, Middlesex, England

velocities of waves propagating along the symmetry axes related to a single stiffness component and the density as follows:

$$OA = \left(\frac{c_{11}}{\rho}\right)^{1/2} = \left(\frac{c_{22}}{\rho}\right)^{1/2}$$

$$OB = \left(\frac{c_{33}}{\rho}\right)^{1/2}$$

$$OC = OD = \left(\frac{c_{44}}{\rho}\right)^{1/2} = \left(\frac{c_{55}}{\rho}\right)^{1/2} \quad (3)$$

$$OE = \left(\frac{c_{66}}{\rho}\right)^{1/2}$$

Measurement of material density and of the velocities of longitudinal and transverse waves travelling along and perpendicular to the fibre axis will therefore allow four of the five stiffness components to be calculated. The velocity at some intermediate angle to the fibre axis is related to several stiffness components including c_{13} so that, from a knowledge of one other velocity, c_{13} may be deduced. Although only five velocity measurements are therefore necessary to characterise the elastic properties, greater accuracy may be achieved from a series of measurements as indicated in [2].

An experimental technique for obtaining these velocities has been described in detail [3] and is based upon the measurement of the time difference produced when the specimen is placed in the path of ultrasonic pulses passing between two transducers, one transmitting the other receiving, in a water bath. The time difference is related to wave velocity by the velocity of sound in water and the path length travelled in the specimen. A complimentary technique [4] relies on the measurement of the amplitude of waves reflected from the surface of the specimen to give critical angles of refraction from which wave velocities may also be deduced. Accuracy can be high for the components given in Equation (3) but c_{13}, being less sensitive to wave velocity, is obtained with less certainty [5].

4 COMPOSITES WITH OTHER SYMMETRIES

In principle, by making sufficient measurements so that the whole of the velocity surface may be plotted, it is possible to determine the symmetry and symmetry axes of any specimen. This would involve a large amount of experimental work and analysis which could be greatly reduced if some prior information were known about the symmetry axes. Measurements made in a plane of symmetry demonstrate the direction of symmetry axes, such as a direction of preferred fibre orientation, and the degree of anisotropy within the plane. Fig. 2 illustrates measurements of q-L and q-T wave velocities made in a plane normal to the direction of preferred orientation in a material containing oriented chopped fibres. The data show that the specimen possesses orthorhombic symmetry and indicate the axes of symmetry and hence the plane in which fibres are predominantly misoriented. From measurements made in each symmetry plane, the nine independent stiffness components for materials with this symmetry may be readily deduced. Although materials having lower symmetries may also be studied, the accuracy achieved in the additional components is low as discussed in the following paper.

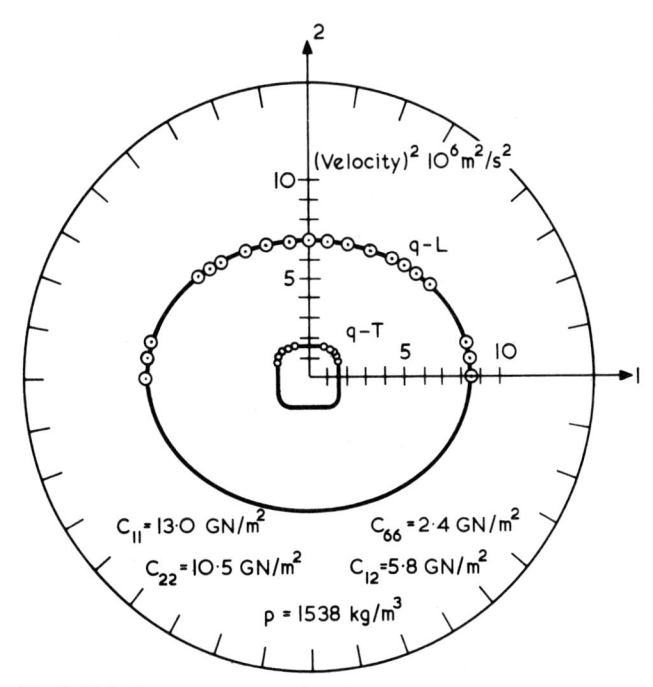

Fig. 2 Velocity measurements in a plane normal to the direction of preferred orientation in a short fibre composite

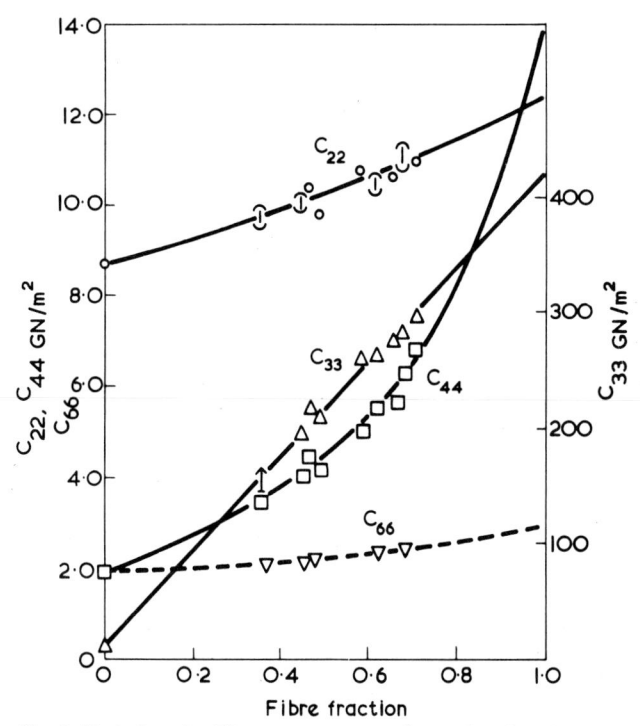

Fig. 3 Variation of stiffness components of type 1 carbon fibre reinforced epoxy with fibre fraction

5 ANALYSIS OF RESULTS ON CONTINUOUS FIBRE MATERIAL

Fig. 3 illustrates the variation with fibre fraction of some of the stiffness components for uniaxial, continuous carbon fibre reinforced epoxy. Specimens were found to be accurately transversely isotropic. The smooth curves are derived from expressions developed by Hashin and Rosen [6] where fibre stiffnesses have been chosen to give the best fit to experimental data. It is apparent that the longitudinal stiffness, c_{33}, is very sensitive to fibre fraction. If properties of the fibre and matrix are known, a knowledge of wave velocity in this direction will, therefore, give an accurate estimate for fibre fraction. The transverse stiffness, c_{22}, is far less dependent upon fibre fraction. The accuracy in velocity measurement for this component is sensitive to the quality of the specimen studied and it is probable that defects within the specimens give rise to the recorded scatter in data. Defects act as scattering centres which give rise to frequency dependent attenuation of the transmitted pulse. Since the pulse is composed of a range of frequency components, the shape is distorted shifting in time the zero cross-over positions which are used to measure delay time. This lowers the measured velocity value even below that expected from the material when allowance has been made for the influence of defects on matrix stiffness. In addition to the reduction in amplitude, observation of the change in pulse width gives a simple indication of the extent to which a specimen is defected and the validity of wave velocity measurements. Analysis of the spectrum of the pulse promises to give information not only on the extent but also on the type, size and distribution of defects. Research along these lines is under progress at NPL.

Because of the small beam diameter, typically one centimetre, and the dependence of wave velocity upon fibre fraction and the presence of defects, measurements made at different positions but along parallel directions can be interpreted to indicate the homogeneity of a specimen or component.

6 ANALYSIS OF RESULTS ON SHORT FIBRE SPECIMENS

For good quality material containing a uniform concentration of fibres, ultrasonic wave velocity values can be interpreted to give an assessment of the degree of orientation in short fibre composites. The interpretation requires a theory which relates composite properties to the misorientation of fibres. The following simple theory is based upon that used to explain mechanical anisotropy in oriented crystalline polymers. A short fibre composite is assumed to consist of a collection of fibre units consisting of a single fibre surrounded by a column of matrix. These units are considered to be transversely isotropic and to have identical properties which are the same as those for a unidirectional continuous fibre composite with the same phase concentration. This assumption allows unit properties to be easily obtained experimentally but takes no account of any dependence of modulus upon length. In composite systems containing stiff fibres in a compliant matrix a length correction may be necessary. The units are allowed to vary in size so that they can be bonded rigidly together to form a void free composite. Fig. 4 shows the misorientation of a fibre unit with respect to reference axes for the composite. The stiffness components of a unit expressed in terms of reference axes for the composite, c'_{ijkl}, and symmetry axes for the unit, c_{mnop}, are related by the transformation:

$$c'_{ijkl} = a_{im} a_{jn} a_{ko} a_{lp} c_{mnop} \quad (4)$$

where the a_{im} expressions etc are direction cosines between the two sets of axes and are functions of θ and ϕ, $a(\theta, \phi)$. Using an assumption of uniform strain which considers that, when the composite is under load, the strain field in each unit is the same as that in the composite, the stiffness components for the composite, $\bar{c}_{ijkl}$, may be obtained by averaging the rotated stiffnesses for the units thus:

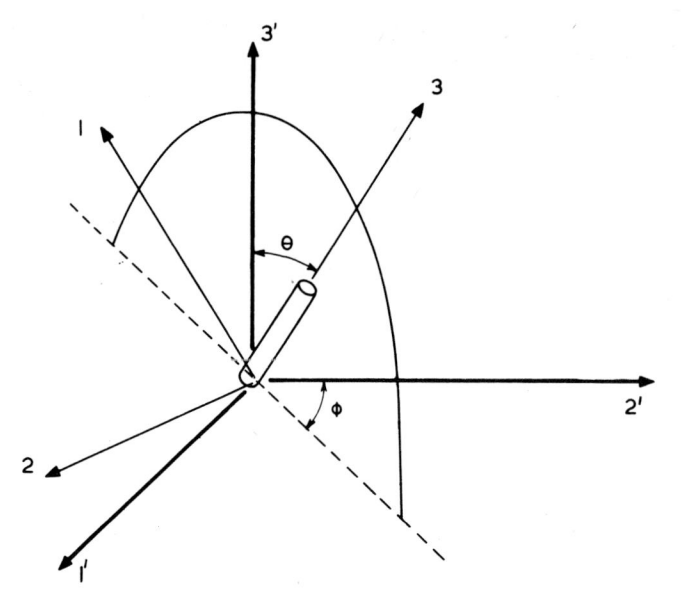

Fig. 4 Orientation of fibre unit axes with respect to reference axes for the composite

$$\bar{c}_{ijkl} = \overline{a_{im} a_{jn} a_{ko} a_{lp}} \, c_{mnop} \quad (5)$$

where the bar over the direction cosines implies averaging over all orientations of the cell axes. The composite properties so deduced give an upper limit to expected values [7] which must lie between this limit and a lower bound given by an assumption of uniform stress where composite properties are derived by addition of compliance components for the units thus:

$$\bar{s}_{ijkl} = \overline{a_{im} a_{jn} a_{ko} a_{lp}} \, s_{mnop} \quad (6)$$

The averages of the orientations are defined by the distribution of fibres, $p(\theta, \phi)$, so that the average value for any $a(\theta, \phi)$ is

$$\overline{a(\theta, \phi)} = \int_0^{2\pi} \int_0^{\pi/2} a(\theta, \phi) \sin\theta \, d\theta \, d\phi \quad (7)$$

$p(\theta, \phi) \sin\theta \, d\theta \, d\phi$ gives the fraction of fibre units lying within the solid angle $\sin\theta \, d\theta \, d\phi$. The magnitudes of the orientation functions, $\overline{a(\theta, \phi)}$, serve to characterise the fibre orientation and these may be deduced from Equations (5) or (6) with a knowledge of the elastic properties of the unit and the composite. This analysis is simplified in the following two cases.

6.1 Case 1

Fibres are randomly oriented about a direction of preferred orientation. The composite then has hexagonal symmetry with five independent stiffness components. The following equations show how some of these are related to unit properties under the assumption of uniform strain, from Equation (5).

$$\bar{c}_{33} = \overline{\cos^4\theta} \, c_{33} + \overline{\sin^4\theta} \, c_{11} + \overline{\cos^2\theta \sin^2\theta} \, (2c_{13} + 4c_{44})$$

$$\bar{c}_{11} = \frac{3}{8}\overline{\sin^4\theta} \, c_{33} + \frac{1}{8}(3\overline{\cos^4\theta} + 2\overline{\cos^2\theta} + 3) c_{11} + \frac{1}{4}(3\overline{\cos^2\theta \sin^2\theta} + \overline{\sin^2\theta})(c_{13} + 2c_{44})$$

$$\overline{c_{44}} = \frac{1}{4}(2\overline{\cos^2\theta\sin^2\theta} + \overline{\sin^2\theta})c_{11} - \frac{1}{4}\overline{\sin^2\theta}\,c_{12} +$$
$$+ \frac{1}{2}\overline{\cos^2\theta\sin^2\theta}\,(c_{33} - 2c_{13}) + \frac{1}{2}(\overline{\cos^4\theta} +$$
$$+ \overline{\sin^4\theta} - 2\overline{\cos^2\theta\sin^2\theta} + \overline{\cos^2\theta})c_{44}$$

where $\overline{\sin^2\theta} = 1 - \overline{\cos^2\theta}$

and $\overline{\sin^4\theta} = 1 - 2\overline{\cos^2\theta} + \overline{\cos^4\theta}$

and $\overline{\cos^4\theta} = \int_0^{\pi/2} \cos^4\theta\, p(\theta) \sin\theta\, d\theta$, etc.

Similar expressions exist derived from Equation (6) for compliance components under the assumption of uniform stress.

6.2 Case 2

Fibres are mis-oriented in the $1'-3'$ plane only giving a composite of orthorhombic symmetry possessing nine discrete stiffness components. Equations (5) then reduce to:

$$\overline{c}_{33} = \overline{\cos^4\theta}\,c_{33} + \overline{\sin^4\theta}\,c_{11} + \overline{\cos^2\theta\sin^2\theta}\,(2c_{13} + 4c_{44})$$

$$\overline{c}_{11} = \overline{\sin^4\theta}\,c_{33} + \overline{\cos^4\theta}\,c_{11} + \overline{\cos^2\theta\sin^2\theta}\,(2c_{13} + 4c_{44})$$

$$\overline{c}_{22} = c_{11}$$
(8)
$$\overline{c}_{44} = \overline{\cos^2\theta}\,c_{44} + \overline{\sin^2\theta}\,c_{66}$$

$$\overline{c}_{55} = c_{44} + \overline{\cos^2\theta\sin^2\theta}\,(c_{11} + c_{33} - 2c_{13} - 4c_{44})$$

$$c_{66} = \overline{\sin^2\theta}\,c_{44} + \overline{\cos^2\theta}\,c_{66}$$

where, in this case, $\overline{\cos^4\theta} = \int_0^{\pi/2} \cos^4\theta\, p(\theta) d\theta$, etc. Under the uniform stress assumption similar expressions again exist in terms of compliance components.

In both cases, the fibre orientation is characterised by the magnitude of the orientation functions $\overline{\cos^4\theta}$ and $\overline{\cos^2\theta}$ and these may be deduced from a knowledge of the properties of the fibre unit and from measured values of at least two elastic moduli for the composite. The unit properties can be obtained either by interpolation of experimental data, as shown for carbon fibre reinforced epoxy in Fig. 3, or, if the fibre and matrix properties are known, from one of the theories giving composite properties in terms of phase values.

It is usually possible to measure a number of stiffness components of any composite system, any two of which can be used to deduce the orientation functions. The derived values, however, are likley to be dependent upon which components are chosen, partly because some components will be more sensitive to orientation but, in particular, because the assumptions of the model may be more valid for some components than for others. In order to give some indication of the reliability of the deduced orientation functions, data have been analysed on some specimens containing oriented, short, type 1 carbon fibres in epoxy resin supplied by the ERDE. Ultrasonic measurements made in a plane transverse to the direction of preferred fibre orientation, see Fig. 2, indicate that the fibres are predom-

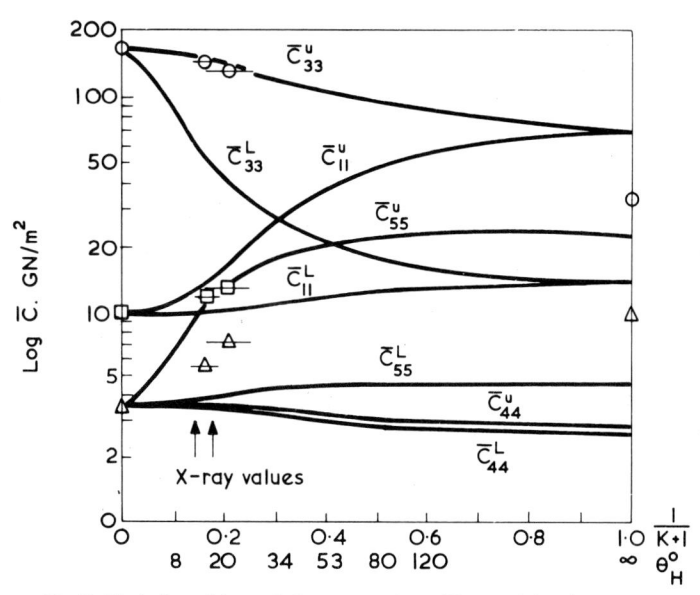

Fig. 5 Variation of bounds for composite stiffness with orientation for fibres distributed according to the function $p(\theta) = Ae^{-k\theta}$

inantly misoriented in one plane as considered in Case 2. Fibre distribution functions within this plane were obtained from x-ray measurements by the deconvolution of crystallite distribution functions in the misoriented material with the function for aligned, continuous fibre material. Although there is some uncertainty in the precise shape of the derived distributions, especially at low values, a function of the form $p(\theta) = Ae^{-k\theta}$ was found to describe, quite accurately, the distribution of fibre axes with angle θ to the direction of preferred orientation. The adoption of a function of this form allows the orientation to be described by the single parameter, k, which is simply related to the half width of the distribution, θ_H, by $k = 2\ln 2/\theta_H$ and enables bounds for the composite stiffness to be evaluated in terms of k from Equations (8) and the equivalent ones in compliance. Fibre unit properties were obtained from the curves in Fig. 3. The variation of these bounds with orientation is shown in Fig. 5. where the orientation is expressed as $1/(K+1)$ which varies between 0 and 1 for unidirectional and random fibre orientations, repectively. Curves are shown for components $\overline{c}_{33}, \overline{c}_{11}, \overline{c}_{55}$ and $\overline{c}_{44}$ since these can be measured with high accuracy.

Experimental data for three materials is given for components $\overline{c}_{33}, \overline{c}_{11}$ and $\overline{c}_{55}$, since it is apparent that these are highly sensitive to orientation. The horizontal lines indicate the range of orientations allowed for each of the aligned materials assuming that experimental values cannot lie outside the bounds. For these materials, since data for $\overline{c}_{11}$ and $\overline{c}_{33}$ fall close to the upper bounds, which have defined the limits of orientation, the size of these limits is small, especially for the more highly oriented material. This indicates that orientation functions deduced from uniform strain predictions for $\overline{c}_{11}$ and $\overline{c}_{33}$ may be considered valid for material containing high fibre alignment but probably become less reliable as the alignment decreases. Data for $\overline{c}_{55}$ fall close to neither bound and are not, therefore, useful for the assessment of orientation. It can be seen that x-ray values for orientation fall within the limits of the ultrasonic predictions but, since the function $p(\theta) = Ae^{-k\theta}$ is not an exact representation of the distribution of fibre axes within the specimens, it is not possible to make precise conclusions on the quantitative comparison of the predictions from both techniques. Materials with near random fibre orientations have moduli close to the geometric mean of the two bounds so that in this range only a qualitative estimate of orientation is possible. If only qualitative estimates are required, these may be obtained very simply from the ratio $\bar{c}_{33}/c_{33}$. The component $\bar{c}_{33}$ is most sensitive to orientation and may be measured rapidly and accurately whilst c_{33} is also easy to derive accurately since it is simply related to known fibre and matrix properties. Accordingly, a knowledge of the variation in $\bar{c}_{33}$ from one position to another in a specimen or component containing a uniform fibre concentration will indicate the extent to which the material is inhomogeneous owing to variations in fibre orientation.

REFERENCES

1. **Musgrave, M.J.R.** 'Crystal Acoustics', Holden Day (1970).
2. **Dean, G.D,** and **Lockett, F.J.** 'Determinations of the mechanical properties of fibre composites by ultrasonic techniques', *ASTM STP* 521 (1973) pp 326–346
3. **Markham, M.F.** 'Measurement of the elastic constants of fibre composites by ultrasonics', *Composites* 1 (1970) pp 145–149
4. **Elliot, J.G.** 'An investigation of ultrasonic goniometry methods applied to carbon fibre composite materials', *Harwell Report* NDT/64 (1973)
5. **Dean, G.D.** and **Turner, P.A.** 'The elastic properties of carbon fibres and their composites', *Composites* 4 (1973) pp 174–180
6. **Hashin, Z.** and **Rosen, B.W.** 'The elastic moduli of fibre reinforced materials', *J Appl Mech* 31 (1964) pp 223–232
7. **Hill, R.** 'Elastic properties of reinforced solids', *J Mech Phys Solids* 11 (1963) p 357

QUESTIONS

1 By C.M.R. Dunn (Research Department, ICI Plastics Division, Welwyn Garden City, Herts)

Have you used beam attenuation to estimate anisotropy?

Reply by G. Dean

No. We have evidence that indicates that attenuation increases with misorientation in short fibre composites presumably because wavelengths are comparable with fibre lengths. However, attenuation used as a parameter for assessing orientation, would not be nearly as sensitive as longitudinal wave velocity and its measurement would require consideration of the path length travelled and the elimination of attenuation be reflection at the surfaces of the specimen and by scattering from other centres such as voids.

2 By Z.A. Raouf (UMIST, P.O. Box 88, Manchester)

Can I ask what frequencies were used?

Reply by G. Dean

Five MHz pulses enable accurate measurements to be made on good quality material and wavelengths are greater than fibre diameters and smaller than specimen diemnsions. Two MHz pulses have been used with slightly voided specimens to reduce pulse distortion caused by scattering and hence increase accuracy.

Anisotropy arising from processing of reinforced plastics

K. THOMAS and D.E. MEYER

Existing NPL equipment for making ultrasonic measurements on plastics has been further developed to permit rapid surveying of industrial components. Examples are given for dough-moulding compound and for glass fibre-filled Deroton (polytetramethylene-terephthalate) of the information which can be obtained. Particular attention has been paid to what can be done completely non-destructively, without cutting sections from the component. Necessary theory and computer programmes for analysis have been developed. It is believed that the technique is valuable for quality control, and because it facilitates determination of the results of processing, it enables improvements in processing to be made.

1 INTRODUCTION

Commercial methods of processing fibre-reinforced plastics often produce anisotropy in a product because the fibres tend to align with the direction of flow of material. This would apply, for example, when considering extrusion of dough-moulding compound (dmc) into a mould, injection-moulding of a thermoplastic, or solid-state forming of a thermoplastic [1]. Ideally the processing might be arranged to align the fibres close to some chosen design direction, so that the reinforcing properties of the fibres were used most effectively. This of course is done when using expensive carbon fibres; but in the case of glass fibre-filled materials the benefits of alignment may not presently justify the extra cost of special processing and of designing with anisotropic material. Thus it has been suggested by Gotham et al that the appropriate design procedure with injection-moulded glass fibre-filled polypropylene is to treat the material as isotropic; and to use for design data the minimum properties of a moulding which models typical processing conditions [2].* This immediately raises the question of what direction gives minimum properties (not necessarily that normal to the apparent flow direction) and naturally these minimum properties are lower than those of random-orientation isotropic material.

The approach to the problem of obtaining minimum property data taken by the above workers was to use fairly large specimens from the moulding, so doing a limited degree of averaging of properties, and to cut these parallel and normal to the evident major flow direction. This is clearly reasonable, because mechanical testing must be held within economic bounds. It should however be noted that cases have been found for *unfilled* polymers where elastic modulus (100s tensile creep) is least in a direction at 45° to the direction of molecular alignment by flow [3] or drawing [4]. A more detailed knowledge of the variation of anisotropy over a moulded article, besides providing greater understanding of conditions, would provide more confidence in the reliability of limited mechanical testing. Perhaps more important is the possibility of then altering processing conditions in order to obtain either less anisotropy, or increased preferred orientation.

Anisotropy measurements can now be readily made, using ultrasonic wave velocity to determine elastic constants. The principles of these measurements have been dealt with elsewhere — eg [1]. Put briefly, where density is known, they yield the elastic constants of polymer sections in the range 2–20mm thick, provided that there are effectively parallel entry and exit faces for the ultrasonic wave. For a full determination of the constants it is best to make longitudinal and transverse (shear) wave velocity measurements over the fullest possible angular range in three orthogonal planes [1,5]. This is easily achieved by cutting suitable specimens from the article, preferably a cube of side about 6mm; but the usefulness of the technique will be much increased wherever truly non-destructive tests on uncut articles are possible. We have therefore explored what can be done in this respect.

The results to be presented are examples for commercially moulded dough-moulding compound, containing 50–55% by weight chalk and 20% glass fibre (BIP material); and for ICI injection-moulded Deroton (TGA 50), which is an engineering thermoplastic having 25% by weight glass fibre in polytetramethylene-terephthalate (PTMT).

2 EQUIPMENT

The ultrasonic measuring equipment used has been based on that developed by Markham [6,7]. We have however carried the development further for making rapid measurements on large components. These need to be placed in a controlled

---

Division of Materials Applications, National Physical Laboratory, Teddington, Middlesex

* Dunn and Turner (this conference) give a more detailed evaluation of the position and propose criteria resulting from fuller testing on mouldings of varied geometry (pp 113-119)

temperature water tank accommodating also the 2.5MHz or 5MHz ultrasonic transducers. We have constructed a tank of internal size 610 × 610 × 406mm (24 × 24 × 16in) and a measurement rig where the component remains fixed whilst the transducers are moved (Fig. 1). The electronic equipment used to time the wave (commercially available from Videoson Ltd) has been supplemented by a unit which digitises all the necessary measurements and reference data, and presents them at a BS 4421 interface (Fig. 2). We use this to provide punched tape for record and computer calculation, or connect the instrument on-line to a computer so that results can be immediately analysed for rapid assessment of a new component. The necessary computer pro-programmes have been developed, and particular attention has been paid to avoiding errors in what otherwise could be complicated measurements by adopting a standard procedure for defining the orientation of the wave in the measured component.

3 ANALYSIS OF RESULTS

It is appropriate to summarise what can now be achieved in the determination of elastic constants. The basic assumption is that the dissipative polymeric material measured is linear viscoelastic for the small strains involved ($\sim 10^{-5}$). The elastic constants can then be exhibited in a 6 × 6 matrix, and it is usual to appeal to thermodynamic theories to show that this matrix is symmetric, reducing the number of independent elastic constants to 21 (see Section 5). This reduction has not been proved conclusively to hold for viscoelastic material [8,9], but since there has been no experimental evidence in the low strain region to indicate asymmetry [9, 10], we may reasonably adhere to it. From this point we can apply the theory developed by Musgrave to relate elastic constants to ultrasonic wave velocities [11, 12]**. The general cubic equation to be satisfied for the longitudinal-type and two transverse-type waves which may travel in a given direction (l, m, n) with individual velocity v is

$$(A - \rho v^2)(B - \rho v^2)(C - \rho v^2) - (\beta\gamma)^2(A - \rho v^2) - (\gamma a)^2(B - \rho v^2) - (a\beta)^2(C - \rho v^2) + 2(a\beta)(\beta\gamma)(\gamma a)$$
$$= 0$$

where ρ is the density of the material, and $A, B, C, a\beta, \beta\gamma, \gamma a$ are functions of l, m, n and of the 21 elastic constants. The relationships are considerably simplified when there is some extra symmetry of the material arising from processing, or from for example regular linear alignment of fibres in an isotropic matrix. In particular, if the material has orthotropic symmetry with three mutually perpendicular planes of symmetry, then there are only nine elastic constants. By measuring wave velocities in the three symmetry planes, it is possible to perform an exact optimization by computer on say 100 data points to calculate these nine constants. This approach differs from the more usual one of making just 9 (or a few more) measurements, in that it gives a more thorough check of the fit of data to the assumed symmetry.

There is a complication in the optimisation analysis caused by the fact that two transverse waves can be activated in the

---

** For a viscoelastic material which formally has complex elastic constants, direct use of measured velocities has to be justified by factors for imaginary components being negligible [13].

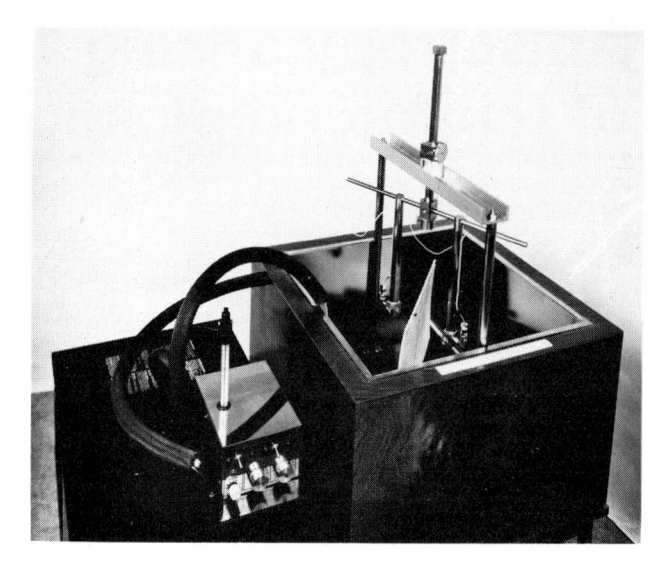

Fig. 1 Constant temperature water tank and rig for ultrasonic measurement of large components. Scale given by foot rule on the side of the tank

Fig. 2 Ultrasonic tank with electronic timing equipment, digitising unit (uppermost), and punched tape output

measured material, although in practice only one is formed. In the case of orthotropic symmetry, the simplifications permit us to exclude mathematically from the analysis the unmeasured wave, and so obtain exact optimisation. It is possible to do this also allowing an extra two (total 11) elastic constants. Beyond this we can extend the analysis on an approximate basis to calculate 15 or the full 21 constants; the process used is better the nearer the symmetry is to being orthotropic, ie the smaller the 12 additional constants.

In the results to be reported it will be seen that these constants are small, and the degree of fit of measured data to calculated solution will be demonstrated. We consider such a solution to be a valuable check on whether the planes of symmetry have been correctly assumed, and on the likely errors associated with using a more restricted set of constants for design purposes.

It may be noted that for purposes of qualitative assessment of anisotropy, the modulus in a particular direction is given roughly by the quantity (material density) $\times V^2$, where V is the velocity of the longitudinal or transverse type wave propagated in that direction. This relation is exact for isotropic material.

4 RESULTS FOR DOUGH-MOULDING COMPOUND

Two industrially moulded articles were measured: one a 15 mm diameter rod of length 100mm, made by a transfer moulding process with the deliberate intention of aligning glass fibres along the axis; the other part of an experimental flange including a brass insert.

When sectioned and polished, the rod (Fig. 3) which defines a set of reference axes) appeared to have a high degree of fibre alignment along the axis, although further examination revealed this to be very variable from place to place, with many fibres not so aligned. Using a cork mask to confine the incident ultrasonic wave to a rectangular area of width 1mm, with its length parallel to the rod axis, it was found possible to make measurements in a transverse plane or a diametral plane of the rod without machining the round face (Fig. 4). By using a plot of inverse longitudinal velocity (slowness) against orientation in a diametral plane, such measurements are sufficient to define the maximum possible axial longitudinal modulus, and hence in this case the maximum possible anisotropy. A smaller, more likely, maximum value is also obtained.

Fig. 5, which is a polar diagram of 1000 (longitudinal-type wave velocity, V_L) versus orientation, demonstrates this for the diametral plane (31). The curve through the solid points measured from the curved surface has to be extrapolated to the axial direction 3, and clearly a long extrapolation is required. However analysis shows that the curve may not contain a point of inflection, because this would produce more solutions for the number of waves which might travel in a particular direction in the plane (31) than is theoretically possible. Hence the straight dashed line drawn in Fig. 5 is a limiting possibility for the required

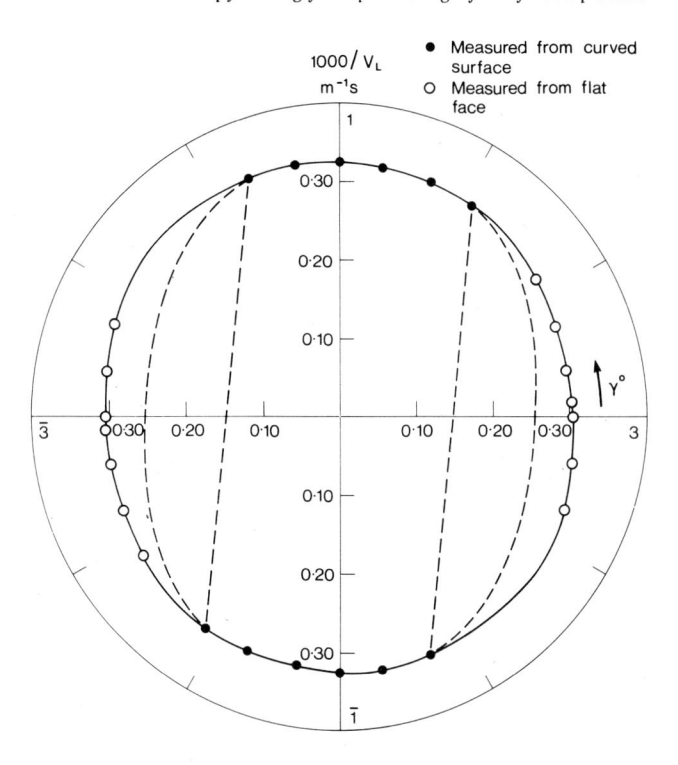

Fig. 5 DMC disc C. Polar plot of 1000/(longitudinal-type wave velocity, V_L) versus wave direction in the diametral plane (31)

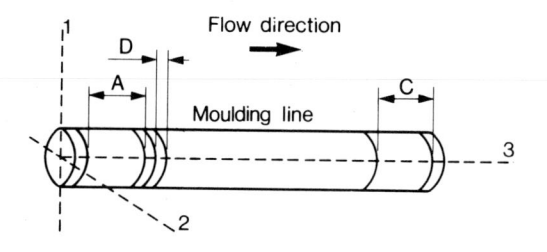

Fig. 3 DMC rod, of length 100mm and diameter 15mm. Reference axes are defined, with direction 3 being the rod axis, and (31) being a diametral plane

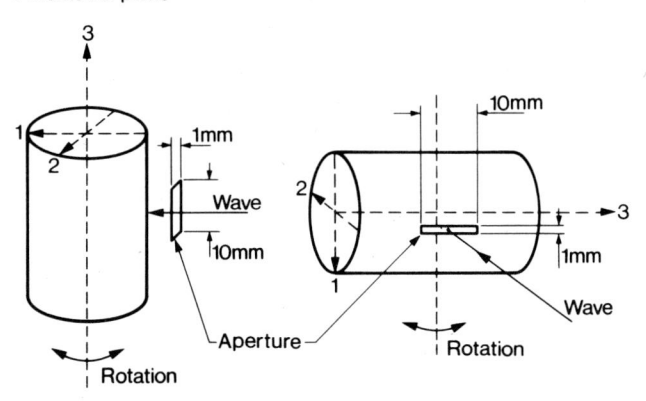

Fig. 4 Measurement of a rod without sectioning, using a cork mask with 1mm aperture to limit the area through which the ultrasonic beam enters the rod. On the left, variation with orientation in a transverse plane (12) is examined; and on the right, a diametral plane (31)

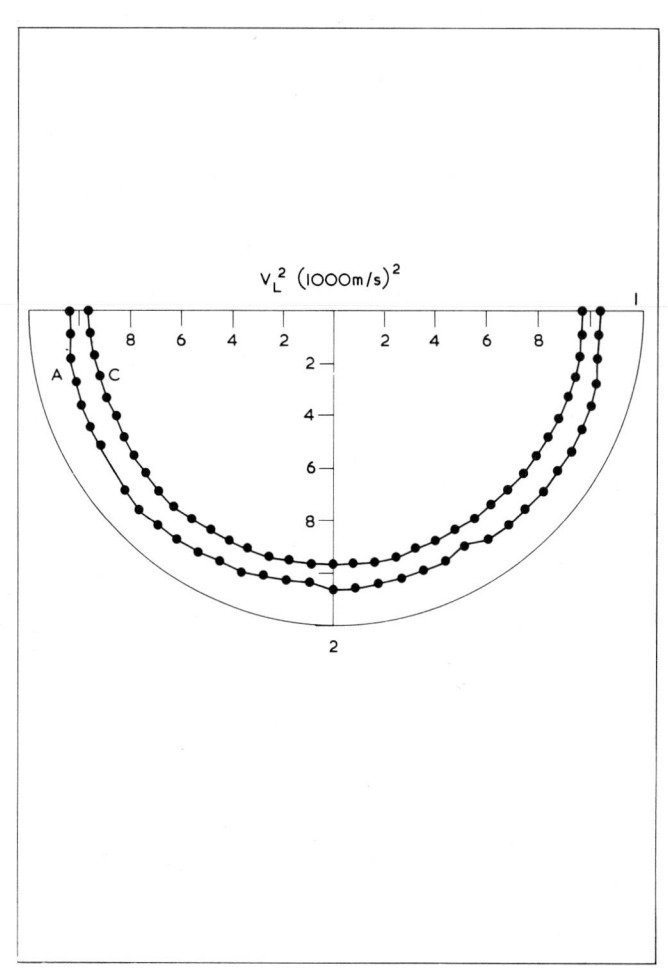

Fig. 6 Polar diagram comparison of V_L^2 values in transverse plane (12) for discs A and C from opposite ends of dmc rod. All values were obtained by measuring through the curved surface, as in the left of Fig. 4. Elastic moduli are roughly proportional to V_L^2, so that disc C has lower moduli. A similar result was found for the diametral planes (23) and (31)

extrapolation; its intersection with the 3-axis defines the minimum possible value of $1000/V_L$, and so the maximum possible value of V_L, in that direction. A more reasonable possible extrapolation has also been drawn. The thickness of the rod and consequent high attenuation of the ultrasonic wave makes this an unfavourable case for extrapolation to obtain the actual value of V_L along the 3-axis. However an example is given in Section 5 (Fig. 12) where the moulding is thinner, measurements can be made without sectioning over a wider angular range, and the extrapolation needed is subject to only a small uncertainty.

For fuller definition of properties, discs were cut from the rod, and measurements between the flat faces of these established the anisotropy. The open circles in Fig. 5 were obtained in this way. The rather unexpected result was obtained that axial longitudinal modulus in the fibre alignment direction is only about 10% greater than the radial modulus. This was a disappointment in one way, for it had been reported that the degree of fibre alignment present had a large effect on the tensile failure behaviour; one could have little hope of correlating a relatively small elastic anisotropy with this behaviour. We were however able to demonstrate a significant 6% fall in elastic moduli (3% fall in wave velocity) along the length of the rod (Fig. 6), accompanied by an improved transmission of a transverse ultrasonic wave and a 2½% fall in density was then found. The cause and effects of this could merit further investigation. Extra measurements were also made at 70°C to indicate the likely degree of variation of moduli with frequency. Only a small change was found (Fig. 7). This is in accord with the finding that the present 5 MHz moduli at 20°C differ very little from values reported for usual 'static' testing [14].

The variations with orientations were $\sim 10\%$, and so it is reasonable to take average values and treat the material as elastically isotropic. On this basis, Table 1 gives the 5 MHz moduli of the rod samples.

Measurements on the flange section (Fig. 8) were aimed at locating the position of regions of local alignment of fibres around the brass insert, which arise from the way the material flows around the insert during filling, and which are found to be the source of mechanical failures. By using a cork mask on the flat surface to define 1.5 mm diameter areas and measuring longitudinal wave velocity through the 12.5–15 mm wall thickness,(Fig. 9), some partial success was obtained, even though the exit face for the ultrasonic wave was curved and not parallel to the entry surface. A region of low modulus through the thickness, which would correspond to fibres tending to align parallel to the surface, was found; optical examination showed local alignment of fibres in this region. However the ultrasonic result could not be confirmed on machining the section to the nominally preferable form with parallel wave entry and exit faces. It was apparent optically that regions with parallel fibres extended for only ~ 2 mm, and then gave way to a region of different orientation. This heterogeneity sufficed to prevent the identification of local oriented regions because of the averaging effect over a pathlength of say 10 mm; but as a corollary, the measurements do yield average properties over this distance. Our conclusions from all the dmc results are as follows:

Table 1 Average moduli for dmc rod

Specimen	Young's modulus E, 10^{10} N/m²	Shear modulus G, 10^{10} N/m²	Bulk modulus K, 10^{10} N/m²	Poisson's ratio
Disc A at 20°C	1.72	0.69	1.16	0.25
Disc C at 20°C	1.61	0.65	1.05	0.25
Disc D at 20°C	1.69	0.67	1.24	0.27
Disc D at 70°C	1.52	0.61	1.07	0.26

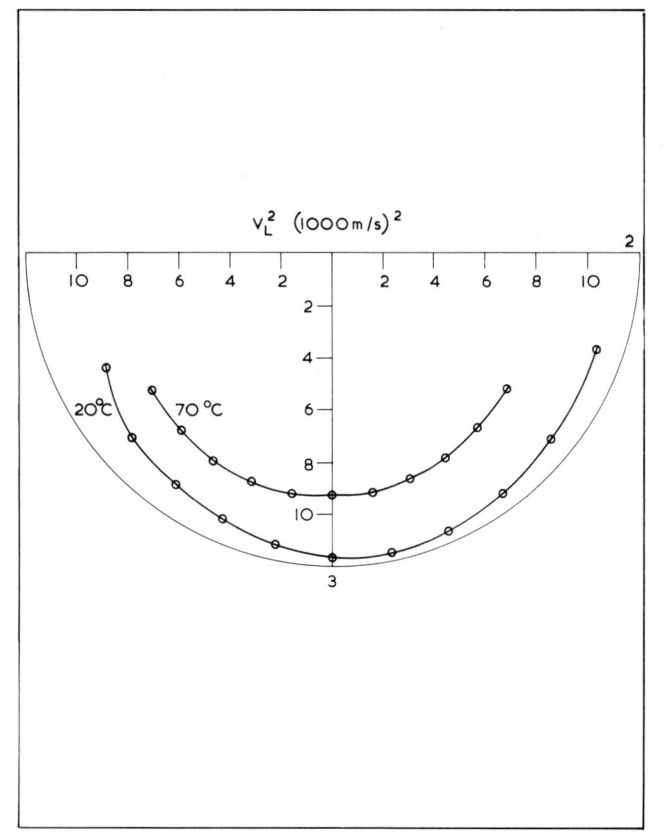

Fig. 7 Polar diagram comparison of V_L^2 values at 20°C and 70°C in diametral plane (23) of thin disc D from dmc rod. All values were obtained using wave entry at the flat face, because attenuation is high at 70°C and a short wave-path through the disc must be used. Moduli are roughly proportional to V_L^2 and so are lower at 70°C

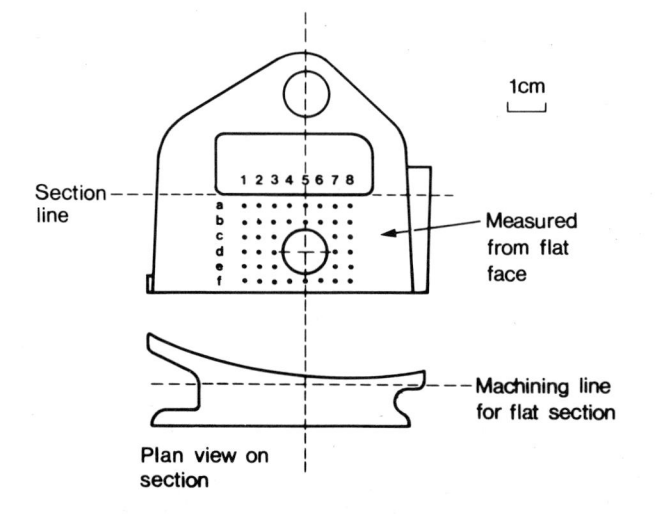

Fig. 8 Part of a dmc flange, showing measuring positions around a circular brass insert, the positions being defined by a cork mask with 1.5 mm diameter holes

Anisotropy arising from processing of reinforced plastics

With curved surface

	1	2	3	4	5	6	7	8
a	10.0	8.2	9.5	9.1	9.3	10.0	9.4	8.8
b	9.0	7.5	8.5	8.8	9.4	9.0	8.9	8.5
c	8.1	8.3	9.0				8.5	8.3
d	8.4	8.4	9.1				8.5	8.0
e	8.3	8.2	8.3				8.1	8.2
f	8.1	8.1	8.5	9.0	8.9	9.0	8.2	8.6

Curved surface removed

	1	2	3	4	5	6	7	8
a	8.6	8.3	8.2	8.1	8.1	8.8	8.2	7.7
b	7.5	7.5	7.6	7.8	8.2	8.1	8.1	7.4
c	7.5	7.8	7.9				8.4	7.6
d	7.5	6.8	7.9				7.5	7.4
e	7.4	7.5	7.4				7.4	7.4
f	8.4	6.9	7.9	8.3	8.5	7.8	8.0	8.4

Fig. 9 Values of V_L^2 through a dmc flange, at positions around a brass insert; units are $(1000 \text{m/s})^2$. The low modulus region outlined was found, but could not be confirmed after machining the flange to give a flat section as in Fig. 8. Optical examination showed local alignment of fibres in the outlined region

(i) 5 MHz ultrasonic wave velocity measurements are possible on dmc samples up to 15 mm thick. However this thickness severely restricts transverse wave measurements, and gives longitudinal moduli which are probably 8% too low due to distortion of the wave form. Allowing for the necessity of averaging inhomogeneity, a pathlength of about 10 mm is preferred for determining average moduli.

(ii) 5 MHz moduli measured at 20°C are about 10% greater than equivalent 'static' moduli.

(iii) The transfer-moulded dmc rod was inhomogeneous, both over distances of a few mm, and from one end to the other. It is suggested that there was a deficiency of filler at one end, giving 6% lower moduli. A transverse plane of the rod was effectively isotropic. There was a strong tendency for fibres to be aligned parallel to the axis, particularly near the surface of the rod, but the degree of alignment produced only 10% increase in modulus along the axis. This variation is insufficient to give real hope of establishing a useful correlation with tensile failure properties.

(iv) Because the dmc casing section was inhomogeneous, the ultrasonic measurements could not detect regions of local orientation which can be observed optically.

(v) The main value of the measurements is:

(a) to give a non destructive measure of elastic modulus and of anisotropy;

(b) to detect general changes in property over a relatively long distance in the direction of flow of material (eg between opposite ends of a 100 mm rod);

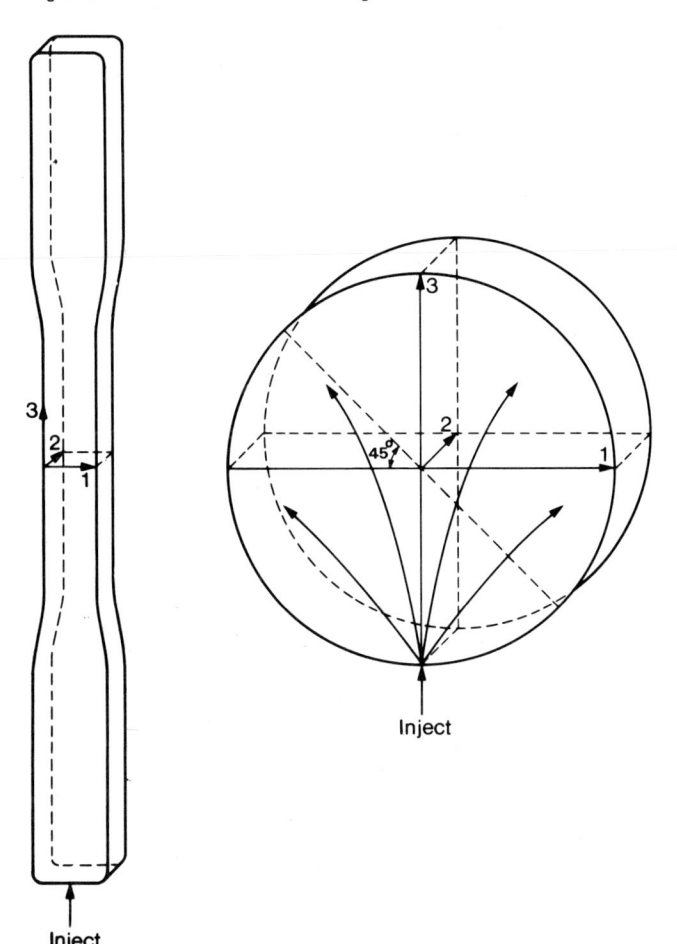

V_T^2 through thickness, at 30° to normal, $10^6 \text{ m}^2 \text{ s}^{-2}$	V_L^2 through thickness, $10^6 \text{ m}^2 \text{ s}^{-2}$	V_L^2 through width, $10^6 \text{ m}^2 \text{ s}^{-2}$
1.27	6.19	
1.22	6.16	
1.22	6.13	
1.22	6.12	
1.26	5.92	
1.27	6.07	
1.21	6.15	
1.19	6.11	6.37
1.19	6.10	6.39
1.19	6.10	6.38
1.19	6.09	6.39
1.19	6.10	6.40
1.19	6.11	6.44
1.19	6.13	6.49
1.19	6.14	
1.22	6.15	
1.26	6.20	
1.24	6.29	
1.25	6.30	

Fig. 10 Deroton injection mouldings modeling typical moulding situations. On the left is a tensile test bar of lath shape; on the right an edge-gated disc. Reference axes are defined, with direction 3 being the main flow direction

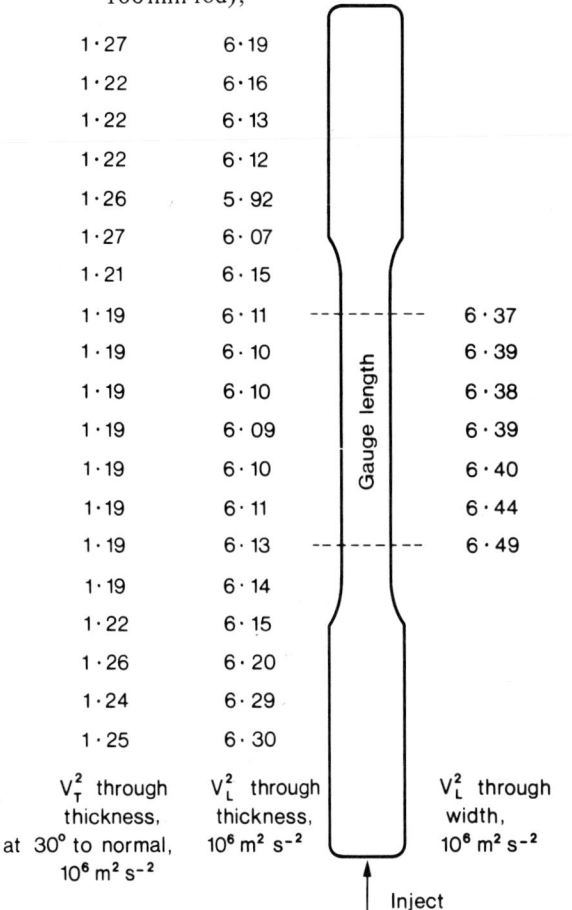

Fig. 11 Check of uniformity of a Deroton lath

COMPOSITES — STANDARDS, TESTING AND DESIGN 135

(c) to form a non destructive quality control test for reproducibility of average properties. This should be particularly valuable with a material known to be liable to wide variations.

5 RESULTS FOR INJECTION-MOULDED GLASS FIBRE-FILLED DEROTON

The samples used were in the form of an end-injected tensile bar, 3 mm thick and 217 mm long, and an edge-injected disc of thickness 3 or 6 mm, and diameter about 100 mm (Fig. 10). These were chosen to model typical situations in practical moulding.

The initial aim with the tensile test bar was to establish the degree of uniformity along the gauge length by measuring the ultrasonic wave velocity in several chosen directions, and Fig. 11 shows that this was very good (V_T is the transverse-type wave velocity). Such a check would be particularly important in the case of creep test specimens. We have also demonstrated the extent to which it is possible to make a completely non destructive analysis of the elastic anisotropy in the gauge length, obtaining the results given in Table 2 for the 2.5 MHz, 20°C elastic constants c at the centre of

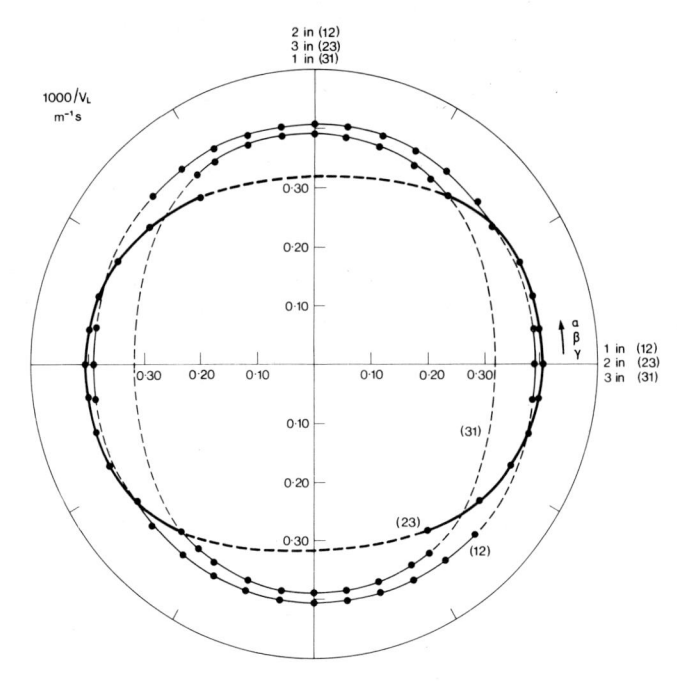

Fig. 12 Centre of Deroton lath. Polar plot of $1000/V_L$ versus wave direction for the planes (12), (23) and (31). The best extrapolation into the inaccessible direction 3 is obtained in the plane (23), shown as a heavy curve. Extrapolation for the plane (31) is also given. Plane (12) requires little extrapolation

the bar (presented in usual matrix notation, with units of $10^9 N/m^2$; the flow direction is designated 3, the width direction is 1, and thickness direction 2 completes a right set of axes). The necessary measurements were made in the planes (12), (23) and (31) shown in Fig. 10, covering as wide an angular range as possible without sectioning. Fig. 12 demonstrates (as already mentioned in Section 4) that a polar plot of 1/(longitudinal velocity, V_L) permits a very good extrapolation into the axial direction 3 of the bar; and the optimisation analysis of data to obtain the set of c values is correspondingly good. This favourable situation occurs because the 3 mm thickness of the moulding permits a wide orientation range to be measured.

The constants c permit stresses σ resulting from various elastic strains ϵ to be calculated using the usual set of six equations

$$\sigma_i = \sum_{\substack{i = 1-6 \\ j = 1-6}} c_{ij} \epsilon_j$$

It is normally more convenient to use the related set of equations:

$$\epsilon_i = \sum_{\substack{i = 1-6 \\ j = 1-6}} s_{ij} \sigma_j$$

The required s values are obtained by inverting the matrix of c values, to obtain the results given in Table 3.

The extra 12 constants in the full solution are small. The only notable change in shape of a plot of (longitudinal velocity, V_L)^2 against orientation due to these extra constants, occurs in the plane (23). Fig. 13 shows the fit of experimental data to the $9-c$ (full curve) and $21-c$ solutions in this plane. The asymmetry represented by the extra c values permits a closer fit of calculated curve to data for the $21-c$ solution.

The small practical effect of the extra c or s values is demonstrated in Fig. 14. The situation considered is the

Table 2 Elastic constants c measured non destructively at the centre of a Deroton lath: units are $10^9 N/m^2$

Matrix notation

c_{11}	c_{12}	c_{13}	c_{14}	c_{15}	c_{16}
c_{12}	c_{22}	c_{23}	c_{24}	c_{25}	c_{26}
c_{13}	c_{23}	c_{33}	c_{34}	c_{35}	c_{36}
c_{14}	c_{24}	c_{34}	c_{44}	c_{45}	c_{46}
c_{15}	c_{25}	c_{35}	c_{45}	c_{55}	c_{56}
c_{16}	c_{26}	c_{36}	c_{46}	c_{56}	c_{66}

Orthotropic solution (9 c values)

9.96	5.94	7.40	0	0	0
5.94	9.31	7.57	0	0	0
7.40	7.57	14.02	0	0	0
0	0	0	1.27	0	0
0	0	0	0	1.75	0
0	0	0	0	0	1.69

Full solution (21 c values)

9.96	5.92	7.40	−0.15	0.03	0.02
5.92	9.28	7.64	−0.09	0.04	0.10
7.40	7.64	14.08	−0.40	0.24	0.08
−0.15	−0.09	−0.40	1.33	−0.10	−0.01
0.03	0.04	0.24	−0.10	1.73	0.15
0.02	0.10	0.08	−0.01	0.15	1.75

result of applying a simple tensile stress σ_3 in the axial direction. It is found for example that the resulting axial extension ϵ_3 is accompanied by a smaller shear strain ϵ_4 in the plane (23), which would not occur if the bar had purely orthotropic symmetry.

Table 3 Inverted elastic constants s for Deroton lath: units are $(10^9 N/m^2)$

Matrix notation

s_{11}	s_{12}	s_{13}	s_{14}	s_{15}	s_{16}
s_{12}	s_{22}	s_{23}	s_{24}	s_{25}	s_{26}
s_{13}	s_{23}	s_{33}	s_{34}	s_{35}	s_{36}
s_{14}	s_{24}	s_{34}	s_{44}	s_{45}	s_{46}
s_{15}	s_{25}	s_{35}	s_{45}	s_{55}	s_{56}
s_{16}	s_{26}	s_{36}	s_{46}	s_{56}	s_{66}

Orthotropic solution (9 c values)

0.188	−0.070	−0.061	0	0	0
−0.070	0.217	−0.081	0	0	0
−0.061	−0.081	0.147	0	0	0
0	0	0	0.787	0	0
0	0	0	0	0.571	0
0	0	0	0	0	0.592

Full solution (21 c values)

0.187	−0.069	−0.061	−0.001	0.006	0.004
−0.069	0.221	−0.084	−0.018	0.008	−0.009
−0.061	−0.084	0.150	0.031	−0.016	0.000
−0.001	−0.018	0.031	0.763	0.040	0.001
0.006	0.008	−0.016	0.040	0.587	−0.050
0.004	−0.009	0.000	0.001	−0.050	0.583

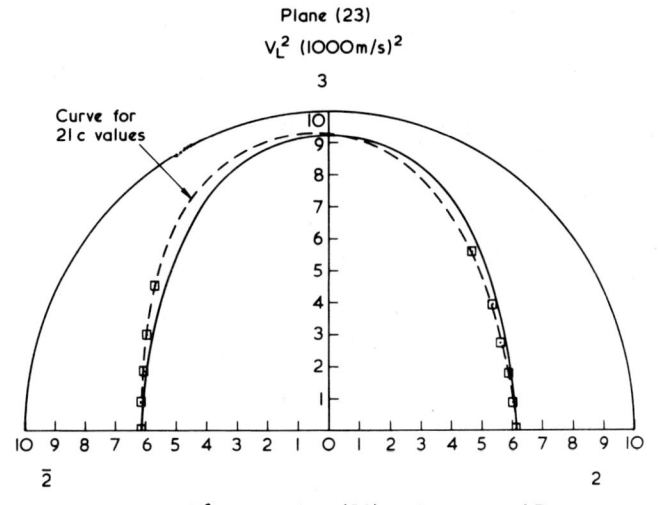

Fig. 13 Values of V_L^2 in the plane (23) at the centre of Deroton lath. The full curve gives the orthotropic solution of Table 2, and the asymmetric dashed curve shows the better fit obtained with the full 21 c values

In the case of the discs, visual examination of the surface indicates that the material fans out from the injection point, and alignment of fibres with the flow direction is anticipated. Measurements on these discs are still in progress. The first and perhaps not unexpected result is that longitudinal modulus in a direction normal to the disc surface does not vary greatly over the disc. However at this stage it seems that variations in shear modulus may be a more sensitive indication of variation in fibre distribution. Purely non destructive measurements are now being made at chosen positions to define shear modulus, and the maximum possible anisotropy of longitudinal modulus in the plane of the disc.

We have for example investigated the variation of V_L^2 with orientation in a diametral plane through the disc which is at an angle of 45° to the injection direction (Fig. 15). Measurements at five positions show little variation in straight-through velocity. Variation with orientation up to 60° from the normal can be measured, and these extreme results are given in Fig. 15. Extrapolation into the direction lying in the plane of the disc gives the same general result, with higher values nearer the injection joint, and the lowest value furthest from this point. It is concluded therefore that in general fibres have a greater tendency for alignment parallel to the surface of the disc, nearer to the injection point. On

$$\epsilon_3 = s_{33} \sigma_3$$
$$\epsilon_4 = s_{43} \sigma_3$$
$$\epsilon_4/\epsilon_3 = s_{43}/s_{33}$$
$$= 0.031/0.150 = 0.21$$

If $\epsilon_3 = 1\%$

$$\epsilon_4 = 0.21\% \equiv 0.12°$$

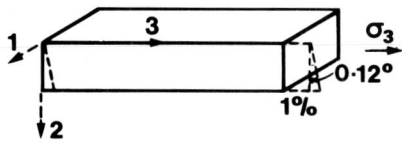

Fig. 14 Effect of extra s values (Table 3) on the calculation of strains ϵ arising from a particular situation of applied stress σ. With stress σ_3 applied, non-zero s_{43} ($= s_{34}$) indicates the presence of an extra shear strain ϵ_4 in the plane (23)

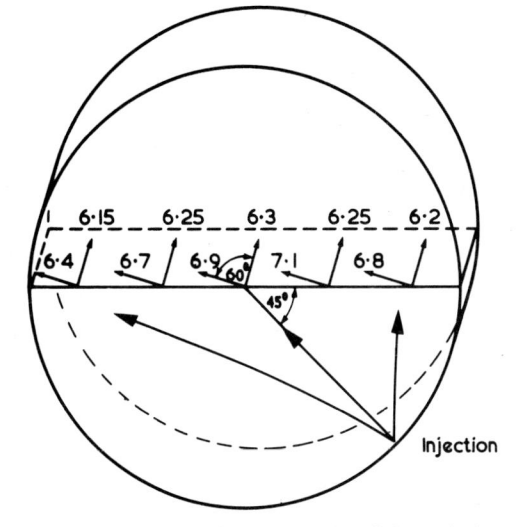

Fig. 15 Measurements for thin Deroton disc of the variation of V_L^2 in a diametral plane which is at an angle of 45° to the injection direction. Units are $(1000 m/s)^2$

the basis that fibres tend to align parallel to the observed flow marks, it had been anticipated that the trend across this plane would be the opposite to that found, with the highest extrapolated values of V_L^2 at the left in Fig. 15. This contradiction between expectation and results demonstrates a complication which has been verified by sectioning, that there is a variation in fibre distribution with depth in the moulding. The ultrasonic measurements give an average result for the volume examined.

Measurements at the centre of the disc, made in the two diametral planes along and normal to the injection direction, are given in Fig. 16. Extrapolation of the results into the flow direction 3 and the lateral direction 1 (which are not directly accessible to non destructive measurement), gives no indication for any notable difference in elastic modulus between these two directions; i.e. there is no notable anisotropy in the plane of the disc at its centre. There is however a difference between planar and straight-through valves.

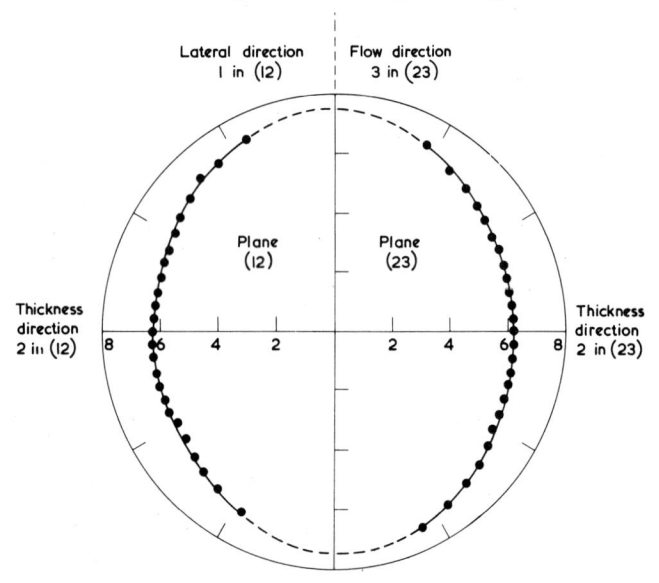

Fig. 16 Measurements at the centre of thin Deroton disc of the variation of V_L^2 in the normal diametral planes (23) and (12) — see Fig. 10. Units are (1000m/s)². Extrapolation in the plane (23) into the flow direction 3 lying in the surface of the disc shows no difference from extrapolation in plane (12) into the lateral direction 1 of the surface.

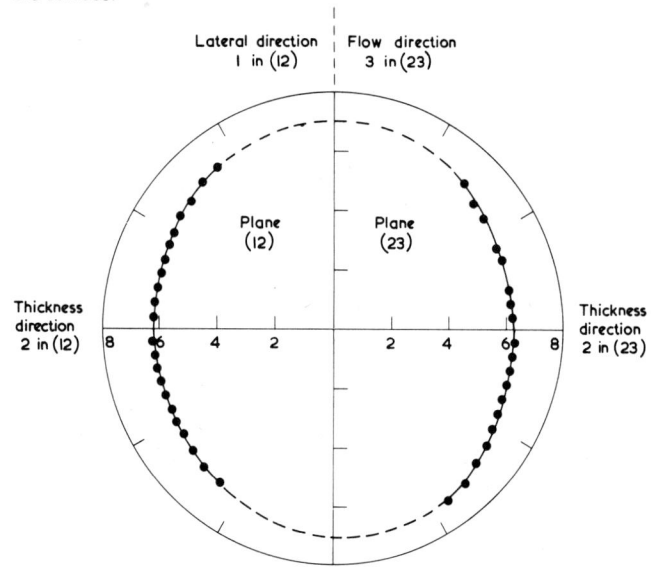

Fig. 17 Measurements at the centre of thick Deroton disc of the variation of V_L^2 in the normal diametral planes (23) and (12). Units are (1000m/s)². Compare with results for thin disc in Fig. 16

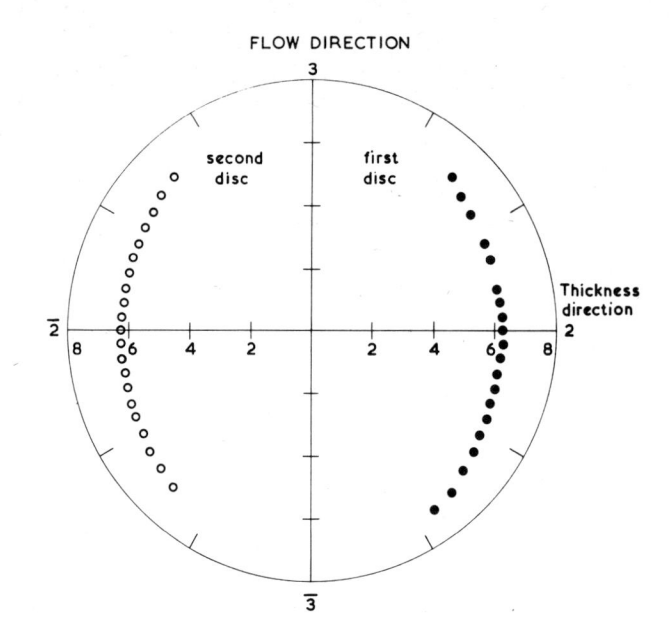

Fig. 18 Measurements at the centre of two thick Deroton discs of the variation of V_L^2 in the diametral plane (23). Units are (1000m/s)². The discs show identical properties in this plane

Further development of the work is to relate results to other processing variables, and as a first step 3 mm and 6 mm thick discs are being compared. Fig. 17 shows results at the centre of a 6 mm thick disc corresponding to those given in Fig. 16 for a 3 mm disc. The orientation range which can be covered is smaller, but again extrapolation indicates no difference between flow and lateral directions. There is however a little less difference between planar and straight-through values, which may be attributed to the greater thickness resulting in more randomisation of fibre direction in the inside of the moulding.

A likely valuable application of ultrasonic measurements is as a ready means of making non destructive quality control tests at selected points in an article. Such measurements are presented for two 6 mm thick discs in Fig. 18, and show identical properties. These measurements in one particular plane are more extensive than would be anticipated for practical use in quality control. The selection of appropriate measurements would be settled by a prior determination of important properties.

This last point is part of a more general consideration in the application of ultrasonic measurements. Looking ahead to the time when anisotropy of injection mouldings could be used in a deliberate fashion, there is a need for information on just which elastic constants will matter most to designers.

We have not in this Section given any comparison between the ultrasonic measurements and the results of usual mechanical testing. Where this comparison has been made, no important discrepancy has been found.

6 SUMMARY

Having developed existing NPL equipment to make possible rapid ultrasonic measurements on industrially processed components, the properties of some glass fibre-reinforced samples typical of industrial mouldings have been explored. This has been done paying particular attention to the extent to which measurements can be made completely non destructively, without cutting the component. In the case of dough-moulding compound, a transfer moulding process which gives fibre alignment produces only a 10% increase in elastic moduli in the alignment direction. Inhomogeneity of this material limits what can be done in detecting regions of local orientation in mouldings. Results for the engineering thermoplastic Deroton give a measure of the anisotropy of injection mouldings in the model forms of a lath or disc, and it has proved possible to determine the elastic moduli of the lath without sectioning. We have not dwelt in this paper on possible errors, but generally these are not serious. It is hoped that sufficient evidence of the possibilities will have been provided to establish the technique as a ready means of quality control and to lead processors and designers to give serious consideration to whether they could take advantage of such information to improve processing.

ACKNOWLEDGEMENTS

We are very grateful for the co-operation of ICI Plastics Division, and of the Company which provided the dmc mouldings, during the course of this work. We wish to thank Mrs A. Woolf of NPL for her valuable contribution to computer analysis of results. The work is published by permission of the Director of the National Physical Laboratory.

REFERENCES

1 Thomas, K., Meyer, D.E., Fleet, E.C. and Abrahams, 'Ultrasonic anisotropy measurements and the mechanical properties of polymeric materials', *J Phys D* **6** (1973) pp 1336–1352 (with plate)

2 Gotham, K.V., Moore, D.R. and Powell, D.C. 'Simplified design procedures for thermoplastic articles', Paper presented at IOP meeting on *'Time dependent stresses and strains in polymers'* London, (February 1974)

3 Ogorkiewicz, R.M. and Weidmann, G.W. 'Anisotropy in polypropylene injection mouldings', *Plastics and Polymers* **40** (1972) pp 337–342

4 Darlington, M.W. and Saunders, D.W. 'Creep in oriented thermoplastics', *J Macromol Sci – Phys* B5–2 (1971) pp 207–218

5 Thomas, K. and Woolf, A. 'Ultrasonic determination of the full elastic constants of anisotropic polymeric material', *IUPAC International Symposium on Macro-molecules* Aberdeen (10–14 September 1973) (Abstracts pp 293–294) paper E12

6 Markham, M.F. 'The NPL ultrasonic tank – its uses in polymer and fibre composite testing', *Ultrasonics for Industry 1969* Conference Papers (Iliffe: Guildford, 1969)

7 Markham, M.F. 'Measurement of the elastic constants of fibre composites by ultrasonics', *Composites* **1** (1970) pp 145–149

8 Rogers, T.G. and Pipkin A.C. 'Asymmetric relaxation and compliance matrices in linear visco-elasticity', *J Appl Maths and Phys* **14** (1963) pp 334–343

9 Halpin, J.C. and Pagano, N.J. 'Observations in linear anisotropic viscoelasticity', *J Comp Mat* **2** (1968) pp 68–80

10 Lockett, F.J. *Nonlinear Viscoelastic Solids* (London and New York: Academic Press, 1972) pp 33 and 57–58

11 Musgrave, M.J.P. 'On the propagation of elastic waves in aeolotropic media. I. General principles', *Proc Roy Soc A* **226** (1954) pp 339-355

12 Musgrave, M.J.P. *Crystal Acoustics* (London: Holden–Day, 1970)

13 Musgrave, M.J.P. 'Elastic waves in anisotropic media', *Progress in Solid Mechanics* **2** (Amsterdam: North Holland, 1961), pp 61–85 (Particularly pp 75–77)

14 Ogorkiewicz, R.M. 'Stiffness and strength of polyester dough moulding compounds', *Composites* **4** (1973) pp 162–166

QUESTIONS

1 On time required to obtain data, by C.M.R. Dunn (Research Department, ICI Plastics Division, Welwyn Garden City, Herts)

How long does it take to obtain a small number of moduli? It would appear that to obtain just a single value in the plane of a sheet a large number of values at different angles are required so that there can be extrapolation. From a commerical point of view the technique would be useful if, in say 1 hour, enough values could be obtained, not to fully characterise the material, but to give a useful guide as to material anisotropy and properties.

Authors' reply

The first step in a measurement is to mount the article in the controlled temperature water bath. The time this takes will depend on what provision has been made for dealing with the particular geometry. It is then necessary to allow the temperature of article and water to stabilise, and variables here are thermal conductivity and thickness of the article, and the required accuracy of measurement; typically 10–15 minutes should be allowed, although this time can be utilised for exploratory measurements.

An individual measurement to obtain the wave velocity in a particular directly accessible direction takes only a few seconds; calculation of velocity with a desk machine occupies about one minute, or less if the machine has a simple programming facility. To extrapolate longitudinal wave velocity results into an inaccessible direction lying in the plane of a sheet requires measurements with angles of incidence of the wave varying from $0°$ up to the maximum possible, say every $2.5°$ or $5°$ up to $30°$ (the angle of refraction in the specimen is greater, up to $60°$, depending on moduli). The results have then to be plotted in a polar diagram of 1/(velocity) versus orientation in the article, to extrapolate to the required direction (see Fig. 12 in the paper for a favourable example). When the general pattern of behaviour in an article, say an edge-gated disc, has been established, then it should be possible to choose a limited number of positions in the article, and a limited number of measurements at each position, to check reproducibility or to characterise the effect of changing a processing variable; and these restricted measurements should be possible within about one hour.

If in contrast the full elastic moduli at some position are required, then full measurements of both longitudinal and transverse waves must be made at the given position; these total about 100 measurements and should be possible, including the required remounting of the article, in less than half a day. The recording and calculation of data becomes tedious and liable to operator error when many measurements are made, and automatic recording and computer calculation of data then becomes important. The time taken to calculate full elastic moduli from velocity data depends on the detail required and also on whether a full range of measurements of both longitudinal and transverse type waves has been possible in three orthogonal planes. The calculation is naturally helped by a prior knowledge of typical moduli.

Measurement of wave velocity is of course affected by any change in thickness of the article. Also any change in density with position directly affects the modulus calculation, but in our experience such changes are usually of only a few percent.

Dynamic testing and performance of unidirectional carbon fibre-carbon composites

L. BOYNE, J. HILL and S.L. SMITH

The dynamic behaviour of unidirectional carbon fibre-carbon composite cantilevers has been measured using a uniformly applied impulse of 100 nS duration time. Dynamic and static properties have been compared for various manufacturing conditions and high speed photography has been used to observe energy absorbing processes and failure mechanisms. The main failure modes have been found to be delamination parallel to the fibre direction and, at high impulse levels, fibre pull-out.

1 INTRODUCTION

The fracture behaviour of unidirectional carbon fibre reinforced carbons (cfrc) is governed by the fact that they consist of a brittle fibre in a brittle matrix. In this type of composite it is possible to design for either a catastrophic failure in longitudinal flexure at a high stress level but with a low work to fracture or a progressive failure mechanism at a lower stress level and a high work to fracture. Examples of the latter mechanisms are longitudinal splitting, matrix–fibre debonding and fibre pull-out. In this paper we describe the behaviour of unidirectional tough cfrc cantilevers of various thicknesses when subjected to a uniformly applied impulse.

2 FABRICATION OF COMPOSITES

The method used for preparing the cfrc has been previously described [1]. Courtaulds HTU carbon fibres were pre-impregnated with a phenolic novolak (hexamine cured) resin, pressed, cured, carbonised and pyrocarbon densified by Fordath, West Bromwich. Two levels of pyrocarbon densification treatment were used:

(i) 100 hours process time and

(ii) 200 hours process time.

Half of the specimens were further heat treated to 2073K for one hour. Table 1 shows the typical composition of the cfrc.

3 IMPULSE GENERATOR

One face of the cantilever is impacted with an electromagnetically driven foil in vacuo. The foil forms the load in a low inductance discharge circuit of a rapid discharge capacitor bank. By varying the amount of energy initially stored in the bank an impulse range of 40–200 N sec M^{-2} is achieved. The elimination of air cushioning effects enables impulse duration times of the order 100 nS to be obtained. This is

Atomic Weapons Research Establishment, Aldermaston, Reading, Berkshire RG7 4PR, England

Table 1 Composition of cfrc

	Pyrocarbon densification time	
	100 h	200 h
Density kg/m^3	1 500	1 520
Carbon fibre v/o	55	55
Resin char v/o	7	7
Pyrocarbon v/o	22	24
Closed porosity v/o	3	3
Open porosity v/o	13	11

generally less than the pressure transit time through the cantilever thickness.

High speed photography is used to record the post-impulse behaviour of the cantilevers.

4 STATIC PROPERTIES AND FAILURE MECHANISM

Table 2 contains the longitudinal flexural strengths and moduli and the work to fracture for the various specimens. Fig. 1 is a typical stress/strain curve showing a pseudo plastic region. In the previous conference Aveston [2] described a possible failure mechanism for unidirectional carbon fibre ceramic composites in a three point bend test which results in an increase in the work to fracture. This involves the sequence tensile failure followed by delamination. This sequence is repeated until complete failure occurs. We have experienced a rather different failure mechanism where partial matrix delamination occurs in the two regions between the central support and the anvil. Fig. 2 shows this effect schematically. By increasing the load complete delamination occurs. This delamination and buckling mechanism can be inhibited by wrapping test specimens in all but the ends and central support point with Sylastic tape. This increases the elastic region of the stress/strain curve, the first yield stress being 1.6 times greater than untaped specimens. The failure mechanism in this case involves some fibre damage and pull-out.

Table 2 Static flexural properties of cfrc

Specimen number	Thickness mm	Densification time hrs	Heat treatment	Flexural strength MN/m^2	Flexural modulus GN/m^2	Work of fracture Kj/m^2
1	1	100	No	1 223	141	46
2	1	100	Yes	—	—	—
3	1	200	No	1 194	141	62
4	1	200	Yes	991	164	59
5	1.78	100	No	1 153	151	63
6	1.78	100	Yes	800	135	38
7	1.78	200	No	874	149	57
8	1.78	200	Yes	371	126	41
9	2.29	100	No	1 015	132	49
10	2.29	100	Yes	981	144	22
11	2.29	200	No	635	199	86
12	2.29	200	Yes	654	157	—

5 DYNAMIC PROPERTIES AND FAILURE MECHANISMS

High speed photography shows that the cantilevers behave completely elastically over the range of impulse levels until onset of failure. The dynamic Young's modulus is within ten per cent of the static value. Beyond the elastic region multiple delamination parallel to the impulsed face is the main failure mechanism. At low impulse levels this takes the form of crack formation (incipient delamination); with increasing impulse complete laminate detachment from the rear surface occurs, in some cases with sufficient momentum to cause failure at the root fibres in tension with fibre pull-out. The degree of incipient delamination in both surviving and detached portions of the cantilevers increases with increasing impulse.

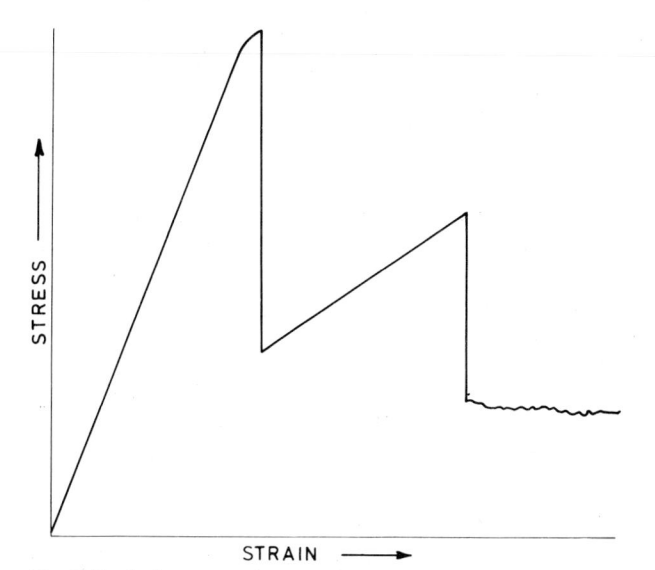

Fig. 1 Typical stress-strain curve for cfrc in three point bend test

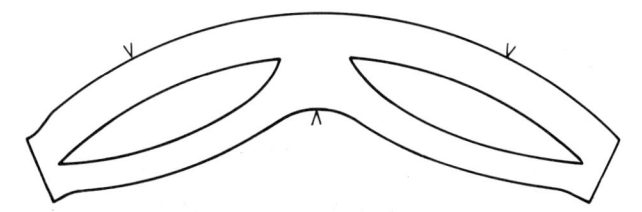

Fig. 2 Failure mechanism of cfrc in three point bend test (delamination and buckling)

Fig. 3 is a series of time lapse photographs showing the response of the cantilevers after impulsing. Figs. 3(i) and 3(ii) show typical response modes of cfrc. Figs. 3(iii) and 3(iv) show that when delamination and buckling are inhibited by attaching wire loops around the cfrc before impulsing then sufficient energy is channelled to the root to cause complete failure. For the non-looped specimen some energy is lost through delamination and despite a loss in stiffness and strength it is capable of supporting ~ 20% of its undamaged static breaking load. The delamination process effectively increases resistance to impulse damage.

This is further illustrated by reference to Table 3. Estimates of the impulse required to take the cantilevers to first fibre failure have been made together with estimates of the impulse required to cause complete root failure knowing the strain energy density required to tensile fracture. In this treatment delamination is ignored. The impulse values so derived are given in Table 3. The results in column 3 compare favourably with the result in Fig. 3(iv) where delamination was inhibited. However for the cfrc under test where delaminations occurred some fraction of the impacted side of the cantilever remained intact at levels up to 9×10^4 Nsec/m^3 (Table 4). This is considerably in excess of failure levels to be expected where no delamination occurs.

The thinnest specimens possess greatest strain energy density to both the elastic limit and complete failure, and consequently would appear from Table 3 to be most resistant to impulse. Unfortunately the modes of failure make it difficult to assess whether this is the case in experiment. There would appear to be no consistent difference between performance of 100 and 200 h densification treated materials for either of the thicker specimens. The 2073K heat treatment however results in lower strain energy densities to complete flexural failure suggesting a lower resistance to impulse. However in experiment the heat treated material appears more resistant to impulse. The 200 h composites also appeared marginally more resistant to impulse over the 100 h composites. It is interesting also that the elastic strain energy density contribution to the impulse for complete failure is 70 and 55% for the non-heat treated and heat treated composites respectively.

6 INTERLAMINAR SHEAR STRENGTH (ILSS)

For the type of failure described above the ILSS is of considerable importance. This property is typically 20MN/m^2

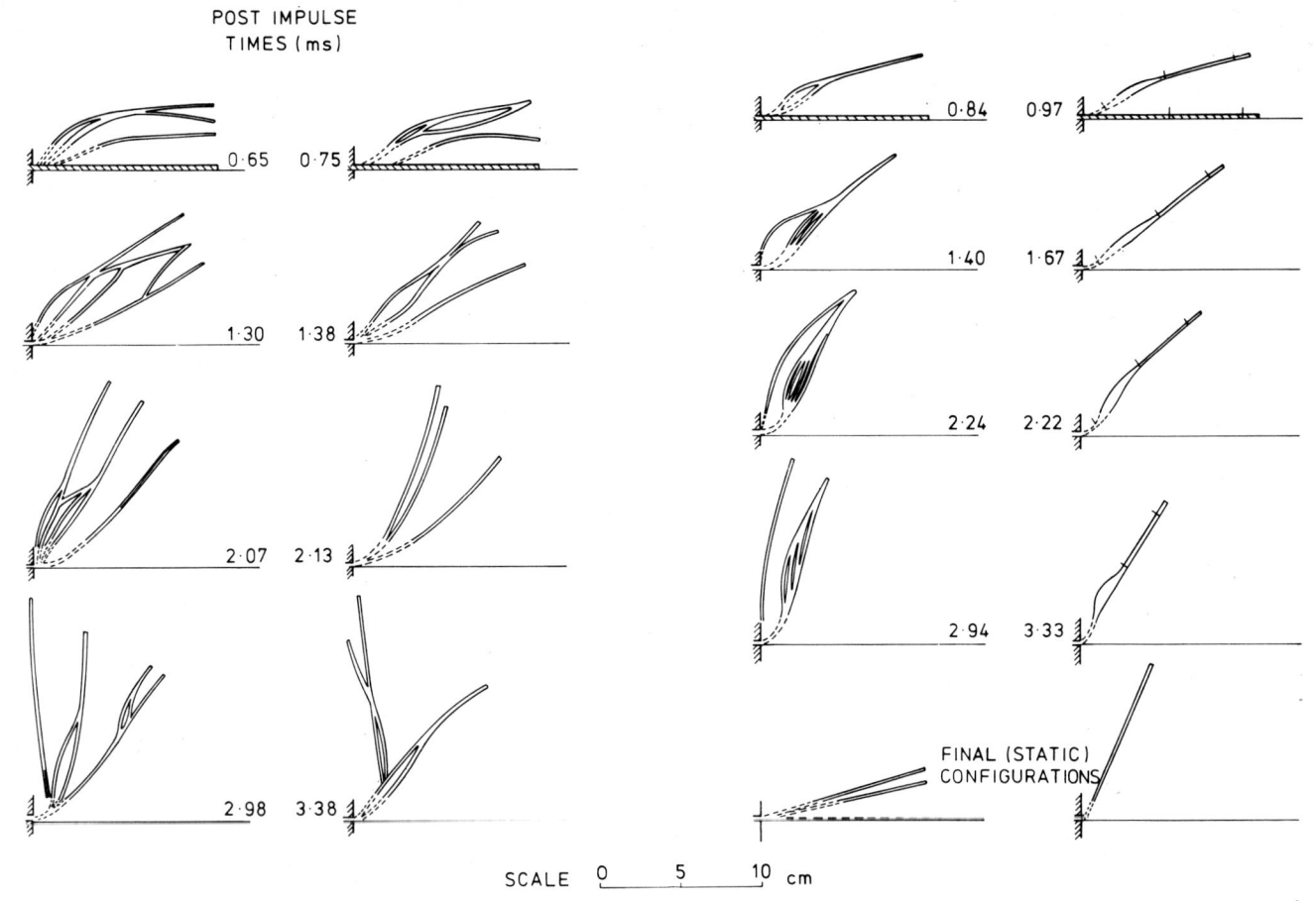

Fig. 3 Time resolved motions of uniformly impulsed cfrc cantilevers: (a) 2.29 mm thick cantilever (200 h heat treated) impulse = 190 N.s/m²; (b) 2.29 mm thick cantilever (200 h) impulse = 135 N.s/m²; (c) 1.78 mm thick cantilever impulse = 95 N.s/m²; (d) 1.78 mm thick cantilever impulse = 91 N.s/m²

Table 3 Estimated impulse required to reach first fibre failure and complete failure assuming no delamination

Specimen number	*Specific impulse to elastic limit $\times 10^4 Ns/m^3$	Specific impulse to cause complete failure at root $\times 10^4 Ns/m^3$
1	3.22	4.13
2	—	—
3	3.20	4.40
4	3.00	4.36
5	2.95	3.63
6	2.10	3.39
7	2.39	3.53
8	2.10	3.39
9	2.29	3.93
10	1.69	3.37
11	1.67	3.10
12	2.75	3.91

* ie impulse per unit thickness (per metre thickness)

in the as-densified state reducing to 17 MN/m², after heat treatment. Taping has the effect of artificially increasing the ILSS and the elastic region of the three point bend test stress strain curve. Similarly wire looping prevents the stored energy from being released during the impulse test thus causing catastrophic failure at the root of the cantilever. The range of ILSS found in this work is critical, below this range the static strength of the cfrc would be unacceptably low whereas above the range the cfrc will become more brittle in their behaviour and lose toughness.

7 CONCLUSION

High speed photography shows that cfrc when subjected to rapid impulses fail by a delamination mechanism. This type of failure is progressive and imparts a degree of toughness to the composites. Such behaviour is also seen in slow bend tests and, therefore, is not dependent on the rate of loading. The thinnest section specimen tested failed at the highest stress in the static test; it presumably was less highly flawed than the thick specimens. Heat treatment has the effect of reducing the static failure stress but increases the dynamic impulse level to complete failure and also reduces the ratio of elastic to post elastic strain energy density to a lower value. It is assumed that fibre movement occurs as the fibre is subjected to higher than its process temperature. The longer pyrocarbon densification process (200 h) marginally increases the density of the cfrc but makes little difference to either static or dynamic performance.

ACKNOWLEDGEMENT

The authors thank the Director of AWRE for permission to publish this paper.

REFERENCES

1 **Hill, J., Thomas, C.R.** and **Walker, E.J.** 'Advanced carbon-carbon composites for structural applications', *2nd International Conference on 'Carbon fibres, their place in technology'*, London (February 1974) paper 19

2 **Aveston, J.** 'The strength and toughness of fibre reinforced ceramics', *NPL Conference 'The properties of fibre composites'* (November 1971)

Table 4 General comparison of cfrc cantilever behaviour under uniform impulse conditions

Applied impulse per unit area per unit depth ($N.S./m^3$)	Untreated		Heat treated	
	100 h	200 h	100 h	200 h
1×10^4	Intact	Intact		
2×10^4	Delamination increasing function of impulse	Delamination increasing function of impulse leading to rear surface layer detachment, volume slightly less than 100 h untreated cantilever at a given impulse level	Intact	Intact
3×10^4	Leading to rear surface layer detachment, volume increasing function of impulse			
4×10^4			Delamination increasing function of impulse leading to rear surface layer detachment, volume increasing with impulse	Delamination and rear surface layer detachment as for 100 h heat treatment cantilever but slightly less severe at a given impulse level
5×10^4				
6×10^4	Root failure	Leads to root failure		
7×10^4				
8×10^4				
9×10^4			Root failure	Root failure

Techniques for measuring the mechanical properties of composite materials

P.D. EWINS

An outline of the problems involved in the measurement of the mechanical properties of composite materials is given, with particular reference to cfrp, together with comments on the short-comings of conventional specimen shapes and test techniques.

The design and development of a coherent set of new specimen shapes is described, and test techniques presented which enable the strength and stiffness properties of a unidirectional laminate to be accurately determined. Also presented are some factors affecting the design of test specimens and techniques for the determination of fatigue and impact performance.

1   INTRODUCTION

An unprecedented boom in the number of new high-performance materials, particularly composite materials, now available and the great potential they offer in many technological fields has resulted in a new and deeper interest in materials testing. Such interest however is far from being academic and is rather born out of necessity; the scientist and research worker, the designer, the quality control engineer and the customer all have a need for materials testing, albeit the information they require may vary considerably.

Let us consider what information can be obtained from the measurement of a particular property and how differing requirements can affect both the way in which a test is carried out and the type of specimen used. This is perhaps best achieved by reference to the apparently simple example of the measurement of tensile strength. In simple terms the test involves pulling a sample of material until it fails, noting the failing load and calculating the failing stress by dividing the failing load by the cross-sectional area at the point of failure. However, two important questions arise from such an exercise and each merits further comment. First, is this the only way in which tensile strength can be measured and, second, in what circumstances is a particular test and the resulting tensile failing stress of interest?

Clearly the answer to the first question is no, for we are all familiar with the flexural test in which tension is a very common failure mode. Of course, the calculation leading to a tensile failing stress is more complicated and includes certain assumptions but nonetheless the test is frequently done. Other types of test can also lead to tensile failure and among these can be cited some forms of hydrostatic testing, and for highly anisotropic materials, compressive testing which can lead to tensile failure in a direction normal to the applied load. In answer to the second question, then if the test is simple and cheap to do and is easily reproducible, it will serve well as a quality control test; in this context the flexural test is often chosen. Generally, however, the failing stress will be of use for design purposes only if the failed region is under pure tension and free from significant stress concentrations, and if environmental conditions can be adequately represented. It is clear that these conditions are likely to be met only by the use of a carefully designed specimen and test technique which can be shown to be theoretically sound. If now we consider the requirements of the scientist carrying out a fundamental investigation into tensile behaviour it is at least possible that he will require yet another specimen shape and test technique, particularly if he is interested in how and where failure first occurs and how such failure propagates. He may, for example, require a test which can be stopped at any moment of time to allow a detailed investigation to take place.

From this brief example it can be seen that several important questions must be answered before we set up a test for any purpose at all and these can be summarised as follows:—

1   For what purpose is the information required?
2   Can environmental conditions be adequately represented?
3   Is the overall technique theoretically sound?
4   Is the material composition known?

The last question has not yet been touched on but it is clearly of prime importance that the composition of the material be known, otherwise a situation could exist in which a wealth of information had been gathered on an 'unknown' material. If such a statement appears obvious, let it not be forgotten that for composite materials this applies not only to the constituent materials but also to the composite as a whole; in this context reinforcement, matrix and void contents are all important.

Because the emphasis of this conference is on design and testing, the majority of tests referred to subsequently will be those used for gathering data for use in structural design. Further, as the paper describes work largely carried out at RAE, Farnborough, the scope will be limited to test techniques and specimens developed for use with carbon fibre reinforced plastics, the composite material in which currently there is most interest. However, it will be readily appreciated that in most instances these will be generally applicable to

---

Royal Aircraft Establishment, Farnborough, Hants

other fibre reinforced plastics such as boron fibre, glass fibre, PRD-49 and whisker reinforced plastics. Their application to other types of composite material will be less direct but the general philosophy will apply.

2 THE NATURE OF THE PROBLEM

As with most composite materials, the difficulties associated with the testing of carbon fibre reinforced plastics (cfrp) arise in the main from its highly anisotropic properties. Presented in Table 1 are typical values for the more important properties of unidirectional cfrp made from the two main types of fibre, high modulus (or type 1) and high strength (or type 2), at a fibre content of 60% by volume. Several of these properties are of importance in the analysis of the behaviour of conventional specimens, particularly the modes of failure, and in the design of new specimen shapes.

The extremely high ratio of tensile or compressive strength in the longitudinal direction to interlaminar shear strength causes two major problems. Firstly, load is normally transferred from the test machine to the specimen by shear and if large loads are required to produce failure at the test section then large shear areas will be required at the specimen ends. Secondly, any waisting of the specimen, necessary to ensure failure in a region of uniform stress away from the ends of the specimen, will have to be very gradual if the shear stresses induced in the waisted region are to be kept below the ultimate shear strength, and unwanted shear failures thereby avoided.

Not only is unidirectional cfrp extremely stiff in the fibre direction, but the stiffness is constant to failure both in tension and compression. Any initial curvature of the specimen or misalignment of the test machine loading mechanism can therefore give rise to significant bending strains and corresponding stresses and result in premature failure of the specimen. This contrasts with the behaviour of most conventional materials where the plastic region offers considerable stress relief and even relatively large bending strains do not cause noticeable reductions in the measured ultimate strengths.

In the transverse direction the properties of unidirectional cfrp, although generally poor, are no less difficult to measure. The stiffness to tensile strength ratio is higher than in the longitudinal direction so that the effects of specimen curvature and machine misalignments have an even more marked effect upon failure.

3 STRENGTH, STIFFNESS AND STRESS/STRAIN DESIGN DATA

In 1971 [1] the Royal Aircraft Establishment presented details of a coherent set of specimens for the measurement of room temperature tensile and compressive properties of unidirectional cfrp, both parallel and normal to the fibre direction, and made reference to the shortcomings of the then conventional specimens. Since that time test techniques for the measurement of shear properties have been studied in detail and, apart from certain limitations, these properties can be measured on the same moulded sheet as tensile and compressive properties.

During the past two years the British Aircraft Corporation (MAD), under contract to Ministry of Defence, have carried out further work on specimen and test technique develop-

Table 1 Typical properties of unidirectional cfrp at a fibre content of 60% by volume

Property	Type 1-S	Type 2-S
Longitudinal tensile strength MN/m^2	1 250	1 550
Longitudinal compressive strength MN/m^2	1 000	1 300
Transverse tensile strength MN/m^2	35	50
Transverse compressive strength MN/m^2	170	200
Interlaminar shear strength MN/m^2	80	110
Longitudinal tensile modulus GN/m^2	210	135
Longitudinal compressive modulus GN/m^2	190	120
Transverse tensile and compressive modulus GN/m^2	7	7
Interlaminar shear modulus GN/m^2	4.5	4.5

ment with the result that currently tensile, compressive and shear data, including stress/strain relationships, can be measured accurately over the temperature range $-40°C$ to $70°C$. All these properties can be determined on the same moulded sheet thus maintaining the desirable coherency.

Thus, an overall technique has been developed which enables unidirectional cfrp to be fully characterised for design purposes. It has been adopted as a preliminary standard by a number of Government Laboratories and the aerospace industry and is gaining wider acceptance both by fibre and pre-preg manufacturers and by various bodies abroad.

Let us now consider briefly each test specimen, the important features governing its design and some reasons why more conventional specimens are unsatisfactory.

3.1 Longitudinal tension

For an evaluation of the tensile properties of metals and plastics several types of specimen are used ranging from flat waisted specimens for use in conjunction with wedge grips or shear pins to circular-section waisted specimens with threaded ends for use with screw fittings. In nearly every case the specimens are highly waisted with large taper angles. None of these shapes is satisfactory for longitudinal tensile testing of unidirectional cfrp as failure occurs either in the ends of the specimen or in the tapered region, in both cases due to excessive shear stresses. The effect of too large a taper angle is clearly shown in a photograph of some failed specimens (Fig. 1). Under an applied tensile load shear cracks have initiated in the taper region and propagated in the fibre direction. Final failure has occurred at a section where the material is weaker, or, more probably, where the stress concentration is large in the vicinity of the end-fitting.

The specimen chosen for use with unidirectional cfrp is shown in Fig. 2. As the design is governed by a number of features which are common to specimens for measuring other material properties, it is worth describing them in some detail. The problem of load input into a material of low shear strength can be approached in several ways but clearly pinned or threaded ends are best avoided because of the associated high stress concentrations. A better solution, and the one chosen here, is to spread the applied load over a large area so that shear stresses are kept below the shear strength. This has been achieved by choosing a thin specimen of rectangular cross-section so that the surface area available for shear load input is large compared with the cross-sectional area. Such a shape also has the advantage of being easily cut from a flat, moulded sheet and allows the use of wedge grips, a preferred method of loading because it is both simple and accurate. Unfortunately, wedge grips acting directly onto the cfrp do not grip well and can cause fibre breakages and associated stress concentrations. However, the use of soft aluminium pads bonded onto the cfrp provides a good gripping surface for the wedge-grips and satisfactorily eliminates potential fibre breakages.

From a knowledge of the maximum load transfer at the ends and the predicted strength of unidirectional cfrp, the maximum specimen thickness at the test section can be calculated if the specimen width is chosen to be constant over the whole length. For the strongest cfrp at 60% fibre volume fraction the maximum allowable thickness is approximately 1.5mm but to allow for fabrication errors, particularly poor bonding between the end-plates and the cfrp, a thickness of 1 mm was chosen.

To allow for the effect of stress concentrations due to the end-fittings, the cross-sectional area at the test section must be less than at the ends. The main point of debate is whether this area reduction is best achieved by a reduction in specimen width, specimen thickness or a combination of the two. The last of these is unattractive, and therefore rejected, because two separate machining operations would be required with a corresponding increase in cost. Of the two remaining possibilities a reduction in the thickness is preferred for two important reasons.

Firstly, from a consideration of the maximum allowable growth rate of either the width or thickness of a specimen of rectangular cross-section to ensure that nowhere is the shear strength exceeded whilst allowing the ultimate tensile stress to be achieved at the minimum section, it has been shown [1] that

$$\frac{dy}{dx} = \frac{F_6}{F_{1T}} \frac{y}{y_0} , \qquad (1)$$

where F_6 is shear strength,

F_{1T} is tensile strength,

y_0 is minimum half thickness (or half width)

and y is local half-thickness (or half width) at a distance x from the point of minimum thickness (or width).

If, further, an interaction between shear and tensile strength is assumed, it has been shown [2] that the expression is modified to

$$\frac{dy}{dx} = \frac{F_6}{F_{1T}} \left(\frac{y^2 - y_0^2}{y_0^2} \right)^{1/2} \qquad (2)$$

In either case it can be seen that generally the maximum rate of growth of either width or thickness depends largely on the shear to tensile strength ratio. It follows, therefore, that for a given cross sectional area reduction, the required taper length and hence the specimen length is considerably less if the reduction is made in the thickness (provided the thickness is significantly smaller than the width).

Secondly, tensile testing carried out using wedge-grips is more likely to result in high out-of-plane bending stresses than in-plane bending stresses. Consequently, a reduction in specimen thickness causing a reduction in out-of-plane stiffness, will minimise the effect of out-of-plane strains on the specimen failing strength. Accordingly, a specimen of constant width tapered only in the thickness was chosen. The stress concentrations due to the end fittings in such a specimen subjected to a pure tensile load have been shown experimentally to be in the range 1.25 to 1.5, so that based on a waisted thickness of 1 mm the minimum specimen thickness before waisting is 1.5 mm; this thickness dimension will subsequently be referred to as the basic thickness. However, as will be shown later the compressive specimens must have a basic thickness of not less than 2mm and to ensure a consistent set of specimens a basic thickness of 2mm was chosen for all specimens.

Although the ideal waisting profile is exponential, there are considerable practical advantages to be gained from making the profile a circular arc. An acceptable approach to this transformation is to replace the exponential by a circle whose radius is such that, for a given basic thickness, waisted thickness and shear to tensile strength ratio, the run-out length produced is the same as that given by integration of Equation (1). This approximation yields a maximum growth rate

Fig. 1 Failures in conventional tensile specimens

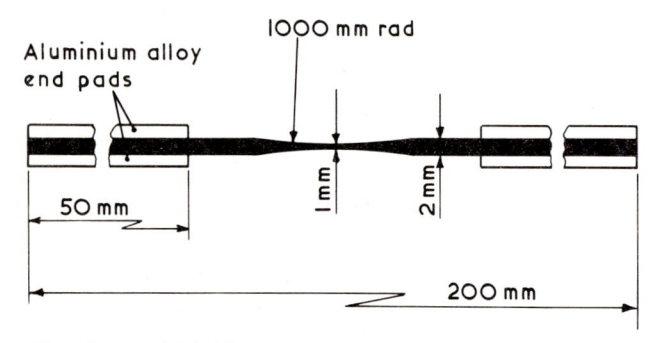

Fig. 2 Longitudinal tensile test specimen

about 5%–10% greater than the allowable rate. For a number of reasons, based mainly on design limitations for shear strength, the maximum tensile to shear strength ratio is taken to be 50:1 and this leads to a waisting radius of about 400 mm.

If the interaction between shear and tensile strength is assumed to exist then a better approximation to the profile is a circular arc whose radius, R, is given by

$$\frac{dy}{dx} = \left(\frac{2\,[y-y_0]}{R}\right)^{1/2} \quad (3)$$

For the same tensile to shear strength ratio this leads to a waisting radius of 1 250 mm. However, experimental evidence seems to confirm that in this type of test there is little shear/tension interaction so, as a conservative compromise, a waisting radius of 1 000 mm was chosen.

It has been calculated [2] using the Lekhnitskii [3] method for an approximate determination of stress concentrations on a hyperbolic profile that the maximum tensile stress concentration on a specimen of this profile gives rise to stresses less than 1.5% greater than that of the minimum section.

The chosen specimen width of 10 mm is a compromise between two conflicting requirements. On the one hand, the width should be sufficiently large to allow a representative cross-section of material to be tested and also allow for the measurement of transverse strains so that Poisson's ratio can be determined. On the other hand, the width should be minimised in order to reduce the effects of in-plane bending strains arising from test machine misalignments.

For the measurement of Young's modulus and stress/strain data up to the failing strain of the material, a parallel gauge length is required and the total specimen length of 200 mm includes an allowance for gauge lengths up to 25 mm.

It is important to note that strain measuring equipment which is attached to the specimen by means of knife edges, most optical equipment for example, should not be used if an accurate measurement of tensile strength is also to be made. Surface damage and stress concentrations beneath the knife edge will cause premature failure of the specimen. Resistance strain gauges bonded to the specimen surface allow simultaneous measurements of stress and longitudinal and transverse strain up to the failing stress of the material. Excellent results have been obtained using 6 mm strain gauges on a 10 mm parallel gauge length and it has been shown that these gauges do not affect the failing stress.

3.2 Longitudinal compression

Measurement of the compressive properties of isotropic materials, particularly metals, has received comparatively little attention in the past and in consequence standard specimen shapes are few. The problems of measuring the true compressive properties of material in sheet form are severe, and certain properties such as ultimate compressive strength are difficult to define. Design procedures have evolved therefore which allow compressive behaviour to be linked to the tensile properties; for instance the ultimate compressive strength of a metal is often taken to be the 0.2% proof stress in tension. The measurement of the compressive properties of unidirectional cfrp is no less difficult and until fairly recently there has been little link between compressive and tensile behaviour. A new specimen shape is clearly required.

Apart from limitations on taper angles at the waisted region which are essentaily the same as for longitudinal tension, there are two main problems to be overcome in the design of a successful specimen. Firstly, the transverse loads set up in the ends of a test specimen under a compressive load, although small, may well exceed the transverse strengths of the material. This will result in failure at the end region in a mode which is typified by a breakdown of the matrix with the fibres splaying out to give a brush-like appearance. Secondly, there is the possibility that an overall buckling failure could occur before the ultimate compressive strength is realised, particularly if the specimen is long and slender.

A specimen shape in which due allowance is made for these problems was developed [4] in 1970 and made use of a circular-section tapered specimen with special steel end caps. These caps served a double purpose in that they prevented end failures in the cfrp and provided adequate restraint against overall buckling of the specimen. This specimen shape is however obviously unsuitable for measurements on flat laminates. Further, both the fabrication and machining have proved to be expensive and this alone would prevent the specimen from gaining wide acceptance. A thin rectangular section specimen has therefore been developed by Purslow and Collings [5] such that test material can be taken from the same standard thickness cfrp laminate as material for other tests and the total cost reduced thereby.

The chosen specimen, shown in Fig. 3, can be considered a combination of the general profile used on the longitudinal tensile specimen and a two-dimensional form of the end caps used on the circular cross-section longitudinal compressive specimen. In order to produce failure in a region remote from end effects the specimen must be waisted and, by similar arguments to those used for longitudinal tension, this is best realised by a reduction in thickness. Further, the taper profile equations developed for tension will apply provided that ultimate compressive strength is used in place of ultimate tensile strength.

To prevent overall buckling before ultimate compressive failure is approached while at the same time waisting the specimen to allow for stress concentrations of up to 1.5 it has been shown [5] that the minimum unwaisted thickness is 3.5 mm. This is for the most adverse case which includes an allowance for a parallel gauge length of 15 mm at the waisted section. Unfortunately, this thickness is not compatible with the optimum thickness of 1.5 mm for the longitu-

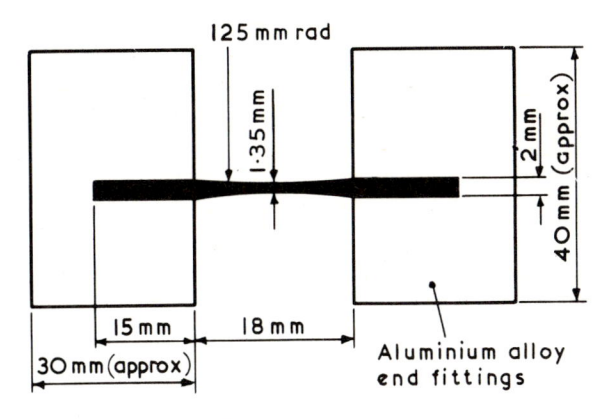

Fig. 3 Longitudinal compressive test specimen

dinal tensile specimen and it was necessary to seek a compromise so that consistency could be maintained.

If the basic thickness were to be reduced to 2 mm it has been shown that in a few instances only would the specimen buckle before the ultimate strength was achieved, but that without a parallel gauge length, which allows a corresponding reduction in specimen overall length, buckling would not occur under any circumstances. On the basis that stress/strain data would be required only infrequently it was decided to choose a basic thickness of 2 mm and a waisted thickness of 1.35 mm but with no provision for a parallel gauge length. A separate unwaisted specimen would enable initial stiffness data to be obtained or, alternatively, a waisted specimen with a parallel gauge length would in most instances yield complete stress/strain data. Where buckling occurred prematurely, data up to about 90% of the ultimate stress could be obtained. A waisting radius of 125 mm was derived on a similar basis to that used for the tensile specimen.

The end-fittings are required to provide a satisfactory method of load input from the test machine to the test piece and to stabilise the specimen against buckling under the compressive load. To satisfy these requirements aluminium alloy blocks are used. A slot of thickness 2 mm is cut across the centre of one of the long edges into which the test piece is bonded. The depth of slot 15 mm was chosen to ensure an appropriate balance between load input by shear at the bonded joint and load input by compression on the end of the test piece; stress concentrations tend to increase with increasing shear input whilst the tendency towards end failure increases with increasing compression on the ends. The opposite edge is machined normal to the slot to ensure that a truly axial load can be applied to the specimen during test, and one remaining edge is machined parallel to the slot to enable accurate alignment of the test piece and end-fittings during assembly.

3.3 Transverse tension

Conventional tensile test specimens are generally as unsatisfactory for a determination of the transverse tensile properties of unidirectional cfrp as they are for measurement of the longitudinal tensile properties but for rather different reasons. In a plane normal to the fibre direction unidirectional cfrp can be considered as isotropic so that, compared with longitudinal tension, stress concentrations tend to be relatively small. Furthermore, the transverse tensile to shear strength ratio is very much lower (about 2:1) than the longitudinal to shear strength ratio (about 50:1) so the limitation on tapers and waisting rates is much less severe.

The main condition affecting the design of the transverse tensile specimen, developed in essence by Mead [6], is the high transverse stiffness to transverse strength ratio which is usually higher than the corresponding ratio in the longitudinal direction. In the presence of any test machine misalignment this ratio can give rise to bending strains which are large compared with the direct strain and with no bending stress relief due to plasticity there is a significant reduction in measured failing load. This effect can be reduced to an acceptable level by an increase in specimen length which leads to a reduction in the curvature caused by misalignment and a corresponding reduction in bending stresses.

Although an increase in length is achieved most easily by the use of a longer cfrp test piece, it is wasteful of material and end-fittings are still required. An alternative method, and one which has proved satisfactory in practice, is the use of soft aluminium alloy extension pieces which serve both to increase the total specimen length and to provide a suitable surface to which the test machine wedge grips can be applied. The extension pieces, which are the same width and thickness as the cfrp test piece, are butt-bonded onto the ends of the cfrp and thin (< 0.5 mm) aluminium straps are bonded across the joints to ensure adequate strength.

A number of the specimen dimensions can be chosen arbitrarily but in order to preserve consistency they are taken to be the same as those of the longitudinal tensile and compressive specimens. Thus, the basic thickness is 2 mm, the width is 10 mm and the waisting radius is 125 mm. Experiments carried out on a number of specimens waisted by different amounts showed that a thickness reduction of approximately 35% (equivalent to a reduction of 0.75 mm on a basic thickness of 2 mm) was sufficient to ensure failure in the waisted region in about 90% of the specimens tested. Although a greater amount of waisting would increase this percentage, it would also increase significantly the possibility of damage during preparation of an already fragile specimen.

The chosen waisted thickness of 1.25 mm is considered to be a satisfactory compromise. To accommodate the taper run-out and allow for a parallel gauge length, a cfrp test piece length of 50 mm was chosen.

The overall specimen length including ends was determined from the arbitrary requirement that, for a given applied load, any induced bending strain shall be less than 1% of the direct strain. For the chosen thicknesses and a maximum test machine misalignment of 0.15 mm (compatible with commercial test equipment), the minimum specimen length satisfying this requirement is 180 mm. With an allowance of 25 mm at each end for gripping in the test machine, the total specimen length is 230 mm. A drawing of the complete specimen as shown is Fig. 4.

As in the previous cases, the effect of strain measurement equipment on the behaviour of the specimen must be taken into account, for both the failing stress and modulus can be affected significantly by the presence of such equipment. Successful results have been achieved using the resistance strain gauges described previously; experiments have shown that these gauges do not affect the failing stress significantly and calculations show their effect on overall stiffness to be less than 2.5%.

3.4 Transverse compression

The design of the transverse compressive specimen is governed largely by a single condition that arises from the high compressive strength to stiffness ratio in the transverse direction which can be several times larger than that in the longitudinal direction (Table 1). In order to maintain the desired consistency amongst specimens, a basic thickness of 2 mm is required and the possibility that a buckling failure might occur before the ultimate compressive strength is realised is therefore high. If the load input into the test piece is partly by shear and partly by end-load, as it is for the longitudinal compressive specimen, then, due to the assumed isotropy of the material in a plane normal to the fibre direction, stress concentrations will be relatively small. Indeed, it has been found experimentally that they are of the order of 1.1–1.2, which corresponds to a waisted thickness of 1.6 mm in order to ensure failure away from the ends. As a result of this small reduction in thickness, the length over which waisting occurs is small. Provided that the specimen ends

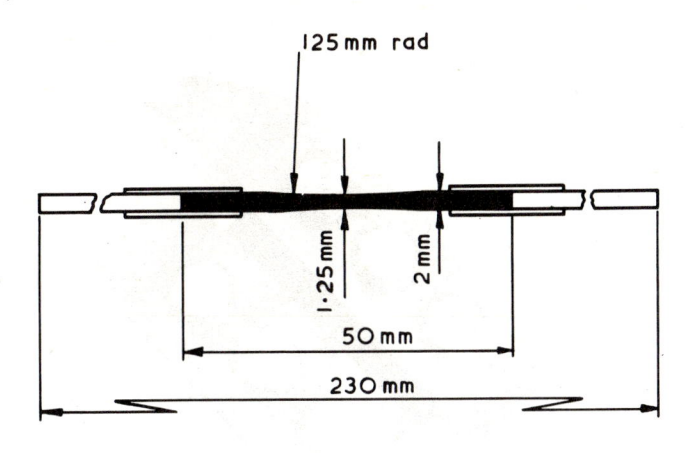

Fig. 4 Transverse tensile test specimen

(Specimen width 10 mm; 125 mm rad; 1·25 mm; 2 mm; 50 mm; 230 mm)

receive adequate support to give an encastré end condition, and that no parallel gauge length is required, a waisting radius of 125 mm will lead to compressive failure before overall buckling in the majority of cases.

The end-fitting developed for the longitudinal compressive specimen was used in this case, since it provides sufficient support to give the necessary encastré end conditions and provides a suitable method of load transfer from the test machine into the cfrp test piece. A diagram of the complete transverse specimen is shown as Fig. 5.

It is a disadvantage that strain measurements cannot be made on this specimen although the initial modulus and some stress/strain data can be determined by the use of an unwaisted specimen. However, an alternative approach to the design of a transverse compressive specimen has been suggested by Dootson [2] in which a plane, unwaisted specimen is used with the load applied solely as compression on the test piece ends. A special loading rig is used so that lateral restraint is applied to the specimen ends in the form of a reaction to the transverse stress caused by Poisson's ratio. This restraint not only prevents premature end failures but also supports the specimen against overall buckling. It is claimed that very little load is transferred into the specimen by shear as a result of the lateral load. The specimen ends are machined flat and normal to the loading direction to a very high tolerance so that excellent axiality of load is obtained. For a specimen free length of 10 mm, stress/strain data up to failure can be obtained and evidence to data suggests that measured failing stresses are no lower than those obtained from the waisted specimen.

Unless and until a specimen better than both of those described is developed it seems likely that both will continue to be used, the choice in a particular situation depending on the data required and other relevant prevailing conditions.

3.5 Shear

For the vast majority of isotropic materials shear properties are generally left unmeasured for reasons which are not hard to find. Firstly, unlike anisotropic materials, the shear modulus, G, of an isotropic material is not an independent property and can be calculated from the measured Young's modulus, E, and Poisson's ratio, ν, according to the formula

$$G = \frac{E}{2(1+\nu)} \qquad (4)$$

Furthermore, this relationship is assumed to hold true in the plastic region also, provided the tangent modulii and Poisson's ratio are used, although the basis for this assumption can be somewhat doubtful.

Secondly, a large number of materials subjected to increasing tensile stress fail in shear on the maximum shear plane where the shear stress is equal to half the tensile stress. For this reason the shear strength is often taken to be half the tensile strength. (There is a combined stress state on the plane of failure so the measured shear strength will be generally less than the true shear strength.)

Finally, because the shear strength of isotropic materials is of the order of half the tensile strength, only relatively rarely is shear strength a limiting factor in structural design. Certainly, shear buckling of panels is often a limiting factor but this is generally an elastic phenomenon for which the relevant elastic properties are known from tests other than shear.

There are situations in which the true shear strength is required and typical of these are the shear strengths of bolts and rivets, and castings and forgings of complex shape. In these contexts, specialist tests have been developed such as single and double overlap tests for bolts and rivets, and torsional tests on rods and thick-walled tubes.

For composites generally, and cfrp in particular, the situation is very different. The shear modulus is an independent property and must be measured, the shear strength (on planes parallel to the fibre direction) is very small compared with longitudinal tensile and compressive strengths and in consequence is frequently a limiting factor in design, and complete shear stress/strain data will be required in some situations. Thus it is clear that satisfactory shear tests need to be developed and standards adopted.

It is not difficult to devise test specimens and techniques which will allow shear data to be obtained. Indeed, Purslow has surveyed [7] recently some twenty-eight tests currently in use, all of which produce shear failure or enable some part of the stress/strain relationship to be measured. The main

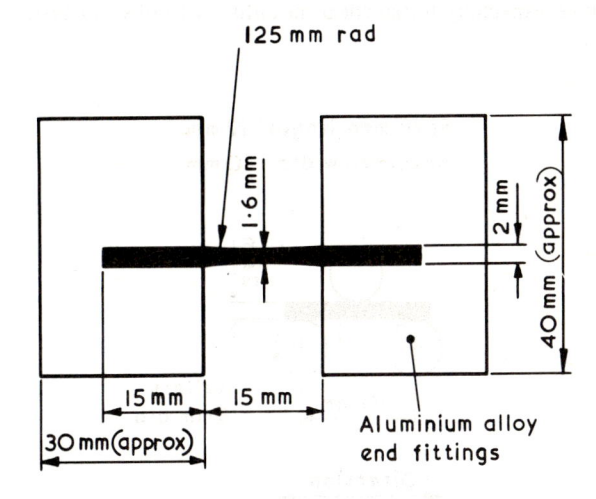

Fig. 5 Transverse compressive test specimen

(Specimen width 10 mm; 125 mm rad; 1·6 mm; 2 mm; 15 mm; 15 mm; 30 mm (approx); 40 mm (approx); Aluminium alloy end fittings)

problem is to develop a test in which there exists a region of pure shear and such that shear failure will occur in that region before failure occurs elsewhere. Obviously, all the methods cannot be reviewed here but it is of interest to note that none satisfies both these conditions completely, although a thin-walled tube with near-ideal end restraints probably comes closest to so doing.

Bearing in mind that for overall consistency specimens should be based on a flat sheet, two tests have been chosen for an overall characterisation of shear; these are short-beam (three-point) shear and rail shear. It should be noted that because unidirectional cfrp can generally be considered isotropic on planes normal to the fibre direction it is acceptable in most instances to characterise shear behaviour by either in-plane measurements (rail shear) or interlaminar measurements (short-beam shear). However, care should be exercised in the use of this assumption because some fabrication techniques, particularly those using preimpregnated tape or sheet, can introduce a significant anisotropy of shear properties; the effect of interlaminar voids, for example, will have a greater effect on interlaminar shear than in-plane shear.

The short-beam shear test, shown diagrammatically in Fig. 6, has gained probably widest acceptance of all shear tests and enables shear strength to be measured quickly and cheaply over a range of temperatures. It also serves as a quality control test in both fibre and preimpregnate manufacture because of the information it yields on fibre/resin bond, which in turn is a measure of both the quality of fibre surface treatment and the general 'condition' of the matrix. Due to the complex stress state and the method of loading however the test is totally unsuitable for the measurement of initial shear modulus ro stress/strain relationships.

The chosen span to depth ratio of 5:1 (ie 10mm between outer loading points and a specimen thickness of 2mm) will ensure shear failure for flexural to shear strength ratios of greater than 15:1 and this encompasses most types of unidirectional cfrp. Kedwood [8] and Sattar and Kellogg [9] have shown that specimen width can cause significant errors in the measured shear strength. In particular, large errors can occur in comparisons of the measured shear strengths of composites in which the fibre moduli differ widely. However, for the chosen width of 10mm, evidence to date suggests that strengths derived from this test are reliable for design purposes.

For a determination of the shear stress/strain data up to failure, especially initial shear modulus, the rail shear test,

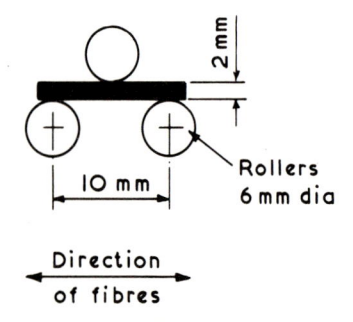

Fig. 6 Short-beam shear test specimen

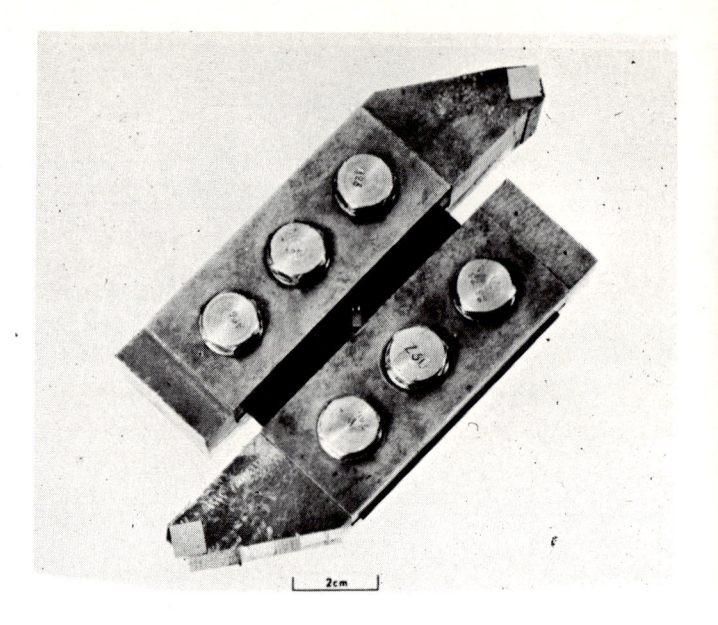

Fig. 7 Rail shear test specimen

shown as Fig. 7, was chosen for a number of reasons. Firstly, the specimen is made from a flat unidirectional sheet and good results have been obtained on specimens of the standard 2 mm thickness. Secondly, a region of near uniform shear stress exists and, provided the specimen is produced with the fibre direction normal to the rails, failure will occur by shear in that region before failure in and around the points of load input. Finally, in the region of uniform shear other stresses are small and generally can be neglected.

Recent experimental work [7] suggests that for similar material shear strengths measured by short-beam and rail shear tests are in good agreement. Furthermore, shear moduli obtained from rail shear tests are in good agreement with those predicted from theory. Although some work is yet to be completed and final dimensions agreed, it seems likely that the rail shear test will be adopted as standard.

3.6 Multi-directional laminates

In general, the problems of measuring the properties of multi-directional material are as complex as those associated with unidirectional material and in some instances, if anything, they are worse. It is fortunate therefore that theoretical approaches enable both elastic and strength properties of multi-directional laminates to be derived from the unidirectional properties of the individual plies. However, to check the accuracy of assumptions that need to be made or are implied in various theories, test techniques for multi-directional laminates are clearly required. A detailed discussion of such methods falls outside the scope of this paper, mainly because specimen development in terms of standardisation is still in its infancy, but it is considered relevant and worthwhile here to give an example of the good agreement that can be obtained between experiment and theory in determining the relationship between unidirectional and multi-directional properties.

Compressive strengths of $\pm \theta$ balanced laminates of cfrp, $0° \leqslant \theta \leqslant 90°$, were predicted using the unidirectional properties measured by the methods described previously and a failure criterion based on maximum strain. A suitable compressive test specimen was designed based on the philosophy used in the design of the longitudinal compressive specimen but waisted in the width instead of the thickness. Clearly, waisting in the thickness is unacceptable for multi-directional laminates because outer plies would be removed in the waisted region thus altering the nature of the material in the test region. Although the end-fittings are common to all specimens, as is the reduction in width at the waisted region of 25%, the distance between end-fittings and the rate of waistings are varied, depending on the value of θ, so that the best compromise is made between allowable waisting rate and tendency towards overall buckling.

Compressive strength tests were carried out for values of θ from 0° to 90° in 15° intervals and the results, which in each case are the mean of 5 tests, are shown in Fig. 8 together with the theoretically determined variation of strength with θ. Good agreement between experiment and theory was obtained not only for failing stresses but also for failing modes. Of course, apart from differences caused by the use of non-optimum specimen shape, exact agreement for $\theta = 0°$ and $\theta = 90°$ is to be expected because these are the longitudinal and transverse values respectively for unidirectional cfrp. It is of interest to note however that at $\theta = 45°$ the compressive test can be considered as an off-axis shear test and the good agreement between experiment and theory demonstrates the similarity between this test and the short beam and rail shear tests. Such similarity helps confirm the accuracy of both rail and shear and short beam shear tests because it has been shown recently that the $\pm 45°$ off-axis shear test can be exact provided the end-restraints conform to particular requirements which are believed to have been met here.

4 FATIGUE

If an overall method for characterisation of the static properties of unidirectional cfrp appears well advanced in terms of specimen design and test standardisation, by comparison the same could not be said of fatigue characterisation. Currently, no standard test specimens or techniques exist although results of fatigue tests have been published for which a host of different specimen shapes and types of loading have been used. It is quite usual with a new material for measurement of static properties to precede those of long-term properties but the apparent lack of progress in the development of specimens and techniques for measuring fatigue performance is due primarily to the complexity of the problems.

The areas of particular concern, from both the materials and structures points of view, have recently been reviewed by Cardrick [10] and it is relevant here to restate briefly some of those which affect materials test methods directly. Firstly, it is important to develop criteria on which fatigue failure can be based. Although a simple fracture criterion has proved adequate for most structural alloys there is some evidence that loss of residual static strength and stiffness, which can occur early in the life, may be limiting factors for composites and that a failure criterion based on these may be required. Recent work [11] suggests that unidirectional cfrp under longitudinal tensile fatigue loading does not exhibit reduced strength or stiffness but the same cannot be said under transverse tensile or shear fatigue loading. In these modes strength and stiffness are matrix dependent and reductions early in the life are to be expected. Under longitudinal compressive fatigue loads reduction might be expected if the compressive properties depend on matrix properties to a significant extent.

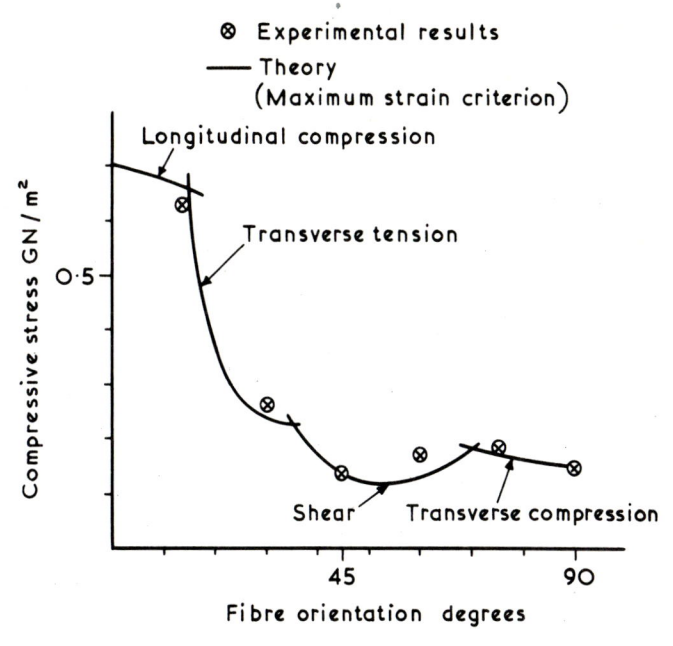

Fig. 8 Compressive strength as a function of ply orientation for balanced laminates

Recent work [12] on the nature of compressive failure tends to suggest that in some circumstances the compressive strength is very dependent on matrix properties so some strength and stiffness reductions should be expected. In multi-directional laminates all loading modes could potentially cause loss of strength and stiffness, the degree depending on such factors as stacking sequence and fibre orientation of the plies.

The unusually high ratio of longitudinal tensile strength (or longitudinal compressive strength) to shear strength of unidirectional cfrp, and to some extent the low values of transverse tensile and compressive strengths, have important implications in specimen design. For a large number of metal alloys, in which the tensile to shear strength ratios are of the order of 2:1 it is usually considered from a damage point of view that a tensile fatigue specimen is loaded only in tension if fluctuating shear stress are everywhere less than about 20 per cent of the tensile value. It is suggested by Cardrick however that if fluctuating shear stresses in unidirectional cfrp are to be disregarded on a similar basis they will have to be kept below as little as 1% of the tensile or compressive values. A similar argument may well apply to secondary transverse stresses.

In addition to these two important factors affecting specimen design and testing techniques, there are a number of others of importance such as the loading frequency, high and low temperatures and creep.

From a consideration of all these factors several important points emerge. Firstly, the design of a satisfactory fatigue specimen must include an allowance for the effect of secondary stresses, particularly shear stresses. It is for this reason that many tensile specimens used so far, both in the UK and USA, are unwaisted and have end-fittings in which shear stresses have been kept very low indeed. Secondly, the fatigue performance may be more important in, say, shear or compression so test specimens and techniques will need to be developed for these types of fatigue loading also. Finally, if as seems probable, failure criteria are to be based on residual strength or stiffness, the chosen specimen should enable strength and stiffness to be measured after a given fatigue test. Of course, some modification to each specimen after fatigue loading is permissible provided no further damage is done.

It is clear that there is considerable scope for research and development in the field of fatigue test specimen design and test procedures. Nonetheless, work is proceeding on the lines outlined above both at RAE and elsewhere and it is to be hoped that standard methods of testing will be adopted in the not too distant future so that comparisons of fatigue performance can be made simply and accurately.

5 IMPACT

Although material properties such as ultimate strengths and stiffness are generally easy to define and arouse little argument among various interested bodies, what is meant by impact depends very much on the use to which impact data is put. Further, impact resistance is not the only way of describing the particular materials characteristic for in the same general context are used other expressions such as fracture toughness, shock resistance and brittleness but again, in most instances, the definitions are loose. Thus it is seen that in the absence of a rigorous definition, the potential for standardisation of a test method to measure 'impact resistance' can exist only within specialised and relatively narrow fields.

As with fatigue testing, it is of prime importance to develop suitable criteria against which the impact performance of a material can be assessed. Most work on cfrp to date has been aimed towards aerospace activity so it is convenient to discuss such criteria in the aerospace context. However, analogous situations will occur in other fields where the development of criteria for impact assessment is equally important.

Bradshaw, Sidey and Dorey [13] considered failure criteria based on three main concepts namely containment, residual strength and stiffness and residual life. They considered also the concept of a threshold damage level in which impact performance is measured in terms of the amount of energy that can be absorbed before damage is initiated. No single type of specimen is satisfactory for determining impact performance according to all these criteria but the pendulum impact test and the ballistic or drop-weight test together are generally adequate.

Among the various pendulum impact tests, the notched Izod test has gained wide use and has been adopted as standard by many organisations. The test is generally well known but a specific specimen geometry has been chosen so that results on different materials and from different sources can be compared. The chosen geometry and overall test method is based on the ASTM standard Izod impact test and is shown in Fig. 9. Although the test is used for determining impact data for a wide range of applications, it is most useful in situations where the requirement is for damage containment, ie situations in which it is required to prevent complete penetration by an incident missile.

Ballistic or drop-weight impact tests have found favour in aerospace research because they give excellent representation of real potential incidents but no standard test for cfrp has yet been developed. Factors such as missile size, shape and velocity and specimen geometry are generally chosen to represent real conditions as closely as possible. However, as

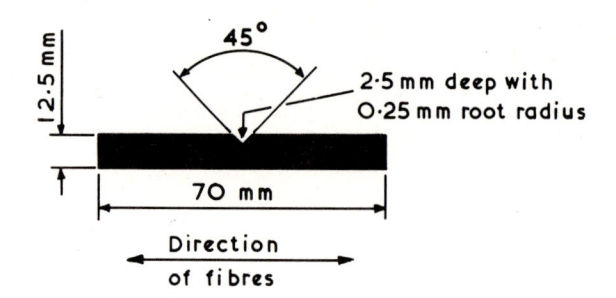

Fig. 9 Notched impact test specimen

this type of test is used both to determine threshold damage levels and in conjunction with failure criteria based on residual strength and stiffness, specimen geometry such as thickness should be chosen so that strength and stiffness test specimens can be abstracted easily and without influencing the subsequent test.

6 CONCLUSIONS

The problems involved in the mechanical testing of most composite materials arise both from their inherent anisotropy and from their unusual combination of properties. These factors influence not only the types of specimen and test technique used but also the definitions of properties, failure criteria and the relative importance of various properties in structural design.

Because of the importance to the aerospace industry as a whole, considerable effort has gone into the design and development of meaningful tests on cfrp but, although a good deal of success can be claimed in the determination of strength and stiffness properties, further effort is required on other no less important properties, such as fatigue, before similar success is achieved. Furthermore, testing on a routine basis over a longish period of time will have to be done before any of these procedures is fully proved and becomes universally accepted as a standard.

Although carried out in a specialised field, it is to be hoped that work described in this paper will be of direct benefit to those involved in the measurement of composite properties for use in other, non-aerospace fields and that it will help direct thought onto the right lines when new specimens and test procedures are required for new and as yet undeveloped composite materials.

ACKNOWLEDGEMENT

The contributions of many colleagues within the Royal Aircraft Establishment are gratefully acknowledged. Crown copyright is reserved.

REFERENCES

1 Ewins, P.D. 'Tensile and compressive test specimens for unidirectional carbon fibre reinforced plastics', *Royal Aircraft Establishment Technical Report* 71217 (1971)

2 Dootson, M. 'The development of unidirectional cfrp test specimens', *British Aircraft Corporation Report* No SQR(P)88 (1973)

3   Lekhnitskii, S.G. 'Anisotropic plates', *Gordon and Breach* (1968) pp 134-139

4   Ewins, P.D. 'A compressive test specimen for unidirectional carbon fibre reinforced plastics', *Aeronautical Research Council, Current Paper* CP1132 (1970)

5   Purslow, D. and Collings, T.A. 'A test specimen for the compressive strength and modulus of unidirectional carbon fibre reinforced plastic laminates', *Royal Aircraft Establishment Technical Report* 72096 (1972)

6   Mead, D.L. 'The strength and stiffness in transverse tension of unidirectional carbon fibre reinforced plastic', *Royal Aircraft Establishment Technical Report* 72129 (1972)

7   Purslow, D. Unpublished work, Royal Aircraft Establishment (1974)

8   Kedwood, K.T. Presentation at IPPS Conference on The Testing of Fibrous Composites Natioanl Physical Laboratory (15 July 1970)

9   Sattar, S.A. and Kellogg, D.H. 'The effect of geometry on the mode of failure of composites in short-beam shear tests', *Composites Materials: Testing and Design ASTM STP* 460 (1969)

10 Cardwick, A.W. 'Fatigue in carbon fibre reinforced plastics – a review of problems', Presented at International Committee on Aeronautical Fatigue (ICAF) Symposium, London (18–20 July 1973)

11 Beaumont, P.W.R. 'The effects of environment on fatigue and crack propagation in carbon fibre reinforced epoxy resin', School of Applied Physics, University of Sussex.

12 Ewins, P.D. and Ham, A.C. 'The nature of compressive failure in unidirectional carbon fibre reinforced plastics', *Royal Aircraft Establishment Technical Report* 73057 (1973)

13 Bradshaw, F.J., Dorey, G. and Sidey, G.R. 'Impact resistance of carbon fibre reinforced plastics', *Royal Aircraft Establishment Report* 72240 (1972)

COMMENTS

1 On the short beam ILSS test, by M.G. Bader (University of Surrey)

I would not wish to underestimate the importance of Mr Mr Ewin's meticulous work on testing composites but I feel that he is a little too satisfied with the short beam interlaminar shear test. My principal objection to this test is that with modern composites (of high interface strength) one just does not get an interlaminar shear failure, but a flexural failure. In this sense the test can be misleading particularly as it is used as one of the principal characterising tests for composites!

My other objection is that when working with composites (cfrp) fabricated from prepreg tape it is extremely difficult to avoid resin rich zones between tapes in the consolidated material. In my experience 'interlaminar' failure usually occurs in one of these layers often a long way from the neutral plane where the shear stress should be at a maximum.

I accept, however, that there is no satisfactory alternative.

Reply by Mr Ewins

Of all the tests used to characterise unidirectional cfrp the short-beam shear test is possibly the least satisfactory; apart from being theoretically rather unsound, failure does not always occur on the neutral plane and the failure mode is not necessarily that of shear. It is perhaps fortunate therefore that the fibre/resin combinations of interest to the aerospace industry do almost invariably lead to shear failure in the short-beam test! However, if failure does not occur by shear but by flexure then only a minimum shear strength can be derived but this in itself may be of some value. (It is, of course, important to state the failure mode.)

Although fabrication techniques developed at RAE successfully eliminate resin-rich zones and interlaminar voids which tend to occur with prepreg material, shear failures away from the neutral plane are sometimes observed. However provided the implications of such failures are recognised, especially in terms of apparent scatter in shear strength values, they can still yield useful information.

In the longer term, complete characterization of the shear properties of cfrp is most likely to be derived from tensile tests on ± 45° balanced laminates, with an admission that some of the desired coherency in the overall characterization has been lost.

2 On the behaviour of the transverse compression specimen, by M. Dootson (British Aircraft Corporation)

I would like to congratulate Mr Ewins on his lucid arguments in favour of the use of a single coherent family of specimens for the characterization of unidirectional cfrp materials.

I feel however that I should comment on the behaviour of th the transverse compression specimen advocated by Mr Ewins. In a careful comparison of a parallel sided specimen designed at BAC with that recommended here, it was noted that even in the parallel sided specimen, a free length of 15 mm led to premature instability. For this reason we reduced our free length to 10 mm and no longer observe this phenomenon. As well as being a convenient specimen for the measurement of modulus in this case, the parallel specimen also yields consistently higher strength values.

Reply by Mr Ewins

The design of a satisfactory specimen for measuring the transverse properties of unidirectional cfrp is difficult, particularly if a coherent set of specimens is required for the complete characterization of the material. If, as is generally the case, there are significant stress concentrations in the end region of a specimen some degree of waisting will be required in order to produce failure away from the ends. Experimental work carried out at RAE showed that, provided buckling did not occur, the measured strengths of waisted specimens were consistently higher than those of unwaisted specimens made from similar material. However, waisting will invariably increase the specimen free length which together with a reduction in thickness due to waisting will reduce the overall stability of the specimen. Thus, where overall stability might be a limiting factor, an unwaisted specimen would prove advantageous.

The specimens proposed by RAE and Mr Dootson represent the two possible solutions to the dilemma. Both have points to commend them and, as was pointed out in the paper (section 3.4), the choice in any situation will depend upon the data required and any other significant factors. If stress/strain information is required, an unwaisted specimen is clearly to be preferred.

3 On surface finish and dimensions of the test specimen, by J.M. Sillwood (NPL Teddington)

I would like to make two points, Firstly, with regard to tensile testing, you seemed to be most concerned that surface scratches, for example from knife edges, should not occur. Would you not agree that one of the advantages of a composite material is its ability to 'stop cracks'?

Secondly, the dimensions of test specimens shown on your slides clearly apply to the particular system chosen for your work and had been arrived at after considerable experience had been aquired. Although I would agree that the general shape of specimens should be similar for other systems I would have thought that exact dimensions could not be arrived at without prior knowledge of the parameters to be measured.

Reply by Mr Ewins

Although in some circumstances, for example, in multi-directional laminates, fibres can arrest cracks in the resin matrix or in adjacent plies, broken fibres themselves can act as crack initiation sites. It is important in materials testing, therefore, that such potential failure initiation sites are eliminated enabling the true mechanical properties to be measured. Of course, a complete material characterization would include an assessment of the effect of surface damage and general discontinuities on the basic properties but we are not here concerned with such effects. The second question involves the well-known paradox which exists in materials characterization and specimen design; material properties need to be known before suitable specimens can be designed to measure those very properties used in the design. In practice however initial values for the properties can be deduced from the fibre and matrix properties or from the measured values using less-than-ideal specimens. Allowances can then be made so that the most adverse combination of properties is used in the design of the coherent set of specimens. Clearly, if these specimens subsequently yield a significantly different set of properties an iterative approach to specimen design can be used but in practice this is unlikely to prove necessary.

QUESTIONS

1 On resin contraction and its effect on testing, by B.R. Watson-Adams (RARDE, Ministry of Defence, Fort Halstend, Sevenoaks, Kent)

Mr Ewins has indicated very clearly the conditions necessary in specimen design and the precautions necessary for testing composites. He has indicated that even void contents are noted — but has any thought been given to the contraction effect of the resin when it cures, and naturally puts the fibres into compression — and how does this effect affect the results obtained in compression and tensile testing?

Reply by Mr Ewins

The interface stresses between fibres and resin, which arise from resin cure and subsequent cooling, may well affect the composite properties particularly the transverse and shear strengths. However, a characterization of the unidirectional composite and the design of the associated test specimens need not take account of these stresses directly. Like voids, they contribute to the properties of the material, which in turn affect specimen design, so in this sense they are taken into account. Of course, fibre type and volume fraction, resin type and cure cycle will all affect the magnitude of the interface stresses and this is just one reason why, like void content, these must be stated as part of the material 'known composition'.

Finally, the effect of resin shrinkage on the overall material properties is not fully understood, partly because it is difficult to separate this effect from those of thermal stresses generally. However, the whole subject is receiving some attention because it is thought to be important in respect of the transverse tensile strain to failure, a property which often limits multi-directional composite performance.

Dynamic flexural properties of composite materials

R.D. ADAMS,[1] D.G.C. BACON,[2] D. SHORT[3] and D. WALTON[4]

Results are presented of flexural vibration tests on a wide range of composites. The damping properties of 0° unidirectional material cannot yet be fully accounted for by theory, but the modulus follows closely the law of mixtures prediction. Off-axis (+ θ), angle-ply (± θ) and generally laminated plates can be analyzed by anisotropic plate theory and the damping and stiffness characteristics successfully predicted from measurements on unidirectional samples of the same material.

1 INTRODUCTION

The main methods of fabrication with composites are either to mould the structure in one piece or to use adhesive joints. Both techniques remove the principal source of vibration damping in structures, that of interfacial friction. Thus, when using composites, the inherent damping properties of the material become important in limiting the amplitude of resonant vibrations and hence the radiated noise and the onset of fatigue. The unit of damping used here is specific damping capacity, ψ, defined as the ratio of the energy dissipated per cycle to the maximum value of the strain energy.

2 EXPERIMENTAL PROGRAMME

Specimens were produced from a variety of glass and carbon fibres using polyester and epoxy resin matrices. Both wet lay up and hot pressing techniques were used.

Beam specimens, usually 200 mm long by 12 mm wide and 2.5 mm thick, were vibrated in their fundamental free-free resonant flexural mode by a coil/magnet pair. The flexural Young's modulus was deduced from the resonant frequency and the damping from the input power at resonance. Extensive precautions were taken to make extraneous losses insignificant compared with the beam damping.

3 DISCUSSION OF RESULTS

3.1 Effect of vibration amplitude

It was generally found for all composites, provided these were properly made, that vibration amplitude did not affect either modulus or damping. Typically, the cyclic loads used were from very low levels up to about 10 per cent of the failure load.

---

(1) Lecturer in Mechanical Engineering, University of Bristol;
(2) Research Engineer, Rolls Royce Motors, Crewe, formerly Research Assistant, Department of Mechanical Engineering, University of Bristol; (3) Senior Lecturer, School of Engineering Science, Plymouth Polytechnic; (4) Research Assistant, Department of Mechanical Engineering, University of Bristol.

3.2 Unidirectional cfrp and gfrp

The conclusion to be drawn from a series of tests on unidirectional material was that the law of mixtures held in predicting the flexural Young's modulus [1–3], whether the variable was volume fraction of fibre or fibre modulus. In addition, Adams and Short [4] also showed that fibre diameter did not affect the predicted modulus. On the whole, it was found that the law of mixtures prediction usually overestimated the experimental values slightly. The discrepancy was principally due to imperfect fibre alignment since only a few degrees misorientation can result in a significant reduction in E (see below). An additional factor in these flexural tests was the effect of shear deflections, particularly with high modulus cfrp materials. Large aspect ratios are required with unidirectional composites since the ratio of E_L/G_{LT} is very much larger than for isotropic materials [2]. (E_L is the longitudinal Young's modulus and G_{LT} is the longitudinal shear modulus.)

It is possible to predict the damping capacity of unidirectional material when stressed in the fibre direction by using the law of mixtures and assuming that all the energy dissipation occurs in the matrix. On this basis, we arrive at the equation

$$\psi_c = \psi_m (1-v) E_m/E_L$$

where the subscripts c and m refer to the composite and matrix respectively, and v is the volume fraction of fibres.

However, it is found that this expression underestimates considerably the measured value of ψ_c, even when considerable effort has been made to eliminate extraneous losses. Basically, there are several contributions to the discrepancy. First, the smaller is the fibre diameter, the larger is the surface area of fibre per unit volume. Adams and Short [4] showed that, for glass fibres of 10, 20, 30 and 50 μm diameter in polyester resin, there was a consistent increase in ψ_c with reduction in fibre diameter. Second, the problem of misalignment is not insignificant as is shown below for angle-ply composites. Third, any structural imperfections such as cracks and debonds lead to interfacial rubbing and hence to additional losses. Finally, although the effect of shear is usually negligible in modulus measurements, this is less true for the damping.

The damping in longitudinal shear, ψ_{LT}, is of the order of 50–100 times larger than the tension/compression component. Thus, although only a few per cent of the energy is stored in shear, this can make a substantial contribution to the total predicted value. Fig. 1 shows that as the aspect ratio of a beam was reduced from 90 to 53, the shear damping contribution was increased. Further, by subtracting the shear damping from the experimental values, the effect of aspect ratio was essentially eliminated.

The difference remaining between the law of mixtures prediction and the 'experimental minus shear' values was mainly due to the combination of misalignment, internal flaws and fibre diameter.

3.2 Off-axis composites (+ θ)

Plates of unidirectional carbon and glass fibre-reinforced plastics were cut to provide beam specimens at various angles from 0° to 90°. From the damping and modulus of unidirectional beams (ψ_L, ψ_T and ψ_{LT}; E_L, E_T and G_{LT}; subscript T denoting properties transverse to the fibre direction) predictions can be made of the flexural damping and Young's modulus.

Since in the free-free flexural test used there were no physical constraints, the prediction for E was that for free flexure, ie such that when the beam bends it is free to twist about its longitudinal axis. The damping prediction was also based on this boundary condition.

Fig. 2 shows typical results for HM-S carbon fibre in Shell DX-209 epoxy resin, and it can be seen that there is good agreement between the predicted and the experimental values.

3.3 Angle-ply composites (± θ)

Angle-ply composites made from HM-S carbon fibres in DX-209 epoxy were produced at ±10°, ±20°, ±30°, ±45° and ±60°. These were made from 10 layers of pre-preg, laid at alternate angles. Essentially, the bending/twisting coupling term is eliminated by the internal restraint of each layer on its neighbours and the flexural modulus is that of the pure flexure prediction (Fig. 2). Similarly, the damping is characteristically different, reaching a peak at a larger angle (45°) than was the case for the off-axis beams.

3.4 General plate

A 250mm square plate, 3mm thick, was hot-pressed from HM-S fibre pre-impregnated with DX-209 epoxy resin, using a lay-up of 0°/−60°/+60°/+60°/−60°/0°. Fig. 3 shows the

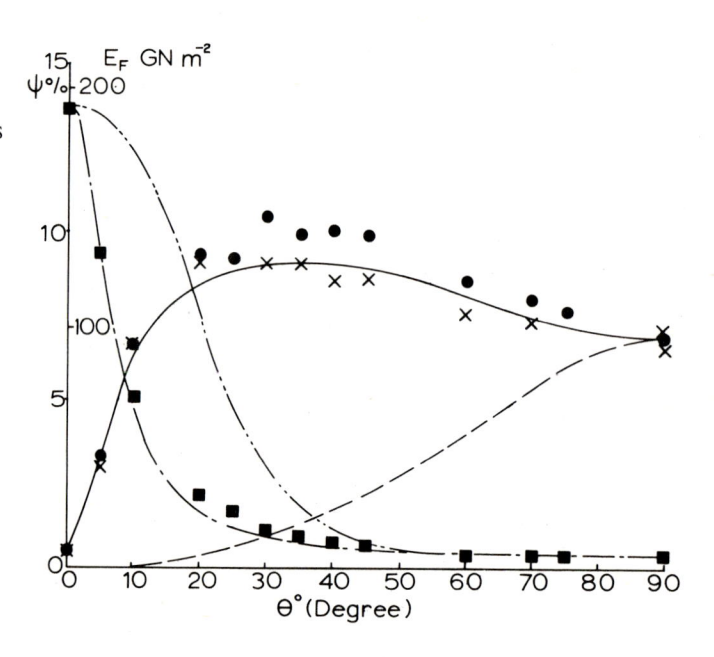

Fig. 2 Variation of flexural Young's modulus, E_F, and damping, ψ_F, with fibre orientation, θ, for HM-S carbon fibre in DX-209 epoxy resin; v = 0.5; ●, × ψ_F plates 1 and 2; ■ E_F average for plates 1 and 2 (values virtually coincident).
Theoretical predictions ——··—— E_{PF} (pure flexure); ——·—— E_{FF} (free flexure); ——— ψ_{FF}

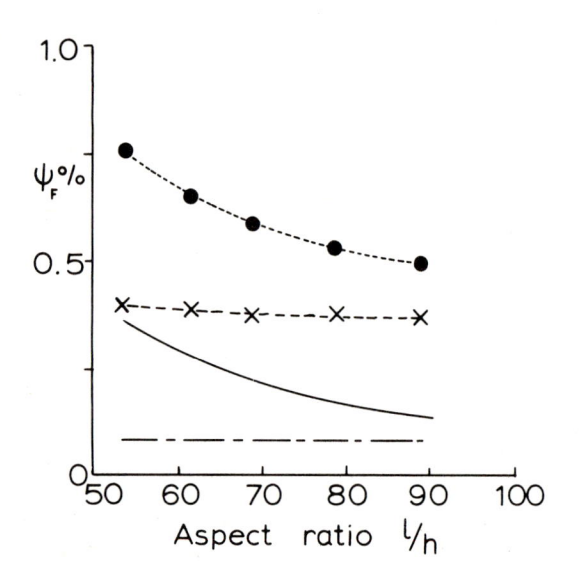

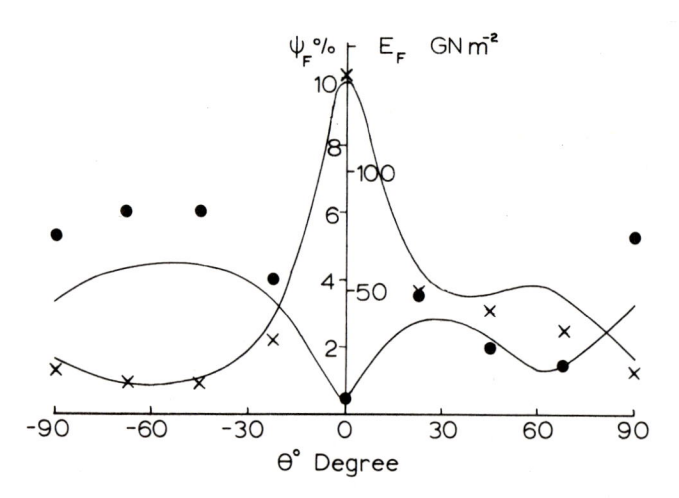

Fig. 1 Variation of damping, ψ_F, with aspect ratio, l/h, for HM-S carbon fibre in DX-209 epoxy resin; v = 0.5; ● measured S.D.C., ——— theoretical shear S.D.C.; ——·—— law of mixtures prediction of S.D.C.; × measured minus theoretical shear S.D.C.

Fig. 3 Variation of flexural Young's modulus, E_F, and damping, ψ_F, with outer layer angle, θ, for a laminated plate (0°/−60°/60°/60°/−60°/0°) from HM-S carbon fibre in DX-209 epoxy resin; v = 0.5, × E_F, ● ψ_F. Free flexure predictions: ——— E_{FF}, ——·—— ψ_{FF}

predicted values based on the method outlined by Adams and Bacon [5]. Beams 13 mm wide by about 200 mm long were cut at various angles to the direction of the outer plies and tested as before in free flexure.

Good correlation was obtained for the experimental and predicted values of the modulus. The damping values tended to be larger than the predictions but the general trend was followed. Subsequent sectioning of this material revealed some dry areas and these most probably contributed to the extra energy dissipation observed, while they would have little effect on the modulus.

4   CONCLUSIONS

In predicting the dynamic flexural properties of unidirectional composites, the effect of shear deformation cannot be ignored, particularly in respect of damping.

From measurements of the properties of unidirectional composite it is possible to predict with reasonable accuracy the damping and moduli of generally-laminated plates.

ACKNOWLEDGEMENTS

Figs. 1 and 2 are reproduced from References [2] and [5]. Much of the work described here was supported by the Science Research Council.

REFERENCES

1   **Adams, R.D**, et al. 'The dynamic properties of unidirectional carbon and glass fiber reinforced plastics in torsion and flexure', *J Comp Mat* **3** (1969) pp 594–603

2   **Adams, R.D.** and **Bacon, D.G.C.** 'The dynamic properties of unidirectional fibre reinforced composites in flexure and torsion', *J Comp Mat* **7** (1973) pp 53–67

3   **Adams, R.D.** and **Bacon, D.G.C.** 'The effect of fibre modulus and surface treatment on the modulus, damping and strength of carbon fibre reinforced plastics', *J Phys D: App Phys* **7** (1974) pp 7–23

4   **Adams, R.D.** and **Short, D.** 'The effect of fibre diameter on the dynamic properties of glass fibre reinforced polyester resin', *J Phys D: App Phys*, **6** (1973) pp 1032–1039

5   **Adams, R.D.** and **Bacon, D.G.C.** 'Effect of fibre orientation and laminate geometry on the dynamic properties of C.F.R.P.', *J Comp Mat*, **7** (1973) pp 402–428

The compression strength of unidirectional cfrp

N.L. HANCOX

It was required to develop a simple compression test which could be used for quality control measurements, preferably with the standard bar specimens already produced for routine testing. Because of these limitations the Celanese compression jig, Fig. 1, was used. This takes a specimen 110 mm × 6.3 mm × 1 mm with aluminium or glass fibre composite end tabs bonded to an approximately 50 mm length of either end. Specimens were made from various types of carbon fibre, laid-up unidirectionally, with a liquid bisphenol A type resin and an anhydride hardner, cured for 2 hours at 120°C.

A series of experiments was carried out in which the unsupported gauge length was varied and the compression strength measured. The results are shown in Fig. 2. It is apparent that the failure stress is independent of gauge length provided this is 6 mm or less. This result will, of course, depend on the thickness of the specimen.

The variation of compression strength with fibre volume loading is shown in Fig. 3. For high strength and high modulus surface treated carbon fibre composites the relationship is linear and compression strengths are similar to tensile ones. Failure invariably occurred by shear at an angle of approximately 45° to the long fibre axis often associated with small amounts of fibre misorientation and cracking. Details of polished sections of specimens are shown in Fig. 4, 5 and 6. For composites made with high modulus untreated material the break was very ragged with a large amount of delamination.

Compression strengths obtained are less than those given by buckling theory and the variation with volume loading linear rather than non-linear. As the strain to failure of the pure resin in compression is considerably greater than that of the carbon fibre in tension (and presumably compression too) it is believed that failure of the composite is caused by the initial failure of the fibres. In accordance with the recent work of Ewins and Ham (1973) it is suggested that longitudinal compression and tensile strengths of carbon fibre and hence carbon fibre composites are governed by the same basic mechanism. With untreated high modulus fibre composites failure is initiated by fibre matrix delamination caused by the poor bonding.

Process Technology Division, AERE Harwell, Didcot, Oxon.

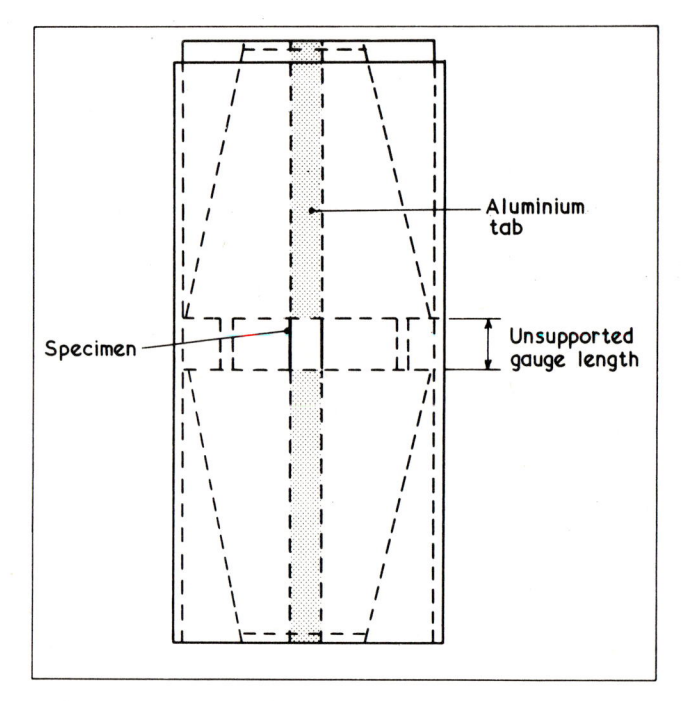

Fig. 1 Compression jig

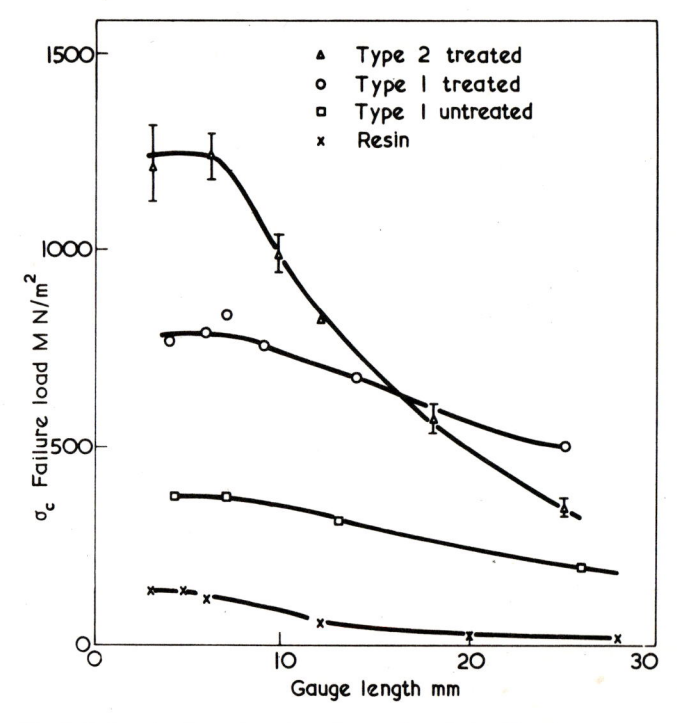

Fig. 2 Failure load as a function of gauge length for 60 volume percent carbon fibre composites

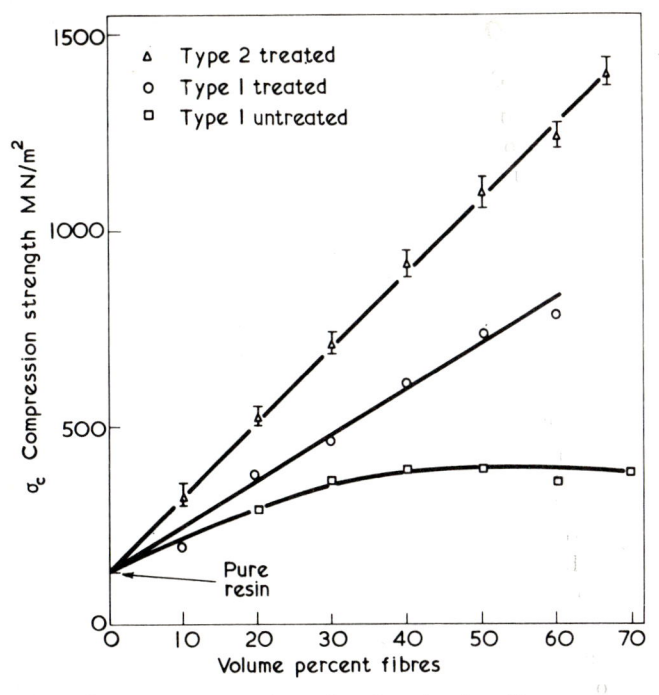

Fig. 3 Compression strength as a function of carbon fibre content

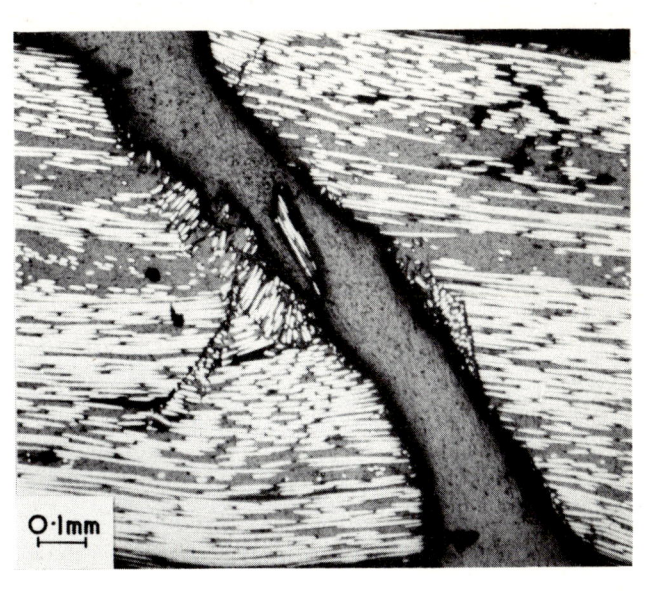

Fig. 5 Section through break in 60 volume percent type 2 carbon fibre composite

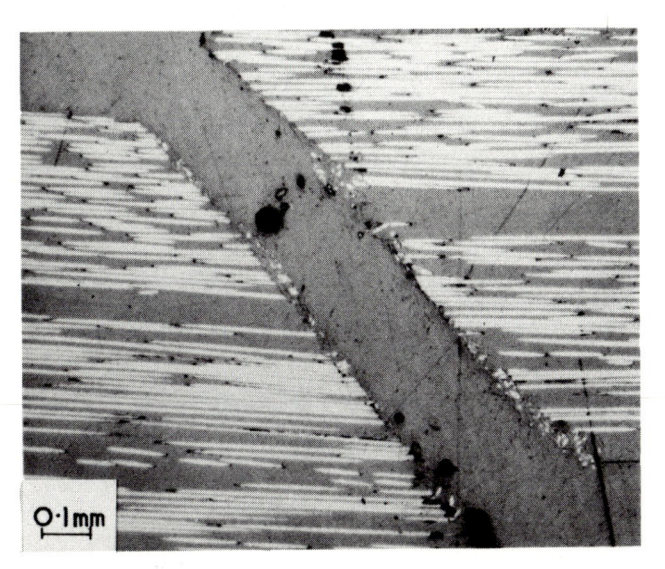

Fig. 4 Section through break in 30 volume percent type 2 carbon fibre composite

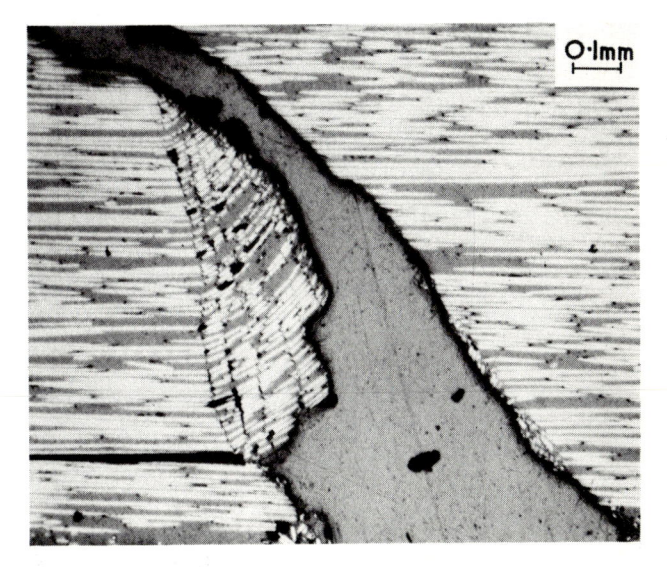

Fig. 6 Section through break in 60 volume percent type 2 carbon fibre composite

Charpy impact strength of uniaxial cfrp

M.G. BADER

1 INTRODUCTION

Several workers, eg [1,2,3], have studied the impact properties of carbon fibre reinforced plastics but their results have not always been compatible. In particular the effects of notches and other aspects of test piece geometry have been confusing. Hancox [1] suggested a correlation between the square of the test piece depth and the energy absorbed, Ellis and Harris [4] suggested a linear relationship whilst Bradshaw et al [5] consider that the length to depth ratio is more significant. It is generally agreed [3,5] that the fracture mode is an important factor and that extensive delamination is associated with high energy absorption, and that low interfacial strength (ILSS) contributes to this situation. The effect of notch depth and sharpness is also confusing; Ellis and Harris [4] have reported that increasing notch sharpness lowers the energy absorbed whilst earlier work [2] suggested the opposite.

In this work test piece geometry and notch dimensions have been systematically varied in a number of typical uniaxial cfrp composites in order to attempt to resolve the above-mentioned uncertainties.

2 MATERIALS

The composites tested were fabricated from surface treated carbon fibres of both high modulus and high strength types in epoxy resin matrices. A wet lay up leaky mould technique was used for fabrication with the exception of one batch of material which was laminated from pre-impregnated tape. The details of the three principal composites used are given in Table 1.

Department of Metallurgy and Materials Technology, University of Surrey, Guildford, Surrey, England

It should be noted that the type I-S fibre gives the lower interlaminar strength and the prepreg system the highest.

In Fig. 1 the result of earlier work [6] with unnotched test pieces in which V_f was the principal variable is depicted. There is a nearly linear relationship between impact energy and V_f but note that the high modulus (type I) composites absorb much less energy than the high strength (type II and type III) materials. It should also be noted that the untreated fibre composites, which have low interfacial strength, absorb more energy than do those made with treated fibres. The HTS-DLS 60 material was only available at one V_f but this shows less energy absorbing capacity than the equivalent type II-S composite which has a somewhat lower interface strength as indicated in Table 1.

3 EXPERIMENTAL

Test pieces were cut from moulded plates or bars approximately 5mm thick to the configuration shown in Fig. 2. Where notches were used these were cut in a face normal to the lay up plane so that inter laminar defects would not unduly influence the delamination behaviour.

Since the Charpy testing machine used had a fixed anvil with supports 30mm apart, variations in test piece geometry were achieved by varing D and d. B was also varied to some extent to maintain the test piece compatible with the available testing ranges.

The Charpy testing machine could be adjusted to give striking velocities of 2, 3 or 4 m/s and striking energies of 1.5, 5, or 15J, but for this work 5J at 3m/s and 15J at 3m/s were found to be most convenient, the majority of the tests being performed at the latter setting.

Table 1 Properties of composites

Composite system		V_f	σ_c GN/m^2	τ_{SB} MN/m^2	Remarks
Fibre	Resin				
Modmor type I-S	Shell 828/HPA/BDMA	0.60	0.75	55	Wet lay-up
Modmor type II-S	Shell 828/HPA/BDMA	0.56	1.4	80	Wet lay-up
Grafil HTS	Ciba DLS 60	0.48–0.55	1.6	90	Prepreg

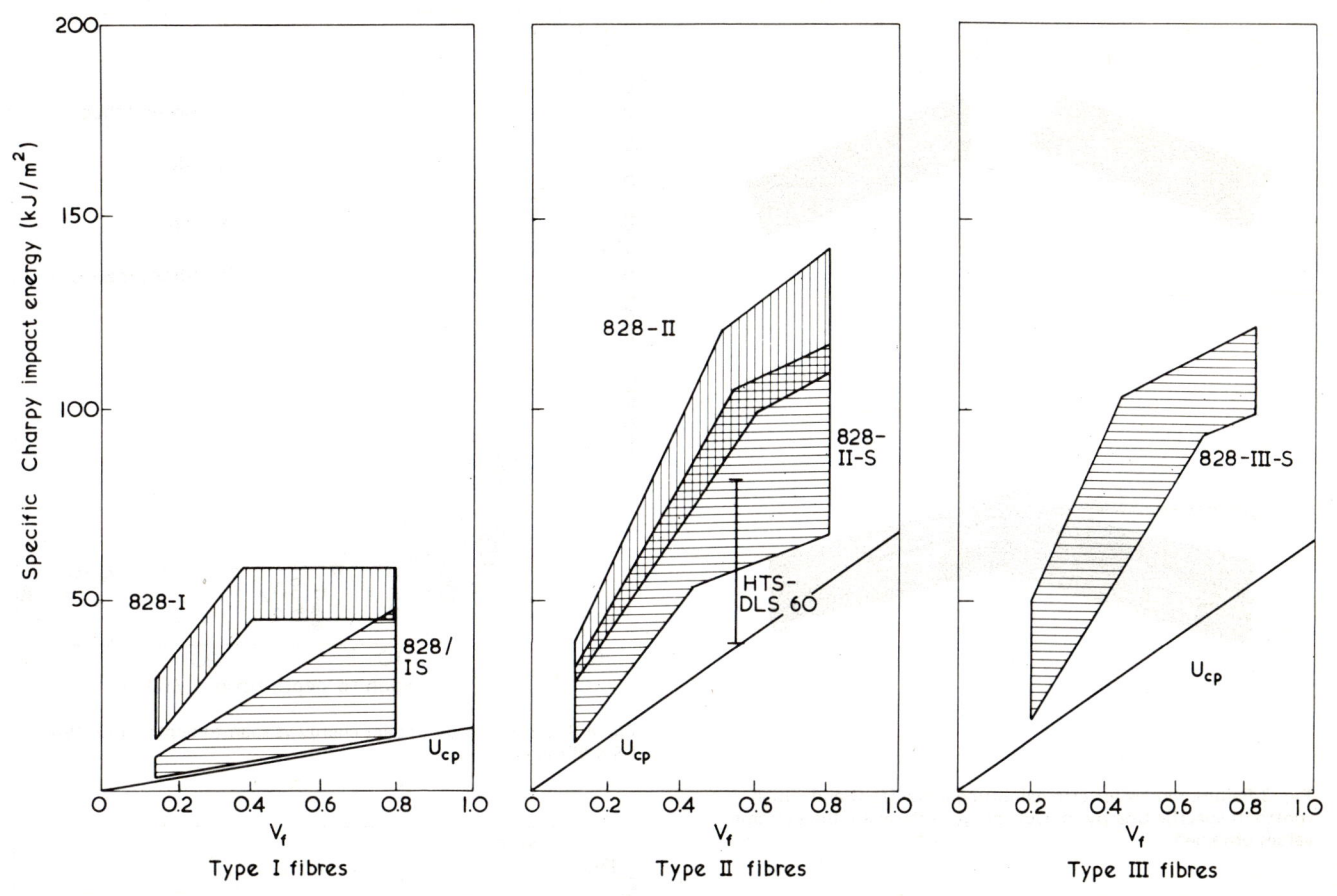

Fig. 1 Specific Charpy impact energy vs fibre volume fraction: uniaxial type I, II, and III carbon fibres in epoxy resin matrices

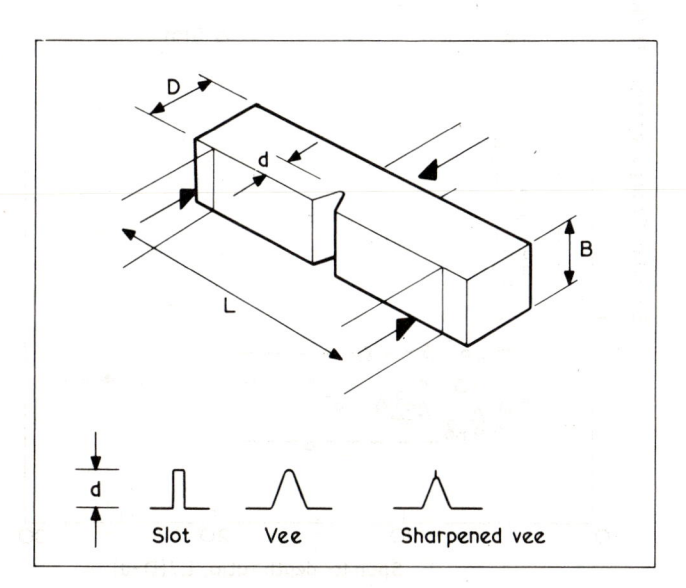

Fig. 2 Charpy test piece and notch configuration

Three alternative notch forms were chosen: a simple slot cut with a diamond tipped slitting wheel, a standard Charpy Vee notch and the latter, sharpened with a cut from scalpel blade.

4 RESULTS AND DISCUSSION

The specific Charpy impact energy (ie total energy abstracted from the penulum divided by the net cross sectional area of the test piece) showed a variation of about four times between the highest and lowest values recorded for each material. The highest values were associated with extensive delamination in both notched and unnotched test series. The lowest values were associated with almost planar smooth fracture surfaces. These two extremes are shown in Fig. 3. At intermediate energies a rougher fracture surface with 'clumped pullout' was produced. In many cases the more brittle failures were initiated on the compressive side of the test piece; even in the case of notched tests. In notched tests there was inevitably a plane of delamination running through the base of the botch.

The specific energies did not correlate well with D, $D-d$, $(D-d)^2$ or d/D as suggested by Hancox [1] and Ellis and Harris [4], although similar behaviour was observed over part of the range of geometries used. The most significant parameter appears to be the effective length to depth ratio $L/(D-d)$. The relationship is shown in Figs. 4, 5 and 6. In each case the specific energy value is uniformly low at high value of $L/(D-d)$ but increased sharply below a critical value of about 5. It was not possible to test at sufficiently low values of $L/(D-d)$ to establish an upper limit but we have observed this in later work on glass reinforced polyester materials.

The lower limiting energy value agrees well with the proposal made earlier [3] correlating Charpy energy with the elastic strain energy absorbing capacity of the fibres. This is the total elastic strain energy in the beam when the failure initiates:

$$U_{CP} = K \times \frac{\sigma_f^2 V_f}{E_f}$$

where K is a factor dependent on the test beam geometry

Fig. 3 Failure modes for cfrp exhibiting a low (top) and a high (bottom) specific Charpy impact energy — these are the extreme values obtained

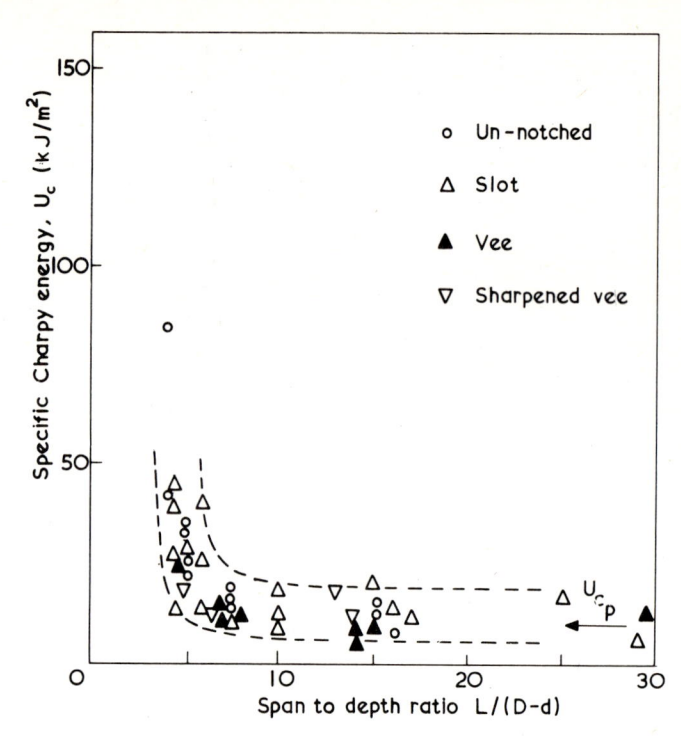

Fig. 4 Specific Charpy impact energy vs span-to-depth ratio: type I-S 828 material

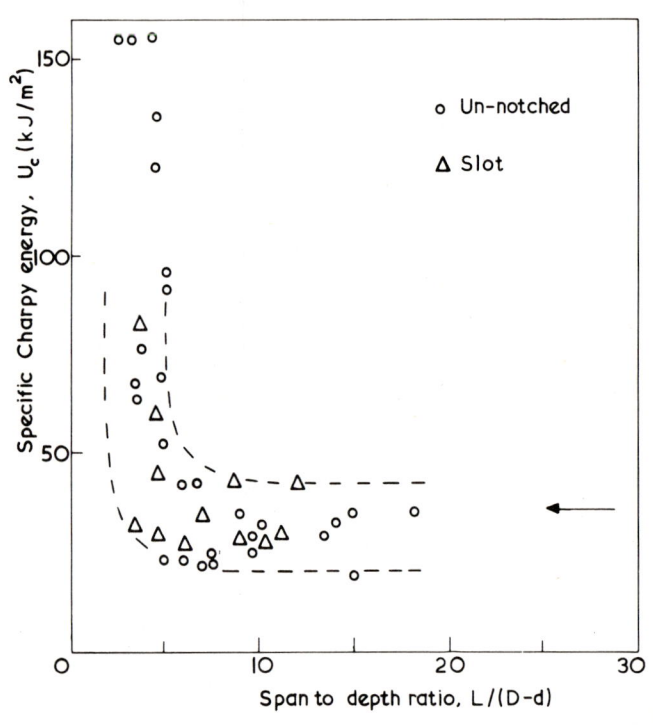

Fig. 5 Specific Charpy impact energy vs span-to-depth ratio: HTS-DLS60 material

and test configuration. The values of U_{CP} for the three materials tested is indicated by the arrows on the right hand side of Figs. 4–6.

The fractures at the high $L/(D-d)$ geometry were smooth and brittle and there was a transition through a rougher fracture to that of extensive delamination through the critical zone for both notched and unnotched tests.

It will be noted from Figs. 4–6 that there is little difference between the notched and unnotched results but that there is possibly a tendency towards higher values for the notched tests at high values of $L/(D-d)$.

This suggests that the materials are not notch sensitive in the normally accepted sense over the range of geometries investigated. Notch sharpness is unimportant since in a uniaxial composite the only significant feature is whether the fibres are continuous or severed by the notch. All notches are therefore of equal severity. The effect of the notch is to act as a shear stress concentration in the plane normal to the notch where the material has the lowest shear strength. On the application of load a shear fracture propogates from the base of the notch. In a short test piece this propogates almost to the outer supports where the stress would be very low. At the centre of the test piece where the bending moment is highest the material above the base of the notch therefore carries little or no stress. The notched test is therefore effectively the same as an unnotched test on a section of dimensions $(D-d)$. The small increase in energy observed on some notched tests arises from the work done prior to the delamination at the base of the notch.

In some earlier work the dimension D has been kept constant together with L and the notch depth d progressively increased. If the intial choice of L and D were such that the test was in the low L/D delamination zone then increasing d would effectively increase $L/(D-d)$ and a lower energy and transition to the brittle fracture mode would be observed. This could easily be confused with true notch embrittlement.

Great care must therefore be exercised when interpreting impact test data since the fracture mode depends on the particular state of stress in the test piece or component. Other things being equal impact energy can be increased by

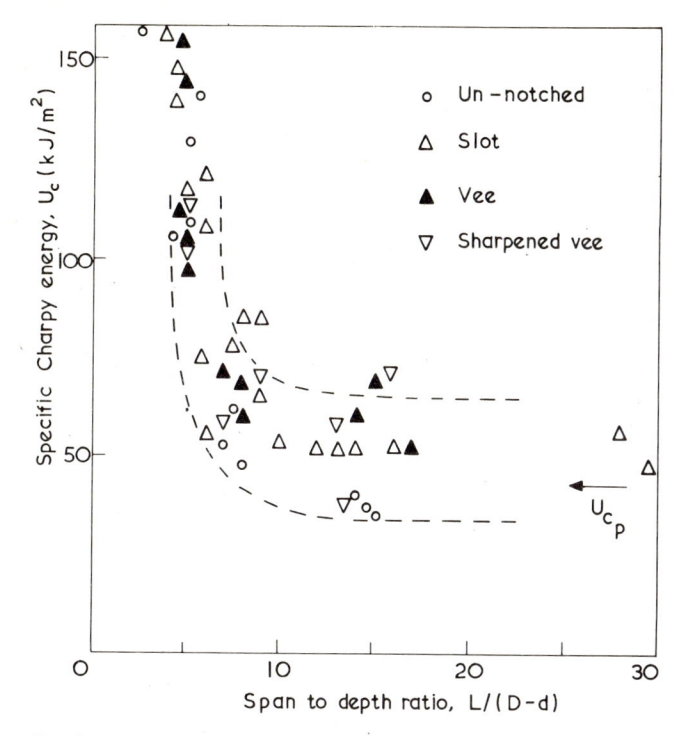

Fig. 6 Specific Charpy impact energy vs span-to-depth ratio: type IIS-828 material

reducing the interface strength until delamination occurs but this is only a function of the relative values of shear stress and principal stresses in the test piece or component, and the optimum material peroperties for one situation will not necessarily apply to another.

5 CONCLUSIONS

The specific Charpy energy for uniaxial cfrp composites is determined principally by the elastic strain energy capacity of the constituent fibres. Highest energies are recorded when the fracture mode is one of extensive delamination and this is influenced by the interfacial shear strength of the composite and by the ratio of principal to shear stresses in the testing configuration. In the Charpy beam test the critical parameter is the length to depth ratio of the beam, and the critical value is about 5 for presently available materials.

In the Charpy beam configuration, notches are not effective in influencing the failure energy or the mode of fracture if allowances are made for their effect in changing the length to depth ratio.

It is extremely important to consider the precise state of stress when attempting to use Charpy test data for design purposes.

REFERENCES

1. **Hancox, N.L.** *Composites* 2 No 1 (1971) pp 41–45
2. **Harris, B., Beaumont, P.W.R.** and **Moncunill de Ferran, E.** *J Mat Sci* 6 (1971) p 238
3. **Bader, M.G., Bailey, J.E.** and **Bell, I.** *J Phys D: App Phys* 6 (1973) p 572
4. **Ellis, C.D.** and **Harris, B.** *J Comp Mat* 7 (1973) p 76
5. **Bradshaw, F.J., Dorey, G.** and **Sidey, G.R.** *RAE Technical Report* No 72240 (1973)
6. **Bailey, J.E., Bader, M.G.** and **Johnson, M.** 'Designing to avoid mechanical failure', *Plastics Institute Conference*, Cranfield (1973) paper 4

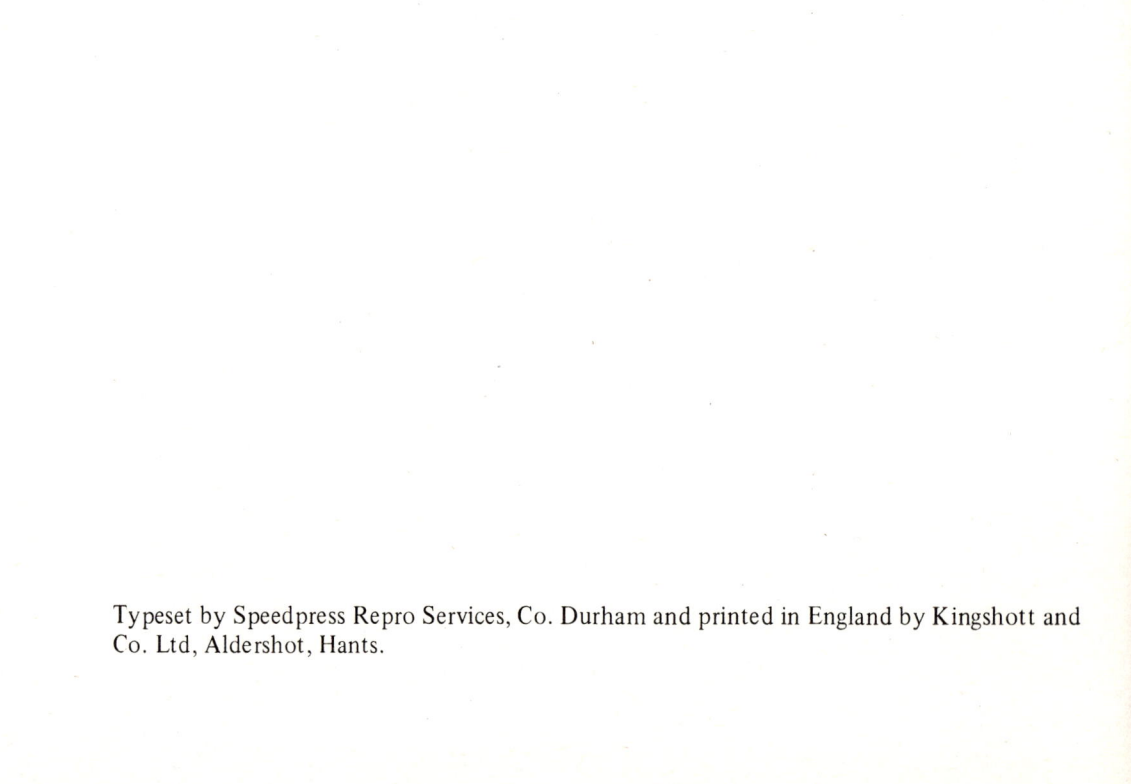

Typeset by Speedpress Repro Services, Co. Durham and printed in England by Kingshott and Co. Ltd, Aldershot, Hants.